TWITTER AND SOCIETY

Steve Jones
General Editor

Vol. 89

The Digital Formations series is part of the Peter Lang Media and Communication list.
Every volume is peer reviewed and meets
the highest quality standards for content and production.

PETER LANG
New York • Washington, D.C./Baltimore • Bern
Frankfurt • Berlin • Brussels • Vienna • Oxford

TWITTER AND SOCIETY

EDITED BY KATRIN WELLER, AXEL BRUNS,
JEAN BURGESS, MERJA MAHRT, & CORNELIUS PUSCHMANN

PETER LANG
New York • Washington, D.C./Baltimore • Bern
Frankfurt • Berlin • Brussels • Vienna • Oxford

Library of Congress Cataloging-in-Publication Data

Twitter and society / edited by Katrin Weller, Axel Bruns,
Jean Burgess, Merja Mahrt, Cornelius Puschmann.
pages cm. — (Digital formations; vol. 89)
Includes bibliographical references and index.
1. Twitter. 2. Online social networks. 3. Internet—Social aspects.
4. Information society. I. Weller, Katrin, editor of compilation.
HM743.T95T85 2   006.7'54—dc23   2013018788
ISBN 978-1-4331-2170-8 (hardcover)
ISBN 978-1-4331-2169-2 (paperback)
ISBN 978-1-4539-1170-9 (e-book)
ISSN 1526-3169

Bibliographic information published by **Die Deutsche Nationalbibliothek**.
Die Deutsche Nationalbibliothek lists this publication in the "Deutsche
Nationalbibliografie"; detailed bibliographic data is available
on the Internet at http://dnb.d-nb.de/.

Cover art:
Klee, Paul (1879–1940): *Twittering Machine (Zwitscher-Maschine)*, 1922.
New York, Museum of Modern Art (MoMA).
Watercolor, and pen and ink on oil transfer drawing on paper, mounted on cardboard.
DIGITAL IMAGE ©2012, The Museum of Modern Art/Scala, Florence.

The paper in this book meets the guidelines for permanence and durability
of the Committee on Production Guidelines for Book Longevity
of the Council of Library Resources.

© 2014 Peter Lang Publishing, Inc., New York
29 Broadway, 18th floor, New York, NY 10006
www.peterlang.com

Printed in the United States of America

Table of Contents

METHODS

Part II: Perspectives and Practices

PERSPECTIVES

PRACTICES

POPULAR CULTURE

Debanalising Twitter
The Transformation of an Object of Study

FOREWORD   Richard Rogers

 three major phases of Twitter development, from stat.us to Twitter, Inc. #socmedhistory

This is an enquiry into how Twitter has been studied since it was launched in 2006 as an ambient friend-following and messaging utility, modelled after dispatch communications. As Jack Dorsey, the Twitter co-founder, phrased it, Twitter also did rather well during disasters and elections, and subsequently became an event-following tool, at once shedding, at least in part, its image as a what-I-had-for-lunch medium. Most recently, Twitter has settled into a data set, one that is of value for Twitter, Inc. and is also archived by the Library of Congress. Each of these objects, described here as Twitter I, Twitter II, and Twitter III, has elicited particular approaches to its study, surveyed below. In the following I take each object in turn, describing the debates and scholarship around them, and provide a framework to situate past, current, and future Twitter research.

INTRODUCTION: TWITTER STUDIES

Founded by Jack Dorsey and associates in San Francisco in 2006, Twitter brought together two subcultures, new media coding culture as well as radio scanner and dispatch enthusiasm. Together they informed what could be called first-generation Twitter (or 'Twitter I'), an urban lifestyle tool for friends to provide each other with updates of their whereabouts and activities (Akcora & Demirbas, 2010). In an early sketch, maintained on Dorsey's dormant Flickr account, the service is called stat.us (see Figure 1). The sketch has two in-built options, "in bed" and "going to park", and the current status is "reading" One is able to watch a user change states, in a sense 'tracking' or following the user's updates like tracking a courier package.

Dorsey's description of the sketch on Flickr also contains the compact name of the service, Twttr, which is in keeping with dispatch and courier messaging protocol. It is a five-digit short code that would comply with the cellular administration of an SMS messaging service, which Twitter is designed to work with. The delivery constraints of text messages provided the rationale for the length of a Twitter message, or tweet, as it has come to be known. With SMS, the message breaks in two after 160 characters, and two messages are sent. It was decided to work within the limits of the one message of 160 characters; 20 were reserved for the name space, and the other 140 characters for the message. The required brevity has spawned growth in URL shortening services, which themselves have grown shorter, from *tinyurl.com* to *bit.ly* and Twitter's own *t.co*.

Twitter's historical roots rely often on Dorsey's own telling. It was conceived as part of a long line of squawk media, dispatch, short messaging, as well as citizen communications services. Dorsey's genealogy of Twitter refers to communications systems for bicycle messengers, truck couriers, emergency services, ambulances, firetrucks, and police. He also mentions GPS, citizen band (CB) radio, as well as Research In Motion's proto-BlackBerry (the RIM 850 interactive pager), for which he wrote a script to batch post to a friend list. Dorsey also recalls a visualisation he made before stat.us that captures the output of radio scanners, and shows on a city map the flows of emergency communication in the city. It demonstrates interest in scanner culture, and has affinity with early locative media art projects. In a two-part interview for the *Los Angeles Times* published in 2009, and in other interviews and public appearances, Dorsey touched on the lineage of the project, at once trying to define Twitter as a new medium in itself, a public instant messaging system. The system was meant to be device and (proprietary) platform independent, thus eschewing the walled

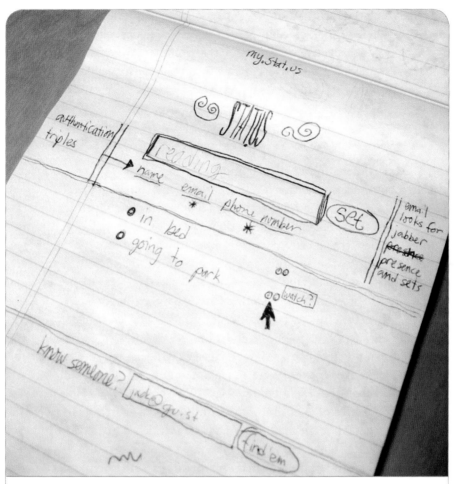

Figure 1: "twttr sketch" by Jack Dorsey, 2000. Source: http://www.flickr.com/photos/
jackdorsey/182613360/ (Dorsey, 2006)

garden model and desiring to be a public information utility. It also is a "new take on the address book", as Dorsey put it. "When I'm visiting New York, I turn on my New York friends just because I'm more interested in their particular interruptions" (Sarno, 2009b). While a universal messaging system, Twitter in that sense was conceived and used also as an ambient, friend-following tool.

One other aspect of the origins story of Twitter is of special interest. The name of the service would try to capture "the physical sensation that you're buzzing your friend's pocket", and after a "name-storming" session resulted in 'twitch' and 'jitter', a dictionary search around *tw* ended with 'twitter' (Sarno, 2009a; Truong, 2011). Twitter means both *bird calls*, as well as "a short burst of inconsequential information" (Sarno, 2009a). The bird which serves as the image for the brand points to the idea of tweets as inconsequential chirpings: a typical early perspective on Twitter I. But from here we move to Twitter II, which is less about ambient friend-following and inconsequential tweets than about event-following and tweets that matter. Dorsey:

> The whole bird thing: bird chirps sound meaningless to us, but meaning is applied by other birds. The same is true of Twitter: a lot of messages can be seen as completely useless and meaningless, but it's entirely dependent on the recipient. (Sarno, 2009a)

The transformation of dispatch and scanner culture into friend status updates could inform Twitter origins stories (Fincham, 2007; Kidder, 2009). The sketches, founder interviews, as well as early Twitter taglines demonstrate imagined uses that are mundane and everyday, yet also intimate. Up until November 2009, the question Twitter posed to its users was, "What are you doing?" In a sense, the question and answers inform the discourse and early study of Twitter as mundane or banal, on the one hand, and highly personal on the other. Twitter studies and reflections by bloggers have come to describe these functions in terms such as 'ambient intimacy' and 'connected presence'.

There are also stories to be told about unintended uses, or how Twitter was adopted differently from how Dorsey envisaged it. Because friends were to follow status updates of friends, there was no organisation of topics built in; users furnished symbols that caught on, such as the hashtag. The @mention marker is a second example of user innovation. Both have been attributed to having roots in Internet Relay Chat culture (Akcora & Demirbas, 2010). From 2006 to early 2009, Twitter remained virtually the same, though it is difficult to study the evolution of its interface, for it has excluded itself from the Internet archive.

TWITTER I: TOWARDS AN AMBIENT, FRIEND-FOLLOWING MEDIUM

In much Twitter research, the software's origins as an urban, mobile lifestyle tool for friends were largely lost, in a sense, to the etymology of the service name, and the inconsequentiality more generally of tweets. Marketing firm Pear Analytic were among those to study the meaning of tweets, finding them of scant interest (Kelly, 2009). The focus turned to their banality. The BBC news headline about the study read: "Twitter tweets are 40% 'babble'" (BBC News, 2009). The firm manually categorised some 2,000 tweets over a two-week period. As became the norm in Twitter research, they conceived of a series of tweet types, beginning with the senseless: Tweets that were 'pointless babble', that is, of the "I'm eating a sandwich" type. The other categories of tweets were 'conversational', of 'pass-along value', 'self-promotional', and 'spam', where those of pass-along value (and thus of particular informational interest) were estimated at under 9 per cent of the total.

Indeed, characterising tweet types, determining how many of them are of value, and evaluating Twitter as more or less interesting content became the focus of the early studies. Java, Song, Finin, & Tseng (2007) characterised most tweets as "daily chatter", and in a sense, also showed that the other types of tweets were not built into the design and relied on subsequent innovations. "Conversations" on Twitter were beginning to take place, owing to the use by early adopters of the @ symbol for replies to a particular user (Honeycutt & Herring, 2009). "Sharing information" concerned commenting on URLs, which required shortening. The fourth category, "reporting news", also prompted user innovation; the # symbol caught on in Twitter when users reported about the San Diego fires in 2007, with #sandiegofire (Sutton et al., 2008).

How to consider Twitter as substantive (and thus worthy of serious use and study)? Or does it only offer the banal? The daily chatter discussed by Java et al. (2007) was illustrated with a tweet: "Off to get some dinner before everything shuts down" (p. 1). Dorsey himself, in the *Los Angeles Times* interview, joined the conversation about what has become known euphemistically as 'food tweets': "Why would I want to join this stupid useless thing and know what my brother's eating for lunch?" (Sarno, 2009b). Two years later he would come to defend that particular usage: a tweet about breakfast is "extremely meaningful to my mother", he said (Truong, 2011). The preponderance of "food tweets" and the more general "mindless stream" emanating from Twitter were the source of multiple news reports, in an analysis of the coverage of Twitter's first three years (Arceneaux & Schmitz Weiss, 2010). It noted how one of the more significant contributions, from trade magazine *Advertising Age*, questioned the value of tweets:

> The amazing thing is that enough people out there think this mindless stream of ephemera ('I'm eating a tangerine', 'I'm waiting for a plane', 'I want a Big Mac') is interesting enough to serve as the basis for a viable advertising platform. (Arceneaux & Schmitz Weiss, 2010, p. 1271)

What value lies in breakfast and lunch tweets? Twitter may not be about imparting great meaning and serious information, beyond breakfast and lunch eating habits. (One reads rather less about dinner tweets.) But geo-located food tweets may be of interest to those studying the geography of taste and other questions of cultural preference. For example, Edwin Chen (2012), a Twitter data scientist, studied geo-tagged tweets for regional variation in language use in the USA, comparing where people employ the words *soda*, *pop*, and *coke*. The work contributes to a series of similar studies about regional language use, yet substitutes methods of survey and interview with so-called unobtrusive data capture (McConchie, n.d.). Here Twitter becomes de-banalised, and serves as a means to study cultural conditions.

The focus on Twitter as small talk has reoriented the study of new media away from the informational. How to study the social Web, or Web 2.0, now that it has come to dominate over the info-Web, or Web 1.0? People connect socially through small talk, without passing along meaningful information, as Bronislaw Malinowski (1923) described in what he called phatic communion. People communicate in order to relate to one another, to connect, and to establish or maintain a bond. After quoting from Malinowski's classic study, Vincent Miller (2008), in his piece on the phatic culture of social media, referred to this short message: "eating a peanut butter-filled corny dog dipped in queso. mmm-mmmm breakfast" (p. 387). To Miller, Twitter (along with other social media) should be studied as a space where neither the dialogue nor the information exchange is the primary object of scrutiny.

Twitter is also not the space for the study of debate and the deliberative process, as new comment and conversational spaces online have been treated. Information that is worth passing along is in the minority on Twitter, as was found. Twitter and other social media should be analysed as spaces or platforms (the newer term) of so-called 'networked sociality' (Gillespie, 2010; Wittel, 2001). Defending a literary tradition of media, or reintroducing the old and new media divide, Miller (2008) is critical of the ascendency of Twitter and other social media, where "content is not king" (p. 395). The main purpose is keeping in touch. Doing so has its online specificity, somewhat different from the exchanges Malinowski (1923) described: "How do you do?" "Nice day today." "Ah, here you are." Rather, the connection is made both with and without words. "The point of Twitter is the maintenance of connected presence, and to sustain this presence, it is necessarily almost completely devoid of substantive content" (Miller, 2008, p. 396). Apart from small talk, one has connected presence when showing an available state, such as being visibly online in the chat feature of Facebook or on Skype. With social media, people are able to maintain what is referred to as ambient or digital intimacy (Reichelt, 2007).

Twitter is studied not only as banal and phatic. It also could be viewed as shallow media, in the sense that it favours the present, the popular, and the ephemeral. Tweets appear in reverse chronological order, so Twitter has genre characteristics of a blog (albeit with character limits: hence the term *microblog*). It is also part of the real-time Web (or Internet), in the sense that the updates continue to be refreshed. The display of messages has been described as a "stream" (Naaman, Boase, & Lai, 2010). Its privileging of the latest has only grown over time, and it has become more ephemeral. The number of days that old tweets were available was once 20, and subsequently 15 and then 7. In a sense, ephemerality was built in. For its first few years, Twitter contained no search, so stepping back in time meant manual scrolling.

Tweeting banally has its consequences, for one may be unfollowed. Indeed, how to tweet well, authentically, and attractively have become objects of study. Kwak, Chun, & Moon (2011) studied why people unfollow others on Twitter, using a South Korean data set. Those who outpoured (many tweets in a short period), or were dull and tweeted about life's trivial details tended to be unfollowed. Similar findings have been made about unfriending on Facebook, although there are other reasons to be unfriended. Those who post frequently about "unimportant", "polarising", and "inappropriate" subject matters are more likely to leave a Facebook friend than for so-called offline reasons, such as an altercation (Sibona & Walczak, 2011).

Friends on Facebook and followers on Twitter are distinctive, however. Here is how Dorsey put it in an interview, employing the old term 'watching', which was later replaced by 'following':

> The important consideration [is] that on Twitter, you're not watching the person, you're watching what they produce. It's not a social network, so there's no real social pressure inherent in having to call them a "friend" or having to call them a relative, because you're not dealing with them personally, you're dealing with what they've put out there. (Sarno, 2009a)

Indeed, in another study of what they described as the entire Twittersphere of some 41 million user profiles and 106 million tweets in 2009, Kwak, Lee, Park, & Moon (2010) found that Twitter is not particularly 'social'. That is, it does not have the characteristics of a social network, for among other reasons there is low reciprocity in following. This lack of sociality on Twitter prompted the researchers to characterise it as news media, where users broadcast or narrowcast to followers.

Perhaps rather than social circles, Twitter users have audiences. Marwick and boyd (2011) complicated the idea of Twitter as only banal, phatic, or shallow media by introducing the notion of the audience of a user's tweets, and referring to the phenomenon of micro-celebrity. Follower numbers are displayed prominently on one's profile page. Large numbers are status symbols, and one's influence or "klout" (as a popular, third-party metric is called) is measurable. Like A-list bloggers, there are A-list Twitter users, but also influencers on smaller scales. Marwick and boyd (2011) discussed the notion of the networked audience, which has elements of the writer's audience (readers, in the minds of the writers) and the broadcast audience (quantities of viewers, for advertising), but new traits, too. For example, Twitter users may ask their followers questions. They may read their audience's inter-exchanges in the web of followers around theirs.

TWITTER II: TOWARDS A NEWS MEDIUM FOR EVENT-FOLLOWING

In November 2009, Twitter's tagline changed. The question Twitter users were asked had been "What are you doing?" It became "What's happening?" To David Crystal, the linguist and author of *Txting: The gr8 db8*, the change signified a move from an ego to a reporting machine (Tate, 2009). Twitter studies were still focussed on the ego machine. Indeed, it has been found that "80% of Twitter Users Are All About Me", as the *Mashable* headline read just prior to the tagline change (Van Grove, 2009). In studying 350 users, Naaman et al. (2010) made more fine-grained the scholarly characterisations of tweets, and many of the nine types they derived concentrate on what one could call 'me-tweets'. In their tweet type classification, note that the banal has been subdivided into many kinds, and that there is really only one tweet type—"information sharing"—that could be considered 'news'.

Twitter's tagline change could be interpreted as an internal shift, as well as a nudge for both users and researchers to consider information sharing tweets. Another co-founder of Twitter, Biz Stone, discussed Twitter's new purpose when the trending topics feature was introduced in April 2009. It is a state-of-affairs machine, or "discovery engine for finding out what is happening right now" (Stone, 2009). Dorsey, whose vision for Twitter usage always appeared to be more in the area of ambient intimacy, did aver that the service did "well at: natural disasters, man-made disasters, events, conferences, presidential elections", or what he calls "massively shared experiences" (Sarno, 2009b). For it to be a machine for media events (as massively shared experiences are sometimes called), and for it to take that role from television, an argument should be made about its significance in a specific event.

The South by Southwest conferences in 2007 and 2008 in Austin, Texas, established Twitter as an event backchannel, a kind of gossip machine for commenting on what one thinks of speakers' talks (McCarthy & Calore, 2008). Twitter use at conferences would be standardised with a conference hashtag, which attendees would use and watch; speakers would be ranked according to mention frequency, showing how each trends (Ebner & Reinhardt, 2009; Ebner et al., 2010). Early Twitter studies often listed the events when Twitter was considered impactful: the San Diego fires (as mentioned); the Sichuan earthquake in May 2008; the Mumbai terrorist attacks in November 2008; James Karl Buck's arrest in Egypt in 2008; and the Hudson River landing of a US Airways jet in January 2009, a story which broke on Twitter (Arceneaux & Schmitz Weiss, 2010).

Andrew Sullivan, the American A-list political blogger, made an allusion to Twitter as a revolutionary machine, when he wrote in reference to the street demonstrations in Iran after the presidential elections of June 2009 that "The Revolution Will Be Twittered", as opposed to televised (Sullivan, 2009). That headline appeared on 13 June, a day after the elections, and on 15 June, Ari Berman (2009), blogging at the revered, left-of-centre publication *The Nation*, entitled his posting, "Iran's Twitter Revolution". Evgeny Morozov, at the time working on his book *The Net Delusion: The Dark Side of Internet Freedom*, strove to debunk the idea of Twitter as revolutionary machine (Morozov, 2009, 2010). His were arguments informed by the scholarly study of the history of technology, and in particular the critique of viewing machines as driving history. Leo Marx and Merrit Roe Smith (1994) summed up such a line of thought with examples: "'the automobile created suburbia.'. . . 'The mechanical cotton-picker set off the migration of southern black farm workers to northern cities.' 'The Pill produced the sexual revolution'" (p. xi).

Discursively, Twitter was being fit into a lineage of revolutionary technologies, like the Xerox photocopier and the fax machine from Soviet times, or the mobile phone and text messaging in the colour revolutions. Morozov critiqued Clay Shirky, whose *Here Comes Everybody* engenders optimism about social media as a democratising force, in a sense being a version of machines driving history. Morozov: "'Tehran's "collective action cascade" of 2009 feels like Leipzig 1989,' tweeted Clay Shirky, new media's favorite cheerleader" (2009, pp. 10–11). Shirky and Morozov would come to debate one another, and in the exchange, Morozov (2010) introduced a phenomenon accounting for why Twitter cannot drive the revolution: "authoritarian governments—those in Belarus, China and Moldova are good examples—are increasingly relying on what is known as 'event-based internet filtering,' whereby they turn off mobile coverage". Jack

Dorsey described Twitter as a service doing well with events, where "a lot of [the] people are not sitting in front of a laptop screen—they're typing from their phone" (Sarno, 2009b). But that is when it is not shut down. Morozov's arguments are also informed by new media user studies, and especially the myth of user-generated content, where very few are responsible for the great majority of content (van Dijck, 2009). In Iran, those very few responsible for Twitter content Morozov finds nearly irrelevant:

> Pro-Western, technology-friendly and iPod-carrying young people . . . are the . . . most frequent users of Twitter. They are a tiny and, most important, extremely untypical segment of the Iranian population (the number of Twitter users in Iran—a country of more than seventy million people—was estimated at less than twenty thousand before the protests). Whatever they do with Twitter may have little relevance to the rest of the country, including the masses marching in the streets". (Morozov, 2009, p. 12)

Morozov also did not appreciate what Sullivan and Berman saw (or digitally witnessed) in Twitter (Morozov, 2009). In a variation on Andrew Keen's argumentation about the decline of quality in journalism and in letters more generally because of the Web, Morozov reported that the traditional media are not in Iran, and can no longer afford to report there. Instead, we must rely on nameless bloggers and other online reporters. So, "what Andrew Sullivan is 'seeing' might be radically different from what is actually happening" (Morozov, 2009, p. 11). No longer only the ambient intimacy machine, Twitter was becoming a news source, replacing old media (however regrettably) when information was shared from the ground. Berman (2009) wrote: "some absolutely riveting and thrilling reporting has been done over Twitter".

Refashioning Twitter as new object of study (what I refer to as 'Twitter II'), researchers took up the project of de-banalising Twitter by identifying new tweet types, and a new purpose, similar to Dorsey's discussion of where and when Twitter has done well (events, disasters, and elections). Tweet characterisation would become rather different from making distinctions about the multiple forms of banality (plus information sharing). Researchers also used the markers in the tweets to create significant collections (hashtags), and to order them (retweets) so as to tell the story of the events on the ground and online. Tweet collections by researchers also caught the attention of Twitter, Inc., which at once banned their sharing, and announced that all tweets would be made available in an archive at the Library of Congress. I return to studying Twitter as archived object in the conclusion.

In Twitter Studies II, the research framework would move away from the implications of ambient intimacy to the value of accounts from the ground and from online for event-following. In the critical study of Twitter as quality source, there are the questions of accuracy and professionalism in reporting of which Evgeny Morozov wrote. The larger question, more in the realm of political science, is also the issue of the significance of Twitter for the so-called revolution. The call for revolution, or at least for shouting from the rooftops, was reported by Andrew Sullivan in his 13 June 2009 blog posting: "ALL internet & mobile networks are cut. We ask everyone in Tehran to go onto their rooftops and shout ALAHO AKBAR in protest #IranElection" (Sullivan, 2009). The revolutionary tweet, as it might be called, was posted by @MirHossein Mousavi; Mousavi lost the election to the standing president, Mahmoud Ahmadinejad, a day earlier. Street protests erupted. Mousavi's Twitter account posted the news that the Internet and mobile coverage were down, and that people should take to the rooftops. Here we note Morozov's admonition about relying on Twitter, social media, and the Internet for uprisings. Note, too, the hashtag contained in the tweet, #IranElection. It became a means to follow the action, and also one to demarcate a set of tweets in order to study the events, and the content of the 'Twitter revolution', or at least the Iran election crisis both online and on the ground. The first study of Iran election-related tweets appeared some two weeks after the election, on 26 June 2009, subtitled "The First Eighteen Days" (Hwang, 2009). It criticised the use of the term 'Twitter revolution', joining many others whom the authors list, including Evgeny Morozov and Clay Shirky. It also outlined a technique to make a tweet collection, using multiple hashtags related to #iranelection, or those other hashtags that appear in tweets containing #iranelection. Further, it adds to the data the results of key word queries in Twitter search (then a new feature). Relying on a single hashtag, #iranelection, misses much of the discourse:

> The number of tweets using hashtags other than #iranelection amount to 1,166,765 messages, or 57.6% of the total set accumulated in our study (a significant portion of the discourse that other studies ignore when focusing solely on #iranelection). (Hwang, 2009, p. 3)

Generally, the researchers concentrated on the characteristics not of the revolution but of the conversation, as they call it, with a description of the users and their relative contributions, including activity measures. They also discuss influential users and contents retweeted most frequently, pointing to a method to order tweets for the purposes of evaluating Twitter users' contributions to event-following.

How to employ retweets in order to debanalise Twitter? Indeed, two research projects, at Rensselaer Polytechnic Institute (RPI) and the Digital Methods

Initiative (DMI) at the University of Amsterdam (where I contributed), examined in some detail the tweets which used the hashtag #iranelection (Gaffney, 2010; Rogers, Jansen, Stevenson, & Weltevrede, 2009). Both examined how to make use of "retweets of interest" (Gaffney, 2010, p. 6). As Berman (2009) pointed out in his "Twitter Revolution" blog posting on 15 June 2009, the accounts of events from Twitter, however sourced, were compelling, and prompted the DMI project to consider how to transform the 'retweets of interest' into a story of the events of June 2009 (Rogers et al., 2009). Can Twitter be made into a storytelling machine that recounts the events on the ground and on Twitter? The result of our efforts, "For the ppl of Iran—#iranelection RT" is a collection of some 650,000 tweets containing the hashtag #iranelection, from 10 to 30 June 2009. The top three retweets per day were captured and ordered by retweet count (see Figure 2). All sets of retweets were placed in chronological order, as opposed to the reverse chronological order of Twitter and blogs more generally. The story of the events unfolds through retweets over the course of the twenty days: Mir-Hossein Mousavi holds an emergency press conference; the voter turn-out is 80%; Mousavi's website and Facebook page are blocked; police are using pepper spray; Mousavi is under house arrest, and declares he is prepared for martyrdom; Neda is dead; there is a riot in Baharestan Square; Bon Jovi sings "Stand by Me" (in Farsi) in support; Ahmadinejad is confirmed the winner, and a last tweet in the collection reads "light a candle for those who have died". In the retweets, one takes note of many of the main storylines discussed above that mitigate Twitter's role in the 'revolution', and detail is offered about how the users reacted. There is the suspicion of infiltration and the call for an act of solidarity to change one's user location to Iran. The Internet is filtered,

Figure 2: Segment of 'For the ppl of Iran—#iranelection RT.' Top 3 retweets per day, of tweets with #iranelection hashtag, 10–30 June 2009, in chronological order. Source: Rettiwt.net, Digital Methods Initiative, Amsterdam, 2009

and subsequently proxies and anonymisers are offered. There is violence in the streets, and first aid as well as digital witnessing pointers are given. In a sense, it is a space which aids events—as other researchers have found in their studies of natural disasters (Bruns & Liang, 2012).

CONCLUSION—TWITTER III: TOWARDS (ARCHIVED) DATA SET AND ANTICIPATORY MEDIUM

Once considered a source of "pointless babble" about one's lunch and a backchannel for interacting at an event (while speakers held presentations and listeners remarked), Twitter increasingly has come to be studied as an emergency communication channel in times of disasters and other major events, as well as an event-following and aid machine for revolution and uprising in the Middle East and beyond. More recently it has settled into a data set, from which researchers have made collections, and one to be archived and made available by the U.S. Library of Congress. Twitter III is thus being studied as data, which requires both contractual access as well as technical infrastructure to take in the tweets, store them, and analyse them. Twitter has an array of access points (so-called firehoses and sprinklers from its own API), intermediary commercial collection vessels (Gnip and DataSift), and analytical tools which are often used for Web data analysis more generally (such as the network visualisation software, Gephi). Twitter is particularly attractive for research, owing to the relative ease with which tweets are gathered and collections are made, as well as the in-built means of analysis, including retweets for significant tweets, hashtags for subject matter categorisation, @replies as well as followers-followees for network analysis, and shortened URLs for reference analysis. Given its character limit and the fact that each tweet in a collection is relatively the same length, it also lends itself well to textual analysis, including co-word analysis (Marres & Weltevrede, 2013). Additional avenues of Twitter analysis have recently opened that take up the invitation made by Biz Stone (and Twitter more generally) to follow meaningfully what is happening, for example by making a list of subject matter or domain knowledge experts (or concatenating and/or triangulating those of others) so as to capture their tweets, and study the evolution of an issue area according to "professional communities of practice" (Turoff & Hiltz, 2009).

There are issues with Twitter as a data provision machine. Twitter was conceived (by Dorsey and associates) as ephemeral, whose users, if we take Dorsey as an avant-garde case, are not thought to be "obsessive about going all the way back in time and catching every single message that people have updated about" (Sarno, 2009b). Owing to issues of scale as well as resources, there are limited quantities of tweets available per user, per hashtag, etc., without special access privileges. As with other Internet or new media data sets, one is often required to be employed by or within the walls of the corporate research lab in order to have access to larger data sets, including longitudinal ones. For example, it was the Twitter data scientist Edwin Chen who conducted the study on regional variation in the use of words such as *soda*, *pop*, and *coke*. As boyd and Crawford (2011) pointed out in their influential paper concerned with big data science,

> During his keynote talk at the International Conference on Weblogs and Social Media (ICWSM) in Barcelona on 19 July 2011, Jimmy Lin—a researcher at Twitter—discouraged researchers from pursuing lines of inquiry that internal Twitter researchers could do better given their preferential access to Twitter data. (p. 13)

Moreover, Twitter, Inc. trades in the so-called data market, and its evolving terms of service and dealings with third parties are increasingly distinguishing between 'good' Twitter data and the black market for data, which could be construed as the research collections made by other means and shared (Puschmann & Burgess, 2013).

Twitter, however, is to be archived by the Library of Congress, and made available for research purposes. As is now customary in Twitter studies, someone brought up the value of sandwich tweets: the first comment posted on the FAQ page of the Library of Congress's Twitter project reads, sarcastically, "it's critical the future generations know what flavor burrito I had for lunch" (Raymond, 2010). Of interest here are the implications of studying Twitter, once it becomes an archived object. The archived tweets under study will be at least six months old, which creates a gap in longitudinal work between the number of days' worth of tweets available currently via Twitter, and those aged six months or more. Of greater interest, perhaps, will be the difference in query and storage environments between an online Twitter (and its hoses and sprinklers), and the archived Twitter. The Library of Congress has already indicated that Twitter the archived object will no longer be Twitter the online service. As the Library's 2013 White Paper on the Twitter Archive put it, "currently, executing a single search of just the fixed 2006–2010 archive on the Library's systems could take 24 hours" (Library of Congress, 2013, p. 4). Gnip, the social media data supplier and partner with Twitter and the Library of Congress in creating the tweet delivery software for the archive, is separately selling historical tweets, from the very first (by Jack Dorsey) on 21 March 2006 onwards. The Historical PowerTrack API documentation provides insights into Twitter as archived object, and the types of research which are precluded, given certain characteristics of the data. For example, geo-location is not available for tweets prior to 2011, and all tweets older than those have the user's profile information from September 2011 (Gnip, 2013). One avenue of inquiry for Twitter studies is thus the difference between the Library of Congress's services for academic researchers and those of Gnip and others.

Sifting through the enquiries made by researchers to the Library of Congress also provides an opportunity to reflect further on the purpose of studying Twitter. Many include studying what Dorsey described as when Twitter does well: natural and man-made disasters as well as elections. Other proposals highlighted by the Library of Congress are the tracking of flu epidemics on Twitter (recalling Google Flu Trends) and stock market prediction, testing Twitter's capacity as anticipatory medium—which is perhaps a new calling for the platform (Meier, 2013).

ACKNOWLEDGMENTS

Thanks to Anne Helmond and Natalia Sanchez for comments on an earlier version of this piece, presented at the Digital Methods Winter School, Media Studies, University of Amsterdam, January 2013. A longer version is published in the *Proceedings of the WebSci'13 Conference*, 2–4 May, Paris. New York, NY: ACM.

REFERENCES

Akcora, C. G., & Demirbas, M. (2010). *Twitter: Roots, influence, applications* (Technical report). Department of Computing Science and Engineering, State University of New York at Buffalo, NY. Retrieved from http://www.cse.buffalo.edu/tech-reports/2010-03.pdf

Arceneaux, N., & Schmitz Weiss, A. (2010). Seems stupid until you try it: Press coverage of Twitter, 2006–9. *New Media & Society*, 12(8), 1262–1279.

BBC News. (2009, 19 Aug.). Twitter tweets are 40% 'babble.' Retrieved from http://news.bbc. co.uk/2/hi/technology/8204842.stm

Berman, A. (2009, 15 June). Iran's Twitter revolution. Retrieved from http://www.thenation. com/blog/irans-twitter-revolution

boyd, d., & Crawford, K. (2011, September). *Six provocations for Big Data*. Paper presented at the Oxford Internet Institute's 'A Decade in Internet Time: Symposium on the Dynamics of the Internet and Society', Oxford, UK. Retrieved from http://papers.ssrn.com/sol3/papers.cfm?abstract_id=1926431

Bruns, A., & Liang, Y. E., (2012). Tools and methods for capturing Twitter data during natural disasters. *First Monday*, *17*(4). Retrieved from http://firstmonday.org/htbin/cgiwrap/bin/ojs/index.phpl/fm/article/viewArticle/3937/3193

Chen, E. (2012, 6 July). Soda vs. pop with Twitter. Retrieved from http://blog.echen.me/2012/07/06/soda-vs-pop-with-twitter/

Dorsey, J. (2006). twttr sketch [Flickr photo]. Retrieved from http://www.flickr.com/photos/jackdorsey/182613360/

Ebner, M., Mühlburger, H., Schaffert, S., Schiefner, S., Reinhardt, W., & Wheeler, S. (2010). Getting granular on Twitter: Tweets from a conference and their limited usefulness for non-participants. In N. Reynolds & M. Turcsányi-Szabó (Eds.), *Key competencies in the knowledge society* (pp. 102–113). Berlin, Germany: Springer.

Ebner, M., & Reinhardt, W. (2009). Social networking in scientific conferences—Twitter as tool for strengthen a scientific community. In *Proceedings of the 1st International Workshop on Science 2.0 for TEL*. Retrieved from http://de.scribd.com/doc/20363438/Social-networking-in-scientific-conferences-%E2%80%93-Twitter-as-tool-for-strengthen-a-scientific-community

Fincham, B. (2007). Generally speaking people are in it for the cycling and the beer: Bicycle couriers, subculture and enjoyment. *The Sociological Review*, *55*(2), 189–202.

Gaffney, D. (2010). #iranElection: Quantifying online activism. *Proceedings of the WebSci10: Extending the Frontiers of Society On-Line*. 26–27 Apr. 2010, Raleigh, NC. Retrieved from http://journal.webscience.org/295/

Gannes, L. (2011, 1 June). Jack Dorsey is 'The James Franco of the Internet.' Retrieved from http://allthingsd.com/20110601/jack-dorsey-of-square-and-twitter-live-at-d9/

Gillespie, T. (2010). The politics of 'platforms.' *New Media & Society*, *12*(3), 347–364.

GNIP. (2013). API documentation—Historical. Historical PowerTrack overview. *GNIP: Support*. Retrieved from http://support.gnip.com/customer/portal/articles/745561-historical-powertrack-overview

Honeycutt, C., & Herring, S. (2009). Beyond microblogging: Conversation and collaboration via Twitter. In *Proceedings of HICSS '09*. doi: 10.1109/HICSS.2009.89

Humphreys, L. (2010). Historicizing microblogging, In *Proceedings of CHI 2010* (pp. 10–15). April 2010. Atlanta, GA,

Hwang, T. (2009). The Iran election on Twitter: The first eighteen days. *The Web Ecology Project*. Retrieved from http://www.webecologyproject.org/wp-content/uploads/2009/08/WEP-twitterFINAL.pdf

Java, A., Song, X., Finin, T., & Tseng, B. (2007, August). *Why we Twitter: Understanding microblogging usage and communities*. Paper presented at the Joint 9th WEBKDD and 1st SNA-KDD Workshop, San Jose, CA. Retrieved from http://aisl.umbc.edu/resources/369.pdf

Kelly, R. (2009). *Twitter study* (White paper). San Antonio, TX: Pear Analytics. Retrieved from http://www.pearanalytics.com/wp-content/uploads/2012/12/Twitter-Study-August-2009.pdf

Kenrick, C. (2011, February 10). Twitter founder tweets his breakfast menu. *Palo Alto Online*. Retrieved from http://www.paloaltoonline.com/news/show_story.php?id=20012

Kidder, J. L. (2009). Appropriating the city: Space, theory, and bike messengers. *Theory and Society*, *38*(3), 307–328.

Kwak, H., Chun, H., & Moon, S. (2011). Fragile online relationship: A first look at unfollow dynamics in Twitter. In *Proceedings of CHI 2011*. 7–12 May. Vancouver, British Columbia, Canada. Retrieved from http://an.kaist.ac.kr/~haewoon/papers/2011-chi-unfollow.pdf

Kwak, H., Lee, C., Park, H., & Moon, S. (2010). What is Twitter, a social network or a news media? In *Proceedings of WWW 2010*. 26–30 Apr. 2010. Raleigh, NC. Retrieved from http://product.ubion.co.kr/upload20120220142222731/ccres00056/db/_2250_1/embedded/2010-www-twitter.pdf

Library of Congress. (2013, January). *Update on the Twitter Archive at the Library of Congress* (White paper). Retrieved from http://www.loc.gov/today/pr/2013/files/twitter_report_2013jan.pdf

Malinowski, B. (1923). The problem of meaning in primitive languages. In C. K. Ogden & I. A. Richards (Eds.), *The meaning of meaning* (pp. 146–152). London, UK: Routledge & Kegan Paul.

Marres, N., & Weltevrede, E. (2013). Scraping the social: Issues in real-time social research. *Journal of Cultural Economy, 6*(3), 313-335.

Marwick, A., & boyd, d. (2011). I tweet honestly, I tweet passionately: Twitter users, context collapse, and the imagined audience. *New Media & Society, 13*(1), 114–133.

Marx, L., & Roe Smith, M. (1994). Introduction. In L. Marx & M. Roe Smith (Eds.), *Does technology drive history? The dilemma of technological determinism* (pp. ix–xv). Cambridge, MA: MIT Press.

McCarthy, M., & Calore, M. (2008, 9 Mar.). SXSW: Zuckerberg keynote descends into chaos as audience takes over. *Wired*. Retrieved from http://www.wired.com/underwire/2008/03/sxsw-mark-zucke/

McConchie, A. (n.d.). Pop vs. soda. Retrieved from http://www.popvssoda.com

Meier, P. (2013, 5 Feb.). Social media: Pulse of the planet? *Explorers Journal. National Geographic*. Retrieved from http://newswatch.nationalgeographic.com/2013/02/05/social-media-pulse-of-the-planet/

Miller, V. (2008). New media, networking and phatic culture. *Convergence, 14*(4), 387–400.

Morozov, E. (2009, 18 Nov.). How dictators watch us on the Web. *Prospect*. Retrieved from http://www.prospectmagazine.co.uk/magazine/how-dictators-watch-us-on-the-web/

Morozov, E. (2010, 5 Jan.). Why the Internet is failing Iran's activists. *Prospect*. Retrieved from http://www.prospectmagazine.co.uk/magazine/why-the-internet-is-failing-irans-activists/

Naaman, M., Boase, J., & Lai, C. H. (2010, Feb.). *Is it really about me? Message content in social awareness streams*. Paper presented at CSCW 2010 conference, Savannah, GA. Retrieved from http://infolab.stanford.edu/~mor/research/naamanCSCW10.pdf

Puschmann, C., & Burgess, J. (2013). The politics of Twitter data (*HIIG Discussion Paper Series* No. 2013-01). Berlin, Germany: Humboldt Institute for Internet and Society.

Raymond, M. (2010, 28 Apr.). The library and Twitter: An FAQ. Retrieved from http://blogs.loc.gov/loc/2010/04/the-library-and-twitter-an-faq/

Reichelt, L. (2007, March 1). Ambient intimacy. Retrieved from http://www.disambiguity.com/ambient-intimacy/

Rogers, R., Jansen, F., Stevenson, M., & Weltevrede, E. (2009). *Mapping democracy*. Paper presented at Global Information Society Watch 2009, Association for Progressive Communications and Hivos. Retrieved from http://www.giswatch.org/sites/default/files/mappingdemocracy.pdf

Sarno, D. (2009a, 18 Feb.). Twitter creator Jack Dorsey illuminates the site's founding document. Part I. *Los Angeles Times*. Retrieved from http://latimesblogs.latimes.com/technology/2009/02/twitter-creator.html

Sarno, D. (2009b, 19 Feb.). Jack Dorsey on the Twitter ecosystem, journalism and how to reduce reply spam. Part II. *Los Angeles Times*. Retrieved from http://latimesblogs.latimes.com/technology/2009/02/jack-dorsey-on.html

Shirky, C. (2009, 11 Feb.). The Net advantage. *Prospect*. Retrieved from http://www.prospect-magazine.co.uk/magazine/the-net-advantage/

Sibona, C., & Walczak, S. (2011, Jan.). *Unfriending on Facebook: Friend request and online/offline behavior analysis*. Paper presented at 44th Hawaii International Conference on System Sciences (HICSS), Kauai, HI. Retrieved from http://www.computer.org/csdl/proceedings/hicss/2011/4282/00/07-05-17.pdf

Stone, B. (2009, April 30). Twitter search for everyone! Retrieved from http://blog.twitter.com/2009/04/twitter-search-for-everyone.html

Sullivan, A. (2009,13 June). The revolution will be twittered. *The Atlantic*. Retrieved from http://www.theatlantic.com/daily-dish/archive/2009/06/the-revolution-will-be-twittered/200478/

Sutton, J., Leysia, P. & Shklovski, I. (2007). Backchannels on the front lines: Emergent uses of social media in the 2007 southern California wildfires, *Proceedings of the 5th International ISCRAM Conference*, May 2008, Washington, DC.

Tate, R. (2009, 19 Nov.). Twitter's new prompt: A linguist weighs in. *Gawker*. Retrieved from http://gawker.com/5408768/

Truong, S. (2011). Jack Dorsey on the history of Twitter and Square. Retrieved from http://tusb.stanford.edu/2011/05/jack-dorsey-on-the-history-of-twitter-and-square.html

Turoff, M., & Hiltz, S. R. (2009). The future of professional communities of practice. In C. Weinhardt, S. Luckner, & J. Stößer (Eds.), *Designing e-business systems: Markets, services, and networks* (pp. 144–158). Berlin, Germany: Springer.

van Dijck, J. (2009). Users like you? Theorizing agency in user-generated content. *Media Culture and Society, 31*(1), 41–58.

Van Grove, J. (2009, 29 Sep.). STUDY: 80% of Twitter users are all about me. *Mashable*. Retrieved from http://mashable.com/2009/09/29/meformers/

Wittel, A. (2001). Toward a network sociality. *Theory, Culture & Society, 18*(6), 51–76.

Acknowledgments

The production of this volume was generously supported by the Strategic Research Fund of Heinrich Heine University, Düsseldorf; and the ARC Centre of Excellence for Creative Industries and Innovation, Brisbane. Our warmest gratitude to Nicki Hall for her proactive, positive, and painstaking contributions to this complex but rapidly executed project; as well as to Steve Jones and Mary Savigar at Peter Lang, not only for supporting the project, but also for working with us to make it as accessible as possible. We are profoundly grateful to our contributors for sharing their knowledge with such enthusiasm and eloquence, and look forward to their further research in this area with eager anticipation.

Twitter and Society: An Introduction

INTRO-
DUCTION

Katrin Weller, Axel Bruns, Jean Burgess, Merja Mahrt, & Cornelius Puschmann

 welcome to #twitsocbook, welcome to Twitter research! (Pls RT.)

Since its launch in 2006, Twitter has turned from a niche service to a mass phenomenon. By the beginning of 2013, the platform claims to have more than 200 million active users, who "post over 400 million tweets per day" (Twitter, 2013). Its success is spreading globally; Twitter is now available in 33 different languages, and has significantly increased its support for languages that use non-Latin character sets. While Twitter, Inc. has occasionally changed the appearance of the service and added new features—often in reaction to users' developing their own conventions, such as adding '#' in front of important keywords to tag them—the basic idea behind the service has stayed the same: users may post short messages (tweets) of up to 140 characters and follow the updates posted by other users. This leads to the formation of complex follower networks with unidirectional as well as bidirectional connections between individuals, but also between media outlets, NGOs, and other organisations. While originally 'microblogs' were perceived as a new genre of online communication, of which Twitter was just one exemplar, the platform has become synonymous with microblogging in most countries. A notable exception is Sina Weibo, popular in China where Twitter is not available. Other similar platforms have been shut down (e.g., Jaiku), or are being used in slightly different ways (e.g., Tumblr), thus making Twitter a unique service within the social media landscape.

In addition to interpersonal communication, Twitter is increasingly used as a source of real-time information and a place for debate in news, politics, business, and entertainment. Televised sports events such as the FIFA World Cup or the NBA Finals cause massive real-time spikes in global Twitter activity; other entertainment news and events also result in particularly high tweet volumes, be it the death of Michael Jackson, the annual Academy Awards, or the royal wedding of Prince William and Kate Middleton in 2011. Public figures and celebrities from the Pope to Lady Gaga attract enormous numbers of followers, and a photo of Barack and Michelle Obama, posted immediately after Obama's re-election as President of the United States in November 2012, rapidly became the single most retweeted message in the history of Twitter. Disasters such as Hurricane Sandy, and tragedies like the shooting spree of a gunman at the Sandy Hook Elementary School in Connecticut (both in the autumn of 2012) show their immediate aftereffects on the platform, as users report their experiences and search for information, often as events are unfolding—a dynamic that makes Twitter seemingly irresistible to the mass media. Such moments demonstrate how deeply embedded the service has become into the media ecology, and, arguably, into the everyday lives of its users around the world. Increasingly, when noteworthy events occur—both on a global and a local level—there will be Twitter users who share the news.

Beyond the spectacle of major news events, Twitter remains a space for mundane expressiveness and interaction: millions of private users chat with their friends and share photos or URLs via Twitter at any one point, using the service as a journal of their thoughts and everyday activities. This is why Twitter has been bluntly criticised at times for consisting largely of 'pointless babble', 'useless information', or 'phatic communication', but such criticism is simplistic. Rather, the highly personal use by each user as a tool for outreach, spreading information, or connecting to friends is at the very heart of Twitter's utility for individuals and organisations alike, and indeed underpins its very success as a platform for global news media and public communication.

Twitter's dominant uses and norms have been co-created over time, not only by the company, Twitter, Inc., but also by third-party developers and users themselves. Users shape the service through their practices of use, and these activities have led to new forms of communication and new phenomena in participatory culture, for example in the form of Twitter-specific communicative trends and memes. It is therefore as important to investigate Twitter users' everyday activities and their perceptions of publicity, privacy, intimacy, and friendship as they are experienced through and reconfigured by the platform, as it is to study the use of Twitter in the context of major societal themes and events. With this volume, we aim to present both a broad and a detailed picture of the many specific practices through which Twitter is located *in* society, in order to explore the intersections between Twitter *and* society. This not only provides a fascinating insight into how this important social network itself is being used, but also continues a tradition of platform-specific studies—covering blogs (Bruns & Jacobs, 2006), social networking sites (boyd & Ellison, 2007), virtual worlds (Meadows, 2007), search engines (Halavais, 2009; Lewandowski, 2012), Wikipedia (Lih, 2009; Reagle, 2010), YouTube (Burgess & Green, 2009), and many others—which document the social co-construction of new media technologies in the often conflicted interplay between platform users, platform providers, and other stakeholders.

The substantial amount of content generated and shared by Twitter users, from individuals to institutions, also opens up exciting new research possibilities across a variety of disciplines, including media and communication studies, linguistics, sociology, psychology, political science, information and computer science, education, and economics. There remains a significant need for the further development of innovative methods and approaches which are able to deal with such new sources of research data, and for the training of a new generation of scholars who are deeply familiar with such methodological frameworks.

Large datasets can be retrieved from the Twitter Application Programming Interface (API), and can subsequently be mined with a range of specialised tools (programming languages, statistics packages, network analysis frameworks, text and data mining tools). API-based access to Twitter data has contributed to the emergence of a variety of tools and services that promise to measure and compare impact, influence, and audience reach on Twitter—which in turn leads to a growing interest in strategies for maximising such 'impact', and a number of books promising swift success for corporate marketing and political campaigns. However, reliably measuring activity or popularity, or quantifying any other aspects of social media use, is far from trivial, and current approaches are usually neither standardised nor independently verifiable, acting instead as black box analytics frameworks whose outcomes the researcher is asked to trust with blind faith. Several chapters in this collection seek to remedy this situation by establishing common frameworks for Twitter analytics beyond merely quantifying attention, and thereby initiating a conversation about methods in researching Twitter.

Furthermore, the opportunities for advanced Twitter analytics are matched by challenges surrounding the long-term availability of data, research ethics, the interpretation of user-generated information, and the relation of qualitative and quantitative, as well as user-based and content-based research approaches. Such challenges extend well beyond the study of Twitter itself, and are instead shared with the wider field of 'big data' research in the digital humanities which is currently emerging. If the current "computational turn" (Berry, 2011) in our research is to result in what Richard Rogers (2009) has described as "natively digital" methodologies or in "computational social science" (Lazer et al., 2009), then a significant amount of further thought must go into the conceptual, methodological, and ethical frameworks which we apply to such work. In the foreword to the present volume, Richard Rogers introduces the key characteristics of Twitter, its history and usage, and provides a sketch of how Twitter research can keep up with the platform's impressive journey from a frowned-upon niche medium to a global information hub. His foreword presents the many challenges which Twitter research must rise to meet. In their contributions to the main body of the book, our authors respond to these challenges by sharing the diverse insights gained through their own research, across a wide range of disciplines, perspectives, and methodologies, and by raising further questions for future Twitter research.

This collection is divided into four thematic sections. Part I, "Concepts and Methods", presents a selection of theoretical frameworks for the study of Twitter, followed by a range of practical approaches for investigating the platform. It opens with Jan-Hinrik Schmidt's introduction of the concept of "personal publics", which describes the multitude of overlapping, hybrid, public/private spaces that are constituted by each individual user's account and its network of followers. The chapter raises important questions for our understanding of tweeting as a form of communication which can be at once intensely personal and highly public. Axel Bruns and Hallvard Moe follow suit with a reflection on the different layers of communication on the platform, supported by a range of sociotechnical constructs ranging from @replies to hashtags. Their contribution serves as a reminder that Twitter can be used strategically to achieve different levels of publicness and publicity, and provides a framework for defining these levels. Alexander Halavais further explores the social and technological conventions which have given rise to the different formations of interpersonal, publicly personal, or all-out public communication which are possible on Twitter, and traces the processes of co-evolution of the platform and its functionality as they are driven by corporate as well as user activities and interventions. Finally, Cornelius Puschmann and Jean Burgess complete the "Concepts" section by offering a critical reflection on the politics of Twitter data, exploring both Twitter's data policies and the politics of utilising such proprietary and increasingly restricted data sources in research projects. Together, these four chapters form the cornerstones for the conceptualisation of Twitter as a hybrid social network and communications platform on which this collection is based.

"Methods" introduces a number of crucial practical approaches to the study of Twitter, from both qualitative and quantitative perspectives. Devin Gaffney and Cornelius Puschmann discuss the Twitter API, and outline a number of key tools for gathering and processing API data. They also consider the limitations and challenges of API-based work. Axel Bruns and Stefan Stieglitz present a range of key metrics for the quantitative analysis of Twitter activity, and demonstrate their use in practice; these metrics provide a standardised basis for Twitter analytics which improves the reliability and reproducibility of Twitter research. Mike Thelwall outlines the use of time-series-based sentiment analysis for corpora of Twitter data, in order to explore and document the mood of tweeting activity in a given dataset at any one point. Jessica Einspänner, Mark Dang-Anh, and Caja Thimm broaden the perspective from sentiment to computer-assisted content analysis, outlining how automated, semi-automated, and manual analysis approaches may be combined to develop a detailed perspective of the communicative activities captured in Twitter datasets. Alice E. Marwick offers an alternative and strongly qualitative approach to the study of Twitter, employing interviews, ethnographic methods, and close reading of tweeted interactions in order to develop a very detailed, fine-grained picture of who uses Twitter, and of how they use it. Michael Beurskens, finally, considers the legal frameworks within which Twitter researchers operate as they draw on the tweets of a wide range of users, access them through proprietary APIs, and collect them in large corpora. Often overlooked in the day-to-day processes of data analysis, such legal questions are crucial for assuring researchers and protecting their research subjects.

Part II, "Perspectives and Practices", presents a set of thematic and conceptual approaches to the study of Twitter, demonstrating the diverse societal contexts in which Twitter has found application. "Perspectives" explores a range of common aspects of Twitter activity. It begins with Alex Leavitt's chapter on the origins of Twitter memes. Memes as a means to contextualise and label information and participate in discussions have become a substantial part of communicating on Twitter, mainly in the form of hashtags. Leavitt describes several popular memes and their contexts, pointing out that they are both influenced by the users' intentions and the technical environment provided by Twitter, which highlights currently trending topics and may thus enforce already existing memes. But not only topics and hashtags can be used to contextualise and mark information; Rowan Wilken describes Twitter's potential as a locative medium, referring to human desire to assign information to places. Not many tweets include actual geocodes that enable us to trace back the origin of a tweet to an exact longitude and latitude, but users provide information about their locations in different ways, for example in their personal information section or within the tweet itself. By doing so, users may accidentally reveal more personal information than they intended. Michael Zimmer and Nicholas Proferes address controversial issues of privacy on Twitter in their chapter, arguing that far too little is known about whether the users themselves perceive their activities as public. Although Twitter works with very basic privacy settings (a user's profile and all of their tweets are either public or restricted), users may not know that what they are writing is publicly accessible by default.

Miranda Mowbray proceeds by examining a type of Twitter user largely unaffected by issues of privacy: automated Twitter accounts are programs that post messages to Twitter without direct human intervention. These bots are not welcomed by Twitter, Inc. (as the service is intended for human users, according to the company's policies), but not all of them are harmful. While spam may be the most common type of automated tweet, other bots provide useful services or entertain human users (and possibly, one another). The final two chapters in this section address information overload as a perceived adverse effect of using Twitter. Ke Tao, Claudia Hauff, Fabian Abel, and Geert-Jan Houben describe the difficulties of finding very specific information in large volumes of Twitter data. Applying information retrieval theory, they show that individual tweets are a problematic form of document which cannot be easily classified as relevant to a specific search query; new approaches are necessary to make sense of tweets in context. Finally, Thomas Risse, Wim Peters, Pierre Senellart, and Diana Maynard discuss a topic that is relevant for both Twitter users and researchers: the storage and archiving of tweets as a knowledge resource for future generations. In addition to the restrictions imposed by Twitter, Inc.'s Terms of Service, this requires that significant technical challenges be addressed: for example, archiving approaches should also seek to capture the original context of posts by preserving the content of any URLs which are included in the tweets.

The final section of this volume, "Practices", is organised around different forms of social interaction as mediated through Twitter. We have selected a variety of case studies that reflect the richness of usage scenarios and illustrate how users with different backgrounds apply Twitter for their purposes. The section begins with four chapters on Twitter's role in popular culture. These consider, among other themes, the changing practices of fandom and fan interactions. Nancy Baym points out the role of Twitter in audience management as perceived by musicians and other artists. As artists are increasingly dependent on being discovered by and on building relationships with their audiences, they use Twitter to reach out and personalise such relationships. How audience members are enabled to find and interact with each other is also the topic of Stephen Harrington's chapter on tweeting about the television. He discusses the ways in which microblogging during live TV broadcasts transcends the small screen and provides shared experiences beyond the anonymity of a mass medium. Through such practices, Twitter can become a medium for actual fan interaction, where celebrities or media personalities share personal observations with fans, and fans may address them directly in return. This is true in sports as much as in the arts: Tim Highfield's chapter, therefore, focusses on interactions during a particular sporting event, the annual Tour de France. He shows how different groups—cyclists, media, and fans—connect through event-related hashtags and engage with each other. In the following chapter, Axel Bruns, Katrin Weller, and Stephen Harrington move beyond specific sporting events to compare the activities of football clubs over the course of an entire season. Their case studies of the English, German, and Australian leagues reveal substantial differences in tweeting practices across these sports markets. As football clubs can be considered a very specific type of brand, this chapter provides a useful link to a more general investigation of brand communication on Twitter. Here, two different perspectives are explored: Stefan Stieglitz and Nina Krüger investigate the strategies of various major brands in dealing with Twitter during brand crises. Their conclusions from these examples may also provide useful advice for other businesses seeking to make better corporate use of Twitter. Further, Tanya Nitins and Jean Burgess concentrate on the discussion between brands and users that can ensue in Twitter's two-way communicative environment. Some brands successfully create an online space for participation and engagement—in other cases, users are deliberately searching for PR mistakes, or set up parody accounts in order to spread satirical messages.

Twitter has received much attention both within mass media and from communication researchers for its role in political discourse, especially when connected to elections and campaigning. Our three chapters on politics and activism apply diverse approaches to this topic. Axel Maireder and Julian Ausserhofer conduct parallel content analyses of news reports and tweets relating to three different events in Austrian national politics. They show that Twitter discourses develop on a trajectory that is partially independent from mass media reporting. Anders Olof Larsson and Hallvard Moe examine three major elections held in Sweden, Denmark, and Norway in 2011. They focus on the debates surrounding the elections' main hashtags, analysing user activities and different types of tweets. Finally, Johannes Paßmann, Thomas Boeschoten, and Mirko Tobias Schäfer critically investigate whether retweeting establishes a novel kind of gift culture within social media, analysing the Dutch parliamentary Twittersphere in a case study. This also reveals how messages circulate within a specific community. The following chapters address broader perspectives on Twitter in the news ecosystem: Christoph Neuberger, Hanna Jo vom Hofe, and Christian Nuernbergk describe different dimensions of the use of Twitter in journalism, including interactions with readers and viewers, and real-time coverage from the scene of news events. They observe that only in exceptional cases, private individuals report exclusively on public events. In the next chapter, Alfred Hermida notes the role played by an individual user who acted as a central distributor of news about the mass shooting at a cinema in Aurora, Colorado. Hermida illustrates how Twitter can serve as a channel for the distribution of materials from journalists and the mainstream media, especially around breaking news, when rumours and speculation play a crucial role.

These issues are salient in the cases covered by our chapters on crisis communication as well. Social media have helped to involve a larger proportion of the general population in online crisis communication during political and religious uprisings, mass violence, and natural disasters. Axel Bruns and Jean Burgess investigate the use of Twitter during natural disasters in Australia and New Zealand, where the platform helped to share information about the unfolding situation and to coordinate emergency responses. Focussing in particular on the role of visual information in a crisis scenario, Farida Vis, Simon Faulkner, Katy Parry, Yana Manyukhina, and Lisa Evans analyse tweets during the civil unrest in the UK in August 2011 that became known as the London Riots. Analysing the distribution of original photos as well as television screenshots, they make a strong argument for the growing relevance of image-sharing on Twitter. While these case studies serve to underline the point that Twitter has become a subject of research across diverse scientific disciplines, we finally also examine the take-up of Twitter in academia itself—as a tool for scholarly communication. The final two chapters in this section reflect the two sides of the academic coin: research and teaching. Merja Mahrt, Katrin Weller, and Isabella Peters provide a broad overview of how scholars use Twitter for their everyday work, concluding that, for now, the use of Twitter remains rare among scholars in general, although there are some differences across disciplines. In education, Twitter—among other tools—is considered to be a valuable addition for e-learning environments. Timo van Treeck and Martin Ebner analyse two massive open online courses (MOOCs) that integrated Twitter as a communication channel. It appears that both in scholarly communication and in learning environments, retweeting and sharing URLs play an important role as a means of information distribution.

The breadth and diversity of these uses of Twitter in contemporary society document the considerable adoption of Twitter as a platform for everyday and extraordinary, personal, and public communication. The work collected in this volume also showcases the rich insights—not only into Twitter itself, but into society as such—which research in this field is able to generate. Thus, the epilogue to the present collection reflects more generally on Twitter's role in society and its relationship with society. The interplay between the platform, with its technical and political restrictions, and the individuals who make use of this service can inform a wide range of questions on modern societies, largely due to the fact that Twitter makes people's activities, communication, and reactions to outside events publicly accessible at an unprecedented level. The study of Twitter and its uses, therefore, extends well beyond platform studies; rather, it forms part of the broader agenda which Richard Rogers (2009) has outlined: to study society *through* the Internet.

Brisbane / Düsseldorf / Köln / Oxford

May 2013

REFERENCES

Berry, D. (2011). The computational turn: Thinking about the digital humanities. *Culture Machine*, *12*, 1–22. Retrieved from http://www.culturemachine.net/index.php/cm/article/view/440/470

boyd, d. m., & Ellison, N. B. (2007). Social network sites: Definition, history, and scholarship. *Journal of Computer-Mediated Communication, 13*(1), article 11. Retrieved from http://jcmc.indiana.edu/vol13/issue1/boyd.ellison.html

Bruns, A., & Jacobs, J. (Eds.). (2006). *Uses of blogs.* New York, NY: Peter Lang.

Burgess, J., & Green, J. (2009). *YouTube.* Cambridge, UK: Polity Press.

Halavais, A. (2009). *Search engine society.* Cambridge, UK: Polity Press.

Lazer, D., Pentland, A. S., Adamic, L., Aral, S., Barabási, A. L., Brewer, D., . . . Van Alstyne, M. (2009). Computational social science. *Science, 323*(5915), 721–723.

Lewandowski, D. (Ed.). (2012). *Web search engine research.* Bingley, UK: Emerald.

Lih, A. (2009). *The Wikipedia revolution: How a bunch of nobodies created the world's greatest encyclopedia.* New York, NY: Hyperion.

Meadows, M. S. (2007). *I, avatar: The culture and consequences of having a second life.* Berkeley, CA: New Riders.

Reagle, J. M. Jr. (2010). *Good faith collaboration: The culture of Wikipedia.* Cambridge, MA: MIT Press.

Rogers, R. (2009). *The end of the virtual: Digital methods.* Amsterdam, The Netherlands: Vossiuspers UvA. Retrieved from http://www.govcom.org/publications/full_list/oratie_Rogers_2009_preprint.pdf

Twitter. (2013). Who's on Twitter? Retrieved from https://business.twitter.com/whos-twitter

PART 1

Concepts
and Methods

Twitter and the Rise of Personal Publics

1
CHAPTER   Jan-Hinrik Schmidt

 #private or #public? communication on Twitter is both and neither at the same time

Since its first public release in 2006, Twitter has established itself as the leading microblogging platform in most parts of the world. Its widespread adoption and integration with other parts of the digital networked media ecosystem have sparked public debate, pop-cultural responses, and academic research alike. Like other "new media," Twitter is both underdetermined and recombinant (Lievrouw, 2002), making it subject to the interpretative flexibility of the particular social groups involved in developing and appropriating the technology (van Dijck, 2011). Thus, there are many different practices of Twitter use: a teenager in suburban USA will tweet differently from a German professional football team, from a British comedian, and from a political party in Spain. Still, they all participate in a shared media technology with particular functionalities and communicative architecture, so it is worthwhile to examine these characteristics and (some of) their consequences.

In particular, this text will focus on the connections between Twitter practices and changes in our understanding of the public. It starts by describing

Twitter as a communicative space, framed by the three dimensions of software, relations, and rules. Based on these analytical remarks, it is then argued that Twitter contributes to the emergence of a new type of "publicness": the personal public. This concept as well as its consequences for journalism and for our understanding of privacy are discussed, followed by a conclusion which situates the ideas presented here in the overall transformation of mediated communication.

TWITTER AS A COMMUNICATIVE SPACE

The main argument to be developed is that Twitter is providing a particular communicative space which is affording the emergence of a new type of publicness: the "personal public". As argued elsewhere (Schmidt, 2011a, pp. 107–133), personal publics are one of the most important characteristics of the social Web, and as such, are not confined to Twitter. We can observe them most prominently on social network sites such as Facebook, but also on video-sharing platforms or on blogs—but have to note that not all communication based on these media technologies is to be considered a personal public (much the same as not everything printed on paper is to be considered a newspaper, or not everything broadcast on TV is a news show).

Rather, we should consider personal publics as an ideal type of communicative space, defined—and placed in contrast to the "traditional" publics afforded by journalistic mass media—by three elements: in personal publics, information is

1. Being selected and displayed according to criteria of personal relevance (rather than following journalistic news factors),
2. Being addressed to an audience which consists of network ties made explicit (rather than being broadcast to a dispersed, unknown mass audience), and finally, communication in personal publics is
3. Being conducted mainly in a conversational mode (rather than in the one-way mode of "publishing").

What exactly are the elements of Twitter as a communicative space that enable the emergence of personal publics? We can identify them along three analytical dimensions that structure communicative space online—thus framing situated social action within these spaces, without determining it (see Schmidt, 2007, for a similar discussion for blogging): technological features and affordances; social and textual relations; and shared rules.

Twitter is an Internet-based communication technology that allows users to distribute short messages (tweets) of 140 characters or fewer on the World Wide Web or through smartphone apps. Over the last years, various additional features have been included in the backend and the interface, such as the facilities for picture upload and display, or the automatic shortening of URLs to save characters in tweets. (See Chapter 3 by Halavais in this volume for a more detailed analysis of the co-evolution of the Twitter service and its practices.) Through an API (Application Programming Interface), third-party applications which offer additional functionalities can be connected to the service.

But the main affordances which distinguish Twitter from other forms of online distribution of messages such as IRC, email, or discussion boards are the particular ways that articulated relations—the nexus of social ties and textual references, based on code-enabled connections—are used to structure the flow of communication and to filter information. Firstly, Twitter relies on articulated social connections to establish "sender-audience" relationships. While single tweets as well as the collection of past tweets of a particular user are usually publicly accessible through permalinks, the basic concept guiding Twitter use is the idea of "following". Becoming a follower of a user is similar to subscribing to their updates, so their tweets will show up (together with those of the other people you follow) in your timeline, the reverse-chronologically sorted collection of updates. Contrary to social network sites such as Facebook, where social relationships are required to be reciprocal, the follower/followee relationship can (but does not have to) be unilateral (for large-scale studies on the resulting network properties see, for example, Kwak, Lee, Park, & Moon, 2010; Wu, Hofman, Mason, & Watts, 2011). The articulated social relationships are also used to calculate similarities with other users (e.g., Twitter displays how many of my followers also follow them), thus suggesting them as potentially interesting.

Besides this basic social relationship, Twitter communication is further based on textual references made explicit via a combination of communicative practices and software affordances. In order to address or reply to a particular user, the @-sign followed by the account name is used. The equivalent to forwarding a message is the retweet, through which a user distributes a tweet to their own followers while preserving the reference to the original sender (on the importance of retweets as a communicative tool, see also Chapter 2 by Bruns & Moe in this volume). In both cases, communicative references to other Twitter users are not only made visible, but navigable as well: people can follow the @-link or the retweet link to see the context of a conversation or the background of a particular user.

Finally, Twitter affords the formation of relations between users and texts (single tweets as well as whole conversations) through the use of hashtags, which consist of the "#" symbol followed by a word or phrase. Because hashtags are made searchable by the interface, they connect tweets from users who have no preexisting follower/followee relationship. Hashtags are unmoderated, so any user can introduce and use them, giving rise to a wide and uncontrolled variety of hashtags. This results in possible ambiguities in meaning and spelling, but processes of suggestion, imitation, and learning, as well as Twitter's "trending topic" functionality promote a shared use of certain hashtags for current events, cultural expression, or engagement in ongoing conversations.

The particular affordances of Twitter as a software service, together with the social and textual affordances articulated in ongoing use, form a communicative space which is partly stable (e.g., the connections between followers and followees) and partly highly dynamic (e.g., the tweets using a popular hashtag). It differs from other forms of online communication in that there is no "shared location" where users and their contributions become visible (as in a thread within a discussion board, a blog posting or Facebook status update with subsequent comments, or a chatroom). Rather, communication on Twitter is happening in networked, distributed conversations: single tweets forming the basic units and serving as "micro-content" (Dash, 2002) or "nanostories" (Wasik, 2009) are bundled (a) in the constant stream of information within a personal timeline, filtered via social connections made explicit, as well as (b) in the spontaneous and *ad hoc* "hashtag publics" (Bruns & Burgess, 2011), filtered via shared keywords and phrases.

But technological features and emerging networks of people and text alone do not suffice to constitute (and describe) a communicative space. A third structural aspect is necessary—shared routines and expectations about "how to do things", or in this chapter's context: how to use Twitter. They include shared understandings about which topics are appropriate or not for communication (which in turn is related to the issue of privacy, see below), but also more detailed expectations about the presentation, style, or tonality of tweets, as well as about the use of Twitter as part of a larger media ecology. While the opportunities and boundaries set by the Twitter interface (e.g., the limit of 140 characters per tweet) are valid for all users, shared rules might range from rather general norms and expectations to those more particular to certain groups or contexts of use.

The idea of authenticity, for example, is widely shared, and fake accounts are seen as a transgression of communicative expectations (see also Chapter 14 by Mowbray in this volume). Twitter supports this norm not only by providing a mechanism to verify the accounts of politicians or celebrities, it also prohibits impersonation in its own "Twitter Rules," stating: "you may not impersonate others through the Twitter service in a manner that does or is intended to mislead, confuse, or deceive others" (Twitter, 2012d).

This points to a different perspective on Twitter rules, which addresses the power to impose (positive or negative) sanctions. As shared norms and expectations have varying degrees of formality, there are different social agents involved in shaping and enforcing these rules: as a business entity providing Web-based services, Twitter has its own Terms of Service which users have to accept and abide by in order to participate on the platform; failure to do so might lead to the suspension or termination of an account. Additionally, Twitter offers a full set of policies, guidelines, and best-practice documents (Twitter, 2012b) which not only cover impersonation and parody accounts, but also topics such as promoted products (Twitter, 2012c), or the use of tweets in media broadcasts (Twitter, 2012a). Some of them are strongly tied to general legal frameworks, such as copyright, free speech, or the protection of minors, and might, as such, also include other sanctions if breached.

Most of the rules framing the everyday use of Twitter will, however, remain implicit. They might be invoked and contested in the context of misunderstandings, failed communication, or other conflicts between users, when they are made explicit to negotiate and regulate behaviour which has been deemed inappropriate. Thus, knowing how to use Twitter is not restricted to being able to set up an account or use the interface of its website or app. Rather, it also includes implicit knowledge with which users demonstrate that they are "getting" Twitter. Possession of this implicit knowledge about shared routines and expectations becomes a condition of inclusion or exclusion in the "community of practice" of Twitter as a whole, as well as of participating in particular subcultures via Twitter (see Baym, 2010, Ch. 4, pp. 72–98, for a general overview on the role of practice and norms in computer-mediated communities).

PERSONAL PUBLICS ON TWITTER

The previous remarks have described Twitter as a communicative space framed by three structural dimensions of technological affordances, social and textual relationships, and shared rules and expectations. Against this background, we can revisit the idea of personal publics (where information is selected by criteria of personal relevance for a known, networked audience in a conversational mode).

For many users, Twitter is "personal media" (Lüders, 2008), in that they have a large degree of control over what and how they communicate. Contrary to, for example, social media editors for corporate accounts or mainstream media brands on Twitter, they neither have to comply with internal guidelines, PR and corporate communications policies, nor have to adhere to the criteria for newsworthiness which journalists have internalised in their professional education (see Clayman & Reisner, 1998). Rather, both selection and presentation of content to be tweeted can follow criteria of personal relevance. Traditionally, the Twitter interface has mirrored this broad scope of topics to be communicated by just asking "What's happening?" Additionally, the integration of Twitter with other online services (e.g., photo-sharing sites such as Instagram, video platforms such as YouTube, or news sites such as nytimes.com) makes it easy to share activities and content from those sites with one's followers. Thus, Twitter can become a personal hub for sharing a mediated everyday life.

Selecting and presenting information of personal relevance is emerging as a shared rule and expectation. This is assisted by the possibility of addressing particular audiences on Twitter. While mainstream media such as TV, radio, and print distribute information to a wide, unknown, and dispersed mass audience, users on Twitter have at least a latent knowledge of the size and composition of their audience: they can see how many followers they have, and they can—in principle—click on each of their followers' accounts to learn more about the people who have chosen to subscribe to their tweets. This will also make visible the heterogeneity of their audience, as there might be people from a variety of social contexts among the followers (see Marwick & boyd, 2010, for a more detailed discussion of the strategies for dealing with the possibility of collapsing social contexts on Twitter). So, even if two users have audiences of similar size, their compositions themselves will not be the same—rather, every Twitter user has their own particular and unique audience, which forms as an articulated network instead of a dispersed mass.

The third aspect distinguishing personal publics from mass-media publics is their respective communicative mode. Mass-media publics, on the one hand, are based on a mode of publishing or broadcasting, where dedicated senders distribute information without being able to receive feedback through the same technical channel. Personal publics, on the other hand, are characterised by the communicative mode of "conversation," where the strict separation of sender and receiver is blurred. (However, one might, for analytical reasons, still identify sender and receiver in any given communicative episode. On Twitter, the idea of "follower" and "followee" mirrors this distinction of communicative roles.) Accordingly, people expect to be able to retweet or reply to other tweets, or, conversely, to be replied to or retweeted. The software interface, and in particular the various functionalities for displaying and searching for @replies, retweets, and hashtags, supports these practices, and helps users engage in distributed conversations. And although Twitter is based on written communication, many tweets do resemble oral communication in their style and tonality (Tufekci, 2011). Thus, they contribute to the maintenance of a "connected presence" (Licoppe & Smoreda, 2005), enabling people to stay in touch over distance by sharing seemingly mundane and trivial information which nevertheless serves to reassure participants of shared social bonds.

Again, it has to be emphasised that not all communication on Twitter necessarily takes place in personal publics, and that personal publics are not restricted to Twitter. Rather, personal publics should be considered as an ideal type of communicative structure that concurrently complements and modifies other aspects of public communication. Two consequences of the rise of personal publics will be discussed in the remainder of this chapter: their relation to traditional media, and the changes in our understanding of privacy that personal publics bring about.

Twitter has been adopted quickly not only by "regular" users, but also by political activists, parties, and candidates; and by companies, brands, and celebrities (see Marwick & boyd, 2011, as well as the chapters in the second half of this book for a more thorough discussion of practices of Twitter use). They all profit—in different ways—from the alternative ways of addressing and distributing information which Twitter provides, and can circumvent the mechanisms of gatekeeping and journalistic intermediation that characterise traditional mainstream media. In turn, other users can adapt their routines of information management and directly follow interesting sources (such as a celebrity or a politician), instead of having to rely on information about them being filtered and "packaged" by journalists. Thus, users can build their own radar of information sources by selecting and following only those accounts or conversations that (promise to) provide content that is relevant to them.

Professional media are, of course, reacting to this shift in informational practices (see Chapter 26 by Neuberger, vom Hofe, & Nuernbergk, as well as Chapter 27 by Hermida in this book for a more thorough discussion). A growing number of news sites include "tweet this" functions in their stories in order to facilitate the spread of their content, and media brands as well as individual journalists are increasingly present on Twitter themselves. This appropriation of Twitter and its integration into professional journalistic routines is contributing to the three trends Meikle & Young (2012, pp. 47ff.) have identified as the main characteristics of news in convergent media industries: news on Twitter is becoming debundled and linkable (rather than packaged in discrete bundles of news, such as a weekday edition or an 8 p.m. newscast); news involves sharing information among audiences (rather than distributing the information to the audience); and news is becoming conversational (rather than remaining a monologue).

Thus, when building the personalised news radar for their own personal public on Twitter, users might choose to also subscribe to the Twitter account of their favorite newspaper or TV news station, or to a number of them, to get a more diverse set of perspectives on current events. They can share and comment on those news items with their own audience, and even get in touch with journalists to correct errors or suggest related information. This not only changes the mechanisms and expectations of audience participation in journalism (Loosen & Schmidt, 2012), but also turns Twitter into a place where conversation and publication converge. In personal publics, news reporting and instances of professional communication can share the same space with personal musings, phatic communication (Miller, 2008), and social grooming.

This convergence of the public and the personal is already pointing to the second main consequence of the rise of personal publics: they contribute to the shift in our understanding of mediated privacy and publicness (exemplary for the debate on this deep and complex change, see the debate between Ford, 2011, and Jurgenson & Rey, 2012). As users are selecting and sharing information of personal relevance based on the central norm of authenticity with an intended audience composed of articulated social ties, they are making information accessible that might be considered private, such as holiday stories, impressions from family events, one's current location or emotional state, etc. While these might be considered and dismissed as instances of "digital exhibitionism" by some, closer inspection shows that a reconfiguration of the practices and context of everyday impression management and relationship management in extended social worlds is taking place.

As has been argued above, such tweets are becoming part of personalised news streams within articulated networks of strong and weak ties. The decision to tweet or withhold a certain opinion, link, piece of information, etc., will be based on the user's perception of their own audience: how large is it, and how many people from which role contexts are among the followers? Since Twitter use, as other communicative practices, will become routinised over time, usually not every single tweet is scrutinised before sending. Rather, users form a general idea of their followers as an "intended audience" (Schmidt, 2011b) or "imagined audience" (Litt, 2012), which they will use to assess the appropriateness of information. In some situations, users might also address a particular group within their audience, for example, when participating in a hashtag conversation. By selectively disclosing information, either based on the perception of their intended audience or to an explicitly addressed audience, users engage in privacy management.

Characteristics of Twitter as networked digital media, however, complicate these practices of self-disclosure and audience control. Following boyd (2008), we can identify the four aspects of persistence, replicability, scalability, and searchability of digital information which make it difficult to assess the empirical audience—who is actually taking notice of a given tweet?—and almost impossible to constrain the potential audience of those who might, in the near or distant future, have access to it. Thus, personal publics on Twitter challenge users to "maintain equilibrium between a contextual social norm of personal authenticity that encourages information-sharing and phatic communication (the oft-cited 'what I had for breakfast') with the need to keep information private, or at least concealed from certain audiences" (Marwick & boyd, 2011, p. 124).

CONCLUSION

This chapter has discussed Twitter from a sociological point of view, situating its individual use within different structural aspects which both frame and result from this use. It has argued, in particular, that Twitter provides a communicative space which is formed by particular technological features, by emerging social and textual relationships, as well as by shared norms and expectations guiding the use of Twitter. These elements enable the emergence of personal publics, a new kind of publicness which consists of information selected and presented according to personal relevance, shared with an (intended) audience of articulated social ties in a conversational mode.

While the focus of this chapter and the book has been on Twitter, the ideas developed here can arguably be applied to other genres of networked digital media which—in combination with other large-scale, long-term developments such as globalisation and the rise of networks as a central morphology for social organisation (Castells, 2000; Tomlinson, 1999)—contribute to profound changes in contemporary societies. Personal publics afforded by social media are one of the most visible results of shifts in everyday identity management, relationship management, and information management: they allow people to express and work on aspects of their own identity, while maintaining and expanding social connections of different degrees. In addition, they help people manage the abundance of information around them by introducing filter mechanisms which are personal and social at the same time.

To argue, as Keen (2008) has done, that personal publics promote a "cult of the amateur", where trivial babble dominates over thoughtful knowledge of the experts, is to miss the point. We should, rather, acknowledge the potential for inclusion and participation inherent in these new ways of communication, expression, sharing, and socialising. Papacharissi (2010) called this nexus of the individual and the social the "private sphere", in which

> the citizen is alone, but not lonely or isolated. The citizen is connected, and operates in a mode and with political language determined by him or her. Operating from a civically *privé* environment, the citizen enters the public spectrum by negotiating aspects of his/her privacy as necessary, depending on the urgency and relevance of particular situations. (p. 132)

Although not mentioned directly by Papacharissi, we should consider personal publics on Twitter as one of the "places" where this private sphere becomes manifest. Not all of the many different practices of Twitter use will eventually lead to personal publics as defined in the previous remarks. But those which do so provide opportunities for participation and social inclusion, because people communicate and share things that are important to them within an extended network of social ties. In this respect, Twitter is indeed and profoundly social media.

REFERENCES

Baym, N. K. (2010). *Personal connections in the digital age*. Cambridge, UK: Polity Press.

boyd, d. (2008). *Taken out of context: American teen sociality in networked publics* (Unpublished doctoral dissertation). University of California-Berkeley, Berkeley, CA.

Bruns, A., & Burgess, J. (2011, August). The use of Twitter hashtags in the formation of *ad hoc* publics. Paper presented at the European Consortium for Political Research conference. Retrieved from http://eprints.qut.edu.au/46515/

Castells, M. (2000). *The rise of the network society: The information age: Economy, society and culture* (2nd ed.). Oxford, UK: Blackwell.

Clayman, S. E., & Reisner A. (1998). Gatekeeping in action: Editorial conferences and assessments of newsworthiness. *American Sociological Review, 63*(2), 178–199.

Dash, A. (2002, November 13). Introducing the microcontent client. Retrieved from http://dashes.com/anil/

Ford, S. M. (2011). Reconceptualizing the public/private distinction in the age of information technology. *Information, Communication & Society, 14*(4), 550–567.

Jurgenson, N., & Rey, P. J. (2012). Comment on Sarah Ford's 'Reconceptualization of privacy and publicity'. *Information, Communication & Society, 15*(2), 287–293.

Keen, A. (2008). *The cult of the amateur: How blogs, Myspace, YouTube and the rest of today's user-generated media are destroying our economy, our culture, and our values.* London, UK: Nicholas Brealey.

Kwak, H., Lee, C., Park, H., & Moon, S. (2010). What is Twitter, a social network or a news media? In *WWW '10: Proceedings of the 19th International Conference on World Wide Web* (pp. 591–600). New York, NY: ACM. doi: 10.1145/1772690.1772751

Licoppe, C., & Smoreda, Z. (2005). Are social networks technologically embedded? How networks are changing today with changes in communication technology. *Social Networks, 27*(4), 317–335.

Lievrouw, L. A. (2002). Determination and contingency in new media development: Diffusion of innovations and social shaping of technology perspectives. In L. A. Lievrouw & S. M. Livingstone (Eds.), *Handbook of new media: Social shaping and consequences of ICTs* (pp. 183–199). London, UK: Sage.

Litt, E. (2012). Knock, knock. Who's there? The imagined audience. *Journal of Broadcasting & Electronic Media, 56*(3), 330–345.

Loosen, W., & Schmidt, J. (2012). (Re-)discovering the audience: The relationship between journalism and audience in networked digital media. *Information, Communication & Society, 15*(6), 867–887.

Lüders, M. (2008). Conceptualizing personal media. *New Media & Society, 10*(5), 683–702.

Marwick, A., & boyd, d. (2011a). I tweet honestly, I tweet passionately: Twitter users, context collapse, and the imagined audience. *New Media & Society, 13*(1), 114–133. doi: 10.1177/1461444810365313

Marwick, A., & boyd, d. (2011b). To see and be seen: Celebrity practice on Twitter. *Convergence, 17*(2), 139–158.

Meikle, G., & Young, S. (2012). *Media convergence: Networked digital media in everyday life.* Basingstoke, UK: Palgrave Macmillan.

Miller, V. (2008). New media, networking and phatic culture. *Convergence, 14*(4), 387–400.

Papacharissi, Z. (2010). *A private sphere: Democracy in a digital age.* Cambridge, UK: Polity Press.

Schmidt, J. (2007). Blogging practices: An analytical framework. *Journal of Computer-Mediated Communication, 12*(4), article 13. Retrieved from http://jcmc.indiana.edu/vol12/issue4/schmidt.html

Schmidt, J. (2011a). *Das neue Netz: Merkmale, Praktiken und Folgen des Web 2.0* (2nd ed.). Konstanz, Germany: UVK.

Schmidt, J. (2011b). (Micro)blogs: Practices of privacy management. In S. Trepte & L. Reinecke (Eds.), *Privacy online: Perspectives on privacy and self-disclosure in the social web* (pp. 159–173). Heidelberg, Germany: Springer.

Tomlinson, J. (1999). *Globalization and culture.* Chicago, IL: University of Chicago Press.

Tufekci, Z. (2011, 19 May). Why Twitter's oral culture irritates Bill Keller (and why this is an important issue). Retrieved from http://technosociology.org/

Twitter. (2012a). Guidelines for use of tweets in broadcast. Retrieved from https://support.twitter.com/ groups/33-report-abuse-or-policy-violations/topics/149-developer-and-media-guidelines/ articles/114233-guidelines-for-use-of-tweets-in-broadcast

Twitter. (2012b). Report abuse or policy violations. Retrieved from https://support.twitter.com/ groups/33-report-abuse-or-policy-violations

Twitter. (2012c). Trademark policy for promoted products. Retrieved from https://support. twitter.com/groups/33-report-abuse-or-policy-violations/topics/148-policy-information/ articles/20170140-trademark-policy-for-promoted-products

Twitter. (2012d). The Twitter rules. Retrieved from https://support.twitter.com/entries/18311

van Dijck, J. (2011). Tracing Twitter: The rise of a microblogging platform. *International Journal of Media & Cultural Politics, 7*(3), 333–348. doi: 10.1386/macp.7.3.333_1

Wasik, B. (2009). *And then there's this: How stories live and die in viral culture*. New York, NY: Viking.

Wu, S., Hofman, J. M., Mason, W. A., & Watts, D. J. (2011). Who says what to whom on Twitter. In *WWW '11 Proceedings of the 20th International Conference on World Wide Web* (pp. 705–714). New York, NY: ACM. doi: 10.1145/1963405.1963504

Structural Layers of Communication on Twitter

2

CHAPTER   Axel Bruns and Hallvard Moe

.@replies, followers, #hashtags:
tweets reach very different audiences
depending on how they're addressed

Twitter is used for a range of communicative purposes. These extend from personal tweets that address what used to be Twitter's default question, "What's happening?", through one-on-one @reply conversations between close friends and attempts at getting the attention of celebrities and other public actors, to discussions in communities built around specific issues—and back again to broadcast-style statements from well-known individuals and brands to their potentially very large retinue of followers.

These different uses of Twitter are intended for, visible to, and able to reach vastly different subsets of the total Twitter user base. However, in the practical understanding of Twitter users, as well as in the existing body of Twitter research, they—and their overlap and interweaving—are often treated with insufficient clarity, and collapsed simply into a cover-all category of "Twitter use". It becomes necessary, therefore, to untangle these different modes of using Twitter and to define them clearly, in order to provide a basis for the Twitter research presented in this volume as well as for the further work that will follow after it.

In this chapter we propose a conceptual model that defines these different modes of communication. We introduce three key layers of communication on Twitter: the micro level of interpersonal communication, the meso level of follower-followee networks, and the macro level of hashtag-based exchanges; we then show how these layers are interconnected in a variety of ways.

This layered structure of communicative exchanges provides a wider context for existing Twitter research, much of which focusses on specific layers within this framework—most frequently, on hashtag communities operating at the macro level. The broader framework we introduce here serves as a necessary foundation for the development of more sophisticated approaches to the study of Twitter as a communicative system, incorporating such single-layer studies into a more comprehensive, multilayer understanding of Twitter as a communication tool. Extending the existing body of literature, we call for new research approaches which move beyond investigating just one of these three layers.

LAYERS OF COMMUNICATION ON TWITTER

The key modes of communication on Twitter are linked to the specific technological affordances of Twitter as a platform, and can be understood as corresponding to micro, meso, and macro layers of information exchange and user interaction. We start from the default level of Twitter communication, which we will describe as the meso layer.

MESO: FOLLOWER-FOLLOWEE NETWORKS

Among the most fundamental affordances which determine the flow of information on Twitter is the capacity for its users to follow one another—that is, to subscribe to the stream of updates originating from the followed user. Following is not necessarily reciprocal—a user may follow any other user (with the exception of 'private' accounts) without requiring the other user to follow back in return; additionally, other than to follow accounts which have been set to 'private' by their owner, no permission is required to follow another Twitter user.

Once an account has gained followers, the tweets posted by the owner of that account will reach all those users who follow the account—if they actively monitor the tweets originating from their network of followed accounts. This default level of tweet dissemination across the follower-followee network upon which Twitter is fundamentally based constitutes the meso layer of communication. Tweets posted (from non-'private' accounts) are public, and in principle, accessible to anyone using the Twitter search functions or visiting the account's profile page—however, the primary intended audience for standard tweets posted by a regular Twitter user is constituted by the account's followers.

In Schmidt's terminology, introduced in Chapter 1 of this volume, this group of followers is the account owner's "personal public". By analogy, for the majority of Twitter users, it can be argued that tweeting to an imagined audience made up of one's followers is similar to making a public statement to a known group of friends and acquaintances—a speech at a family gathering, a lecture to a class of students. The user addresses a group of at least broadly known others whose numbers are limited, and who may or may not pay attention to the statements made. The analogy breaks down, however, for accounts with very large follower networks—here, the exact make-up of the audience becomes too large to be known, or to be accurately imagined (see Marwick & boyd, 2011). This illustrates that the forms of mediated communication which social media support tend to constitute new models which do not have clear offline equivalents.

MACRO: HASHTAGGED EXCHANGES

Such meso layer communication, whose messages reach some hundreds or thousands of followers on average, arguably constitutes the vast majority of everyday communicative activity on Twitter, but is complemented by particular forms and formats of tweeting that use specific syntax to indicate an intention to extend or narrow the range of addressees. Of these, hashtags (simple keywords preceded by the hash symbol '#') are commonly used to mark a tweet as being relevant to a specific topic and make it more easily discoverable to other users. These are not the only uses of hashtags, however, a point to which we will return below. (For a full discussion of the history of hashtags as a user-defined innovation on Twitter, see Halavais, Chapter 3 of this volume.)

The inclusion of a topical hashtag in a tweet means that the message has the potential to reach well beyond the user's existing number of followers. Hashtags can work as markers of a topic, an issue, or an event—from Justin Bieber through the U.S. presidential election to the earthquake and tsunami which struck Japan (several chapters in the "Practices" section of Part II of this volume address such topical uses of hashtags)—and help to coordinate the exchange of information relevant to such topics. Twitter users are able to directly track such hashtagged tweets, independent of whether the messages originate from accounts they already follow, or from previously unknown Twitter users.

In turn, including a hashtag in one's tweets signals a wish to take part in a wider communicative process, potentially with anyone interested in the same topic. Where used in such a way, hashtags can aid the rapid assembly of *ad hoc* issue publics (Bruns & Burgess, 2011b), especially also in response to breaking news or other sudden developments. Constituted independently of meso-level follower-followee networks, such publics can be more dynamic and ephemeral in their development, but can also solidify into long-standing communities of Twitter users.

The communicative flows which result from the establishment of active hashtag exchanges, at least in the short term, are usually less predictable than those enabled by follower-followee networks—but they are also amongst the most visible phenomena on Twitter, and most accessible to research. At the same time, however, even for well-established hashtags (and perhaps especially for hashtags with a high volume of tweets), it cannot be assumed that all users participating in—posting to—a hashtag public will also follow the full feed of tweets containing the hashtag: Twitter users may simply, speculatively include a hashtag to increase the visibility of their own messages, even if they do not themselves track the hashtagged tweets. The assumption that hashtagging does indeed improve the visibility of tweets cannot always be sustained, therefore: if all users were to use the hashtag simply to mark their own tweets, but did not themselves follow other users' hashtagged tweets, the primary utility of hashtagging would be negated.

This is true especially for what may be classed as non-topical hashtags, which are mainly used as emotive markers (#fail, #win, #facepalm, or #headdesk), but possibly also for popular memes (as explored by Leavitt in Chapter 11 of this volume): given the wide and incongruous variety of the tweets marked as such, it is highly unlikely that many Twitter users will deliberately subscribe to a hashtag feed such as #win, for example. The hashtags which do constitute the macro layer of Twitter communication largely represent the more topical uses of the hashtag syntax, therefore; most non-topical hashtags, by contrast, are used to enhance tweets from the meso layer.

By analogy, then, tweeting to a topical hashtag resembles a speech at a public gathering—a protest rally, an *ad hoc* assembly—of participants who do not necessarily know each other, but have been brought together by a shared theme, interest, or concern. Here, many voices may compete to make themselves heard, and their ability to do so above the fray depends largely on those around them taking up the message and passing it on—on Twitter, by retweeting (a key practice we discuss below).

MICRO: @REPLY CONVERSATIONS

If the hashtag takes communication on Twitter from the meso to the macro layer, then, another communicative convention, which by now has been deeply embedded into the Twitter infrastructure itself, enables users to proceed in the opposite direction: towards the third, micro layer of communication on Twitter. By including an @mention of another user (that is, the addressee's username preceded by the '@' symbol), it becomes possible to highlight a tweet specifically to that user. The Twitter platform and standard Twitter client applications will specifically collect such @mentions and notify the recipient of incoming messages as they are received.

@mentions can be seen, therefore, as attempts to strike up a conversation with another Twitter user; any known Twitter user may be addressed in this way, regardless of whether the addressee is already connected to the sender through the meso layer of follower-followee networks or not. Where @mentions are reciprocated by their recipient, multi-turn exchanges of what can now accurately be described as @replies may eventuate; subject to the limited number of individual @mentions which may be contained in one 140-character tweet, this may involve a small group of participants.

While @mentions and @replies clearly indicate an underlying intention to specifically address one or more other Twitter users, over the total number of the sender's followers, Twitter infrastructure makes this implicit narrowing of communicative focus explicit at least if the tweet *begins* with the @mention of another user: if this is the case, the message is visible in most circumstances only to the sender and addressee, as well as to any users following both accounts. (It will also be visible on the sender's Twitter profile page, however, and in datasets retrieved through the Twitter API.)

@reply conversations constitute a micro-level layer of communicative activity on Twitter, then: though they may be visible to users beyond the actively engaged parties, they are centred around these principal participants first and foremost. Such conversations are analogous to an offline conversation with one or several friends or acquaintances, possibly conducted in the presence of a group of non-participating bystanders. (To ensure that their @reply conversations *are* visible to these non-participants, Twitter users have introduced the .@-syntax: as any tweet which does not begin with @username is visible to all of the sender's followers, prefixing the @reply with '.'—or any other character, in fact—ensures full visibility of the message.) Much as is the case offline, too, to the extent that they are aware of the conversation, these bystanders may always enter it by sending their own @replies.

As with hashtags, however, here, too, it is important to note that not all @mentions are attempts to strike up a conversation—especially where the account referred to in the @mention belongs to a celebrity user, brand, or institution, the @mention may indeed be no more than a third-person mention of that user, by their Twitter handle rather than by their full name, as in "I support @BarackObama". This distinction between explicit interpellation and simple reference is often far from clear, however: an @mention of a celebrity or brand may sometimes also be made in the hope that it does result in an @reply.

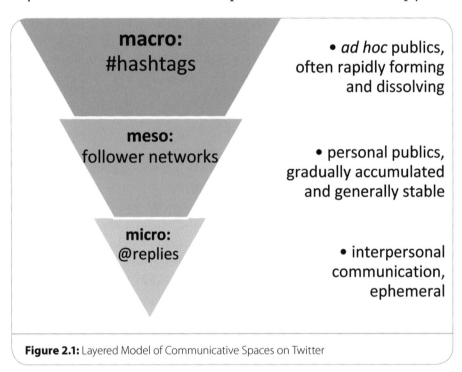

Figure 2.1: Layered Model of Communicative Spaces on Twitter

CROSS-LAYER COMMUNICATION FLOWS

As these descriptions of the three key layers of communication on Twitter already show, the layers do not exist in isolation from one another. While users are likely to envisage a specific set of primary addressees (that is, differently delimited publics—from tight personal networks to broad public assemblies) as they @reply with specific others, tweet general messages, or use hashtags, they will usually be aware that their tweets may also reach users well beyond that initial set of addressees. In the first place, hashtagged tweets as well as @mentions (at least if the tweet does not begin with the @mention itself) will also always be visible to the followers of the message sender, of course: the meso layer serves as a default level of communication on Twitter which it is virtually impossible for users to elude.

But in addition to such inherent interconnections between the layers, determined by the fundamental technological affordances of the Twitter platform, many users also very actively and deliberately transition between the layers. This is self-evident in the use of @replies and hashtags as a means to move from the default meso layer to the more intimate micro layer or the more public macro layer of Twitter communication, but the reverse is also true: so, for example, the syntactic convention of the .@reply enables senders to move from the micro back to the meso layer, while the conscious choice to refrain from adding a known hashtag to an otherwise topical tweet can be regarded as a intentional move from the macro back to the meso layer.

Even direct moves between micro and macro are common: so, for example, an @reply response to a hashtagged tweet transitions the conversation, without a need for the conversation partners to follow one another at the meso level, directly from the broader public space of the hashtag to the one-on-one exchange of @mentions (especially if the @reply does not itself contain the hashtag, and is therefore visible in the first place only to sender and recipient, and any shared followers). Conversely, @replies—or retweets, as we will discuss shortly—which introduce a new hashtag suddenly make the interpersonal conversation visible to the undefined group of Twitter users following the hashtag.

Arguably, it is this flexibility of Twitter as a platform for public communication at various levels of 'public-ness', this versatility of transition between the three major layers of public communication, which serves as the fundament for Twitter's considerable success as a social media service, and makes possible the wide range of uses which the remaining chapters in this collection outline. The triple-layer model (as illustrated in Figure 2.1)—which, it should be noted, evolved through a co-evolutionary process between the platform developers and their users, who introduced the @reply and hashtag conventions (see Halavais, Chapter 3 in this volume)—also constitutes a clear point of distinction from the other global social network, Facebook. The latter offers functionality in the first place for a form of semi-private, personal interactions which are situated somewhere between Twitter's micro and meso layers, and supports macro layer communities only in the context of Facebook pages—but even here, not with the ease of *ad hoc* creation and potential universal reach which Twitter hashtags afford their users.

The most important mechanism for transitioning between the three key layers of communication in Twitter deserves to be discussed separately, however: the retweet (in both its manual forms—e.g., "RT @user [original message]"— and in the form of verbatim 'button retweets'). Retweets—another user-generated communicative convention on Twitter—constitute a mechanism which is inherently designed to move tweets across layer boundaries: Twitter users habitually use them to bring messages from the hashtag level to the attention of their own followers (in the form of manual or 'button' retweets), or even to that of specific recipients, e.g., through manual retweets to which they have added an @mention of the intended addressee: "Hey @recipient, look at this: RT @user [message] #[hashtag]".

If such retweets direct information from the macro to the meso or even micro layer, the reverse is also true: retweets of incoming @replies, or of tweets sent by one of the user's followees, can make these tweets visible to a considerably larger audience if a hashtag is added to the (in this scenario, necessarily manual) retweet. Here, messages from the micro or meso layer are brought to the attention of the macro layer audience by virtue of a newly hashtagged retweet; and even if no new hashtag is included, the retweet of an incoming @reply at least makes that message visible to all the retweeting user's followers, thus transitioning it from the micro to the meso layer.

Finally, even if no new @mentions or hashtags are manually added in the process of retweeting a message—if the retweet is a verbatim 'button' retweet, for example—this passing-along of an incoming message at least fulfils the important function of *horizontally* transitioning the message, even if it remains in the same *vertical* layer of communication on Twitter. What such 'simple' retweets do is to move a message from the specific, meso-layer personal public of the originating user, constituted by that user's Twitter followers, to the meso-layer personal public of the retweeter, thereby reaching a new and almost certainly different group of followers. As much as the *ad hoc* publics which can rapidly gather around hashtags, and operating in concert with them, this horizontal transitioning of messages through the meso-layer follower networks of individual users is responsible for the unprecedented effectiveness of Twitter as a medium for the dissemination of breaking news and rumours.

In this context, it is especially difficult to understand that Twitter and its developers have had a somewhat troubled relationship with the retweeting phenomenon and the functionality underlying it. Early retweeting was entirely manual, but the various Twitter clients gradually automated the process (thereby also standardising the format to the most common "RT @user message" syntax). In late 2009, however, Twitter itself introduced an alternative retweeting mechanism, the 'button retweet' (named after the retweet button which was now displayed next to each message on the Twitter website and in authorised clients), which generated a verbatim, non-editable retweet.

While Twitter co-founder Evan Williams insisted that this new functionality was designed to simplify the retweeting process (Williams, 2009), to avoid the necessity of shortening original messages in order to insert the "RT @user" prefix, and to thus ensure accuracy in retweeting and evade any accidental or deliberate misrepresentation, this streamlined functionality also meant that adding hashtags, @mentions, or any other new material to the retweet was now no longer possible. Button retweets can no longer serve the function of transitioning tweets between the three layers of communication on Twitter, therefore—they can merely transition tweets horizontally. (See also Halavais, Chapter 3 in this volume, on the introduction of button retweets.)

For this reason, many Twitter users continue to use manual retweets; many third-party Twitter clients that had overzealously removed manual retweeting functionality quietly reinstituted it as an alternative option; others never removed it in the first place. Notably, even some of Twitter's own interfaces—at the time of writing, for example, the mobile Twitter websites for iOS devices, but not the Twitter website for desktop computers—once again offer a choice between button and manual retweets, if in a non-standard syntax (cf. Bruns, 2012). This betrays a limited understanding, on behalf of Twitter management and developers, of the wants and needs of the users of the platform, and of the three-layer structure of the key communicative channels which the platform offers—or indeed, a significant divergence in the aspirations which developers and users have for 'their' platform.

CONCLUSION: IMPLICATIONS FOR RESEARCH

The conceptual model for understanding flows of communication and information exchange on Twitter which we have outlined in this chapter has clear implications for how Twitter must be approached by researchers. For obvious practical reasons—hashtags are designed to make tweets more easily discoverable, after all—the majority of extant Twitter research has so far focussed on the macro layer of Twitter communication: on the engagement with breaking news and other topics by participants in hashtag audiences (or, in some cases, hashtag communities, in the narrow sense of the term).

Such work has been able to demonstrate how Twitter users respond almost instantly to natural disasters (Bruns & Burgess, Chapter 28 in this volume; Bruns, Burgess, Crawford, & Shaw, 2012; Mendoza, Poblete, & Castillo, 2010), political unrest (Gaffney, 2010; Lotan, Graeff, Ananny, Gaffney, Pearce, & boyd, 2011; Tonkin, Pfeiffer & Tourte, 2012), celebrity deaths, or other breaking news. It has also been able to illustrate how hashtag activities operate alongside and intersect with the mainstream media coverage of major events, from awards ceremonies (Highfield, Harrington, & Bruns, 2013) and political elections (Bruns & Burgess, 2011a; Larsson & Moe, 2012) through royal weddings to sporting contests. Extant research has also been able to trace how, around some long-standing hashtags, genuine communities of regular participants can form and evolve (e.g., Lindgren & Lundström, 2011; Moe, 2012). In doing so, this research has been able to document the utility of Twitter as a key many-to-many medium which complements, and sometimes even outperforms and supplants, conventional mass media.

However, despite this understandable and often appropriate emphasis on the macro layer, the findings of such studies must always be understood against the background of the greater conceptual model of Twitter communication as we have introduced it here. Hashtag activity in itself does not tell the full story of how Twitter and its users respond to a given event or engage with a given topic. While it may show how many users actively *posted* to the hashtag, it cannot even determine how many others *encountered* subsets of the total volume of hashtagged tweets because one or more of the users they follow were posting or sharing messages from the hashtag feed. Similarly, the volume of follow-on communication (for example in the form of themselves non-hashtagged @replies to hashtagged tweets) usually remains outside the ambit of such studies.

Further, not all topically relevant messages exchanged on Twitter will be marked with an appropriate hashtag; the hashtagged macro level of communication therefore represents only the tip of an iceberg of communicative activity which extends much further down towards the meso and micro levels (and most likely beyond, into private, direct messages). Hashtag studies are able to determine how many hashtagged tweets about a given event or topic were exchanged at any one time—but how many more tweets about the topic, without hashtags, reached only meso-level audiences or engaged with specific @reply recipients at the micro layer?

The bulk of the iceberg is likely to substantially outweigh the tip, in most cases (but is also considerably more difficult to delineate with any degree of exactness): over a period of five days following the March 2011 tsunami on the Japanese east coast, for example, we captured some 790,000 tweets containing the hashtag #tsunami, but close to four times as many tweets simply featuring the *word* 'tsunami'—and even this does not begin to take into account the additional number of topical tweets which happened not to use either hashtag or keyword, but referred to the disaster in other terms or languages.

Correspondingly, studies of Twitter use during election campaigns have shown how key politicians such as major party leaders only show up in hashtag-based datasets when other users tag these leaders' tweets, i.e. when users transition the tweets from the meso to the macro layer of communication through retweeting (e.g., Moe & Larsson 2012). The extent and character of these party leaders' overall tweeting activities largely remains obscured in these studies, therefore.

Methodologically, it is considerably more difficult to move beyond the relatively well-behaved confines of macro-layer hashtag studies. Suggested options include collecting tweets from a pre-defined set of users (e.g., Benney, 2011; Sæbø, 2011; Vergeer, Hermans, & Sams, 2011), or archiving based on keywords (Tumasjan, Sprenger, & Sander, 2010). While the first approach captures communication across the layers *from* a population, it misses any communication *to* the users, as well as retweets of their messages. The latter option, while not being explicitly tied to hashtags, by and large has the same limitations as outlined above. To study public interactions on the meso layer, researchers would need to scrutinise the interactions of all the followers of one or more identified user(s), potentially adding up to a very large number of users to track, and thus exceeding the usage restrictions of the standard Twitter API (necessitating the use of costly third-party services providing access to Twitter data on a larger scale) (but cf. Gaffney & Puschmann, Chapter 5 of this volume). To examine micro-level interactions through @replies, research tools which reliably capture all @reply interactions between two or more identified users must be developed. In turn, the observations made at the micro or meso layer of communication must be integrated again with those at other layers, in order to avoid a repeat of the single-layer problem which exists with hashtag studies.

Finally, the specific communicative context of the phenomena to be studied must also be taken into account. Micro, meso, and macro layers may play considerably different roles depending on the particular groups of Twitter users who use them to communicate, to the point that for users with a very large follower network, the layer order reverses: for a Lady Gaga or Barack Obama, for example, the audience constituted by their followers is likely to be much larger than that made up of the participants and followers of almost any hashtag imaginable. This does not mean that hashtags lose their inherent utility, however; by contrast, a single tweet from such leading Twitter users can be instrumental in publicising the existence of a given hashtag, resulting in a substantial influx of new followers and participants. (This was demonstrated most clearly by the successful, celebrity-centred campaign to publicise the #kony2012 hashtag.)

Such vast follower networks around specific celebrity users already provide their focal accounts with a (meso-layer) Twitter reach which rivals that of the most popular hashtags. Yet, the (macro-layer) audience for hashtags remains less predictable, less unified by shared interest in a specific, leading Twitter user; more multidirectionally interactive; and more changeable. Anyone can subscribe to a hashtag feed, or contribute by posting hashtagged tweets. As the most open and flexible layer of communication on Twitter, then, it makes sense to continue to consider hashtag exchanges the macro level of communicative activity on Twitter.

This threefold conceptual model, stretching across micro, meso, and macro layers of communication, is crucial for an understanding of Twitter both from a practical perspective—from the view of the user attempting to communicate with others through Twitter—and from a scholarly perspective—in order to place observable phenomena on Twitter in the wider context of the full range of communicative activities which take place on the platform. It is important to note here that the model deals only with *public* communication *on Twitter*: in addition to the three layers we have outlined here, there is a further, still lower layer of private communication through direct messaging on the platform itself, as well as through any other forms of private interaction which may be available to any two Twitter users; similarly, there are additional layers of public communication outside of Twitter which, due to the embedding of the Twitter platform into the wider media ecology, are interwoven with communicative processes on Twitter itself.

To fully understand information flows not just *on*, but *through* Twitter as a communicative tool, these outside layers must also be taken into account. During the 2011 south east Queensland floods, for example (cf. Bruns & Burgess, Chapter 28 in this volume), situation updates for the central crisis response steering group were disseminated—hashtagged and in real time—through the Queensland Police Service's (QPS) Twitter account, copied from there to the live tickers of mainstream news channels, posted back to Twitter by viewers of these channels (or retweeted directly from the QPS account), and eventually passed along in person through local neighbourhood networks. Information flows weaved in and out of Twitter, and across the three communicative layers, multiple times. To examine such complex processes of information dissemination only from the perspective of any one layer, or even of any one medium, is to miss an important dimension of their communicative dynamics.

The argument we are making, then, is that while the three layers we have outlined here can be understood in part as determined by the specific technological affordances of Twitter as a platform, they also exist independently of it, and have their equivalents in many other forms of mediated communication. More by chance than by design, and due not least to the considerable influence of Twitter users in guiding their evolution, the communicative mechanisms which Twitter now offers its users are well suited for public communication in a variety of forms: from the comparatively intimate, one-on-one level of @replies through the narrowcast level of personal publics constituted by follower networks to the collective, diffused, many-to-many level of hashtags.

These levels do not simply stem from the underlying technological settings of the Twitter platform, then; rather, in fact, they have co-evolved with it, and sometimes persisted even against the pressures exerted by Twitter's management and developers. Put another way, these different layers of communication precede Twitter itself, and Twitter technology simply gives them concrete, if temporary, form. From this perspective, finally, communicative processes on Twitter also provide us with a glimpse of far more fundamental aspects of human communication.

REFERENCES

Benney, J. (2011). Twitter and legal activism in China. *Communication, Politics & Culture*, *44*(1), 5–20.

Bruns, A. (2012). *Ad hoc* innovation by users of social networks: The case of Twitter. *ZSI Discussion Paper 16*. Retrieved from https://www.zsi.at/object/publication/2186

Bruns, A., & Burgess, J. (2011a). #ausvotes: How Twitter covered the 2010 Australian federal election. *Communication, Politics & Culture, 44*(2), 37–56.

Bruns, A., & Burgess, J. (2011b, 25–27 Aug.). The use of Twitter hashtags in the formation of ad hoc publics. Paper presented at the European Consortium for Political Research conference, Reykjavik, Iceland. Retrieved from http://snurb.info/files/2011/The%20Use%20of%20 Twitter%20Hashtags%20in%20the%20Formation%20of%20Ad%20Hoc%20Publics%20 (final).pdf

Bruns, A., Burgess, J., Crawford, K., & Shaw, F. (2012). *#qldfloods and @QPSMedia: Crisis communication on Twitter in the 2011 south east Queensland floods*. Brisbane, Australia: ARC Centre of Excellence for Creative Industries and Innovation. Retrieved from http://cci. edu.au/floodsreport.pdf

Gaffney, D. (2010). #iranElection: Quantifying online activism. In *Proceedings of the WebSci10: Extending the Frontiers of Society On-Line*. 26–27 Apr. 2010, Raleigh, NC. Retrieved from http://journal.webscience.org/295/

Highfield, T., Harrington, S., & Bruns, A. (2013). Twitter as a technology for audiencing and fandom: The #Eurovision phenomenon. *Information, Communication & Society, 16*(3), 315–339. doi: 10.1080/1369118X.2012.756053

Larsson, A. O., & Moe, H. (2012). Studying political microblogging: Twitter users in the 2010 Swedish election campaign. *New Media & Society, 14*(5), 729–747.

Lindgren, S., & Lundström, R. (2011). Pirate culture and hacktivist mobilization: The cultural and social protocols of #WikiLeaks on Twitter. *New Media & Society 13*(6), 999–1018.

Lotan, G., Graeff, E., Ananny, M., Gaffney, D., Pearce, I., & boyd, d. (2011). The Arab Spring: The revolutions were tweeted: Information flows during the 2011 Tunisian and Egyptian revolutions. *International Journal of Communication, 5*, 1375–1405.

Marwick, A., & boyd, d. (2011). To see and be seen: Celebrity practice on Twitter. *Convergence: The International Journal of Research Into New Media Technologies, 17*(2), 139–158.

Mendoza, M., Poblete, B., & Castillo, C. (2010). *Twitter under crisis: Can we trust what we RT?* Paper presented at the 1st Workshop on Social Media Analytics (SOMA '10), Washington, DC.

Moe, H. (2012). Who participates and how? Twitter as an arena for public debate about the Data Retention Directive in Norway. *International Journal of Communication, 6*(1), 1222–1244.

Moe, H., & Larsson, A. O. (2012). Twitterbruk under valgkampen 2011. *Norsk Medietidsskrift, 19*(2), 151–162.

Sæbø, Ø. (2011). Understanding Twitter use among parliament representatives: A genre analysis. In E. Tambouris, A. Macintosh, & H. de Bruijn (Eds.), *ePart 2011* (pp. 1–12). Heidelberg, Germany: Springer.

Tonkin, E., Pfeiffer, H. D., & Tourte, G. (2012). Twitter, information sharing and the London riots? *Bulletin of the American Society for Information Science and Technology, 38*(2), 49–57. doi: 10.1002/bult.2012.1720380212

Tumasjan, A., Sprenger T. O., & Sander, P. G. (2010, May). *Predicting elections with Twitter: What 140 characters reveal about political sentiment*. Paper presented at the Fourth International AAAI Conference on Weblogs and Social Media, Washington, DC.

Vergeer, M., Hermans, L., & Sams, S. (2011). Is the voter only a tweet away? Micro-blogging during the 2009 European Parliament election campaign in the Netherlands. *First Monday, 16*(8). Retrieved from http://firstmonday.org/htbin/cgiwrap/bin/ojs/index.php/fm/article/view/3540/3026

Williams, E. (2009, 10 Nov.). Why retweet works the way it does. *Evhead*. Retrieved from http://evhead.com/2009/11/why-retweet-works-way-it-does.html

Structure of Twitter
Social and Technical

3
CHAPTER   Alexander Halavais

 Since the early days of twttr, users have been involved in the co-development of its meanings, uses and affordances

Twitter's creation represented a revolution in simplicity. In the early days of Twitter, an explanation of its functionality would likely be met with varying forms of incredulity (Arceneaux & Weiss, 2010): "Is that it?" "Why would anyone do that?" "I don't get it." It was too simple to be easily understood. And yet today, when people want to find out not only what important events are happening in the world, but what their fellow citizens think of those events, in many cases, Twitter is the first place they turn to. Twitter's evolution from a system nearly bereft of formal structure to a complex, networked, social phenomenon is often presented as a process of 'paving cowpaths', iteratively adjusting the design to meet the needs demonstrated by users. While Twitter users were undoubtedly drawn to the system because of its openness and simplicity, they also found that there were needs not met by the formal system. Workarounds emerged that allowed particular communities to connect ideas and people in useful ways. When these patterns were widespread enough, they were often incorporated into the core Twitter system, either by Twitter itself or by other companies making use of its programming interface. The result is a Twitter—in its interface and patterns of use—that was invented in no small part by its users.

But there is a wrinkle in this story. In incorporating these changes, Twitter did more than merely make formal the informal workarounds of its users. These appropriations often displaced social practices that better represented the diversity of users and their needs, replacing them with model uses (and users) imagined by Twitter's developers.

CO-EVOLUTION AND AFFORDANCES

The idea that technologies evolve is hardly a new one, tracing its way back at least to the 19th century and Samuel Butler's 1863 'Darwin Among the Machines' (Dyson, 1997, p. 15). The morphology of a hawk or a handsaw changes over time to meet the requirements of the environment. But that environment is essentially a social one, and is itself often changing. Technologies come to meet the needs of the social groups that use them. We discover technologies as much as invent them, and the social environment in which a technology diffuses influences the way in which it evolves (Pinch & Bijker, 1984). In order to understand the evolution of technologies, we need to follow the relevant groups and the ways in which they employ an innovation or conceive of an artefact (Bijker, 1992; Bijker & Law, 1992).

But the process is never simple or clean, and neither determined by technological constraints nor completely free from them. This is perhaps not as obvious for a piece of machinery; say, an automobile. Kline and Pinch (1996), for example, suggested that there has been a great deal of scholarly work on how the diffusion of the automobile shaped American culture, but "rather less attention has been given to how American society shaped the car" (p. 763). In Twitter's case, the opposite imbalance holds. Much has been made of the ways in which user practices have shaped Twitter, but comparatively less about how Twitter's developers have shaped those practices, or how Twitter has shaped social practices on the Web and more broadly. José van Dijck (2011) tracked how both hardware constraints and the influence of relevant groups resulted in an "interpretive flexibility" that gave way to a more stable, ossified, and commercial service.

Two elements of the architecture of Twitter have made it particularly likely to be influenced by its user community. First, Twitter's interface is extraordinarily simple. Second, it was deliberately open to alternative user interfaces via an application programming interface (API). As a result, the majority of users contribute to the Twittersphere via third-party applications (Johansmeyer, 2009). This makes it difficult to examine the affordances of the technology, as different users understand the interface and capabilities of the system differently (Fragoso, Rebs, & Barth, 2012).

While different Twitter users came to use a variety of devices and applications to read and post tweets, they shared a basic commonality, a text window limited to 140 characters. This visible interface both conceals the structural relationship inherent to the subscription system, and invites posters to be creative with their use of the textual space they are given to work with.

Built into Twitter from the very start was an asymmetric relationship—not 'friends', but 'subscriptions' or 'follows'. For example, as of the time of writing, the singer Katy Perry follows 114 people, but is followed by more than 27 million. On Facebook, with some rare exceptions, the relationships are intentionally reciprocal (Porter, 2009). On Twitter, each user is provided with a conversation that is unique to the mix of subscriptions they have made. Users have no explicit, shared set of connections, and this in many ways shapes the style of conversation and communication that occurs via Twitter.

The push and pull of the designers of Twitter and its user base might be mapped out in the many updates to the service over the years. But among the many changes to Twitter—the ability to mark favourite tweets, or delete your own tweet, or directly message another user, for example—are several that are nearly definitional: they make Twitter what it is. We will examine @replies, retweeets, and hashtags, in particular.

TWITTER BEFORE TWITTER

Twitter did not evolve merely through the interaction of the service itself and its users. Many of the ideas and ideals users brought to Twitter came of its users' interactions with similar technologies. In some cases, Twitter has sought to be integrated with this larger ecosystem. So, for example, the early adoption in September of 2006 of RSS (machine-readable) feeds and permalinks (stable URLs that allow individual tweets to be hyperlinked) no doubt owed some of their impetus to users' familiarity with blogging.

As Fred Stutzman (2007) noted, "the most useful metaphor I've found for describing Twitter is to liken it to a web-based IRC (Internet Relay Chat) client. . . . Twitter is essentially a net-based chatroom filled with your friends" (para. 2,

3). IRC was created in 1988 as a system for chatting online in 'channels'—what might in other contexts be called 'rooms'—and the similarity to Twitter might not be immediately apparent. Nonetheless, both allow the text box to be used to issue commands. In IRC, a '/' preceding an entered phrase indicated that it should not be transmitted to the chat room, but rather should be considered as a request directly to the server. Likewise, users of IRC and other chat systems found the need to reduce ambiguity by addressing a comment to a specific user or users, so that transcripts from IRC often consist of lines beginning with the addressee's name and a colon (Werry, 1996).

IRC continues to be used, as do a number of other synchronous chat systems. Some of them have contained their own innovations, like the threaded synchronous chat system Google Wave. An open protocol called Jabber provides a way to encourage person-to-person chat. And a range of microblogging platforms, including Yammer, Jaiku, and identi.ca, among others, enjoyed varying degrees of success. Since users move between these systems, the practices of one are likely to be replicated on others. Likewise, the designers of many of these platforms are aware of each other's interfaces, and are known to borrow elements (Kincaid, 2010).

@REPLIES

Many have used the term 'microblogging' to describe Twitter (e.g. Java, Song, Finin, & Tseng, 2007), suggesting similarities with the practice of blogging. Although blogs have been called a conversational medium (Efimova & de Moor, 2005), they do not resemble conversations in the traditional sense. Even so, the ability to explicitly link to an addressee means that the back-and-forth turn-taking of a conversation can proceed asynchronously. Twitter's open design means that there is no obvious way to respond to someone in the space. It is not even easy to know whether any other user shares a view of the comment stream you see.

The comment section of large-scale blogs faced a similar problem. Since most blogging platforms provided no easy way of threading comments early on, the commenter wishing to respond to an earlier comment needed to employ a workaround to indicate the target of her response—not unlike the user on IRC. The convention that emerged, and eventually diffused to comment sections of many high-traffic blogs, was to address an earlier comment with @username (Halavais & Martin-Elmer, 2009). Eventually, a number of these blogs changed their comment sections to organise comments visually into threads of comments and rejoinders.

It was perhaps not surprising that a similar convention emerged around Twitter. Twitter may not have been built for conversations, and conversations may not be as popular on Twitter as on microblogging platforms that better support threading (Riemer, Diederich, Richter, & Scifleet, 2011), but it remains clear that the @reply functionality is something users wanted and used. Honeycutt and Herring (2009) noted that Twitter supports conversation and collaboration, and suggested that the system might be changed to better support these social uses.

In fact, the evolution of support for @replies marks a pattern for the creation and adoption of a number of the affordances of the Twitter system, a pattern that would be repeated with other user-led practices. As noted in the official Twitter blog (2007), "at some point, Twitter-ers came up with their own method of directing updates to one another using an @ symbol". Twitter responded to this use by linking these @replies to the addressee's profile page, creating an 'in reply to' link, collecting replies on a separate tab, and eventually integrating a 'reply' link on the Web interface (the 'swoosh') to make replying even easier. The Web interface also uses this new reply metadata to display the chain of comments and replies associated with any selected tweet. They have not gone so far as to fully integrate threading of conversations—no doubt what Honeycutt and Herring (2009), among others, had in mind—but several third-party Twitter clients are capable of this.

The lack of threaded conversation became an early area of contention for users of Twitter. Some users saw the open and disconnected nature of tweeted conversations as an advantage. One commentator noted that among the things he enjoyed about Twitter were "half conversations: '@dude55: you are so totally right on, and I believe what you just said was the most poignant, important, compelling sentence that has ever been posted to the internets.' I sure wish I had a friend named dude55" (Cederholm, 2006). For others, the public conversations held less appeal: "the way people are using Twitter right now, it's rapidly becoming the most inefficient and unusable version of IRC *ever*. Look, people, if you want to chat, then get a chat room. You know?" (Meyer, 2007)

Naturally, at the same time as many of these changes were taking place to help support @replies, the Twitter interface was trying to respond to other demands from its rapidly growing user base, and "tweaking the interface of Twitter on almost a daily basis" (Barber, 2007). Not all of these responses have been well received, and in some cases, users were forced to create their own conventions yet again, to thwart Twitter's new functionality.

As of the time of writing, Twitter suggests that @replies are a subset of @mentions. In each case, the author of a tweet includes the Twitter handle of another user, either at the beginning of the tweet (@reply) or somewhere else in the tweet (@mention). These, along with the direct messaging ability ('D username') introduced in late 2006, provide ways of orienting comments toward a particular user. But the Twitter developers continued to see the visibility of partial conversations (like the @dude55 example above) as a problem needing to be fixed (Twitter, 2008). Thanks to a change in the way the system classified tweets, users began, by default, to not see @replies addressing those they did not follow.

In May of 2009, Twitter did this not only by default, but made it mandatory. In a post on the Twitter blog entitled, 'The Replies Kerfuffle' (Twitter, 2009a), the administrators explained that the change only affected the 3% of users who wanted to see replies to people they were not following. They claimed that it was both a technical and a design issue—they wanted it to be clear who would see the @replies and who would not. But for many users, the transparency in the change was lacking, as was the outcome. After all, other forms of @mentions behaved just as they always had: if you were subscribed, you would see them. The particular case of the @mention being at the start of a tweet (that is, an @reply) was treated differently.

The response from part of the Twitter user base to what the official Twitter blog called "a small change" was immediate. The day after the change, the top trending hashtag on Twitter was #fixreplies (Calore, 2009). Many felt that this was a tone-deaf change that removed an important part of what made Twitter special: the serendipitous discovery of interesting ideas and people. There was also speculation that this new feature was at least in part a response not to user desires, but to a quick fix for an underperforming database that was taxed by the large number of replies that needed to be presented to users.

Just as with the @reply practice itself, it is not entirely clear where the 'dot-at' workaround began. It was clear that to return to the previous functionality, you needed to turn @replies into @mentions. This could be done by inserting any character into the start of a tweet, except a space, and following it with your @reply address. Eventually, the most widely used character was "the simplest, smallest, least annoying punctuation mark", the full stop (Gough, 2009)—for example, ".@halavais Thanks, enjoyed the dirigible cruise!" This gathered these types of posts as @mentions rather than as @replies, though they probably are a bit of both.

RETWEETING

While using the @mention and @reply as a way of threading discussion could find its roots in earlier conversational media, the retweet was, in some sense, unique to Twitter, and based on its model of multiple, user-centric publics (on these personal publics, see also Schmidt, Chapter 1 in this volume). Each person's view of Twitter was at once shaped by the group of other users to which they had subscribed, and by a separate list of users that had subscribed to their own tweets. Retweeting a message represented both an affirmation of the contents of a particular tweet, and a way of spreading a conversation more widely. In one sense, it might be seen as similar to the 'people's microphone' that gained renewed popularity during the Occupy Wall Street protests in New York. To work around restrictions on bullhorns and other forms of amplification, a speaker's words would be repeated by a chorus of those who were within earshot, spreading in waves around Zuccotti Park (Kim, 2011).

But as boyd, Golder, and Lotan (2010) suggested, retweets do more than spread messages more widely—they invite a structure for conversation and comment. In fact, there are a wide range of reasons people choose to retweet a message, and the practices surrounding the retweet are equally various. Though the structure of 'RT @halavais My dog barks some' is the most common formulation, others use 'via' or 'by'. A less frequently seen 'MT' or 'modified tweet' indicates that the original has been shortened or otherwise changed. All of these point to a fairly broad set of uses for retweets that evolved in various parts of the Twittersphere.

After incorporation of the @mention workaround into the Twitter system itself, it seemed clear that other patterns of user behaviour might also be incorporated. As one commentator put it: "in true web 2.0 style the people's usage habits are an input to the design eg. [sic] replies were so heavily used it became a feature, and we just know re-tweets and hashtags will be coming next" (Tropea, 2009). Perhaps recognising the criticism it had received with the switch in how @replies were handled, Twitter rolled out the institutionalisation of retweets more carefully and slowly. In Twitter's implementation, the retweet marker was no longer displayed as part of the textual line itself, instead being indicated in the metadata. This new form of retweet looks identical to the original tweet, but now shows up in another user's stream, and can be seen by their followers. Not only was it missing the familiar 'RT', but also any commentary from the person retweeting. Further, unlike in the old system, users were prevented from (intentionally or unintentionally) retweeting private tweets.

The response was decidedly mixed. While some applauded the new 'button' retweets, those who followed Twitter closely found the official implementation of retweets lacking, suggesting that they might go unused, or might shape the nature of Twitter use in a less than satisfactory way (Grifantini, 2009). The traditional manual retweet (inserting 'RT' into the tweet itself) allowed users to set context, to shape diffusion, and to preserve deleted tweets—in other words, it provided for a wide range of behaviours and expectations that had been built up around the service.

Nonetheless, today the new style of retweet, fully integrated with the official Twitter interface, like the new form of @replies, has become the favoured approach, especially among new users. (Confusingly, at present the Twitter mobile app still provides for 'quoted tweets' as well.) When Twitter administrators first communicated their intention to incorporate retweeting into the interface, they noted that "some of Twitter's best features are emergent—people inventing simple but creative ways to share, discover, and communicate" (Twitter, 2009b). As with @replies, it seems clear that Twitter responded to emergent retweeting practices within the community. And as with @replies, it seems clear that the platform-level solution to the problem only partially reflected the intentions and desires of a diverse user community.

#HASHTAGS

Hashtags represent a way of indicating textually keywords or phrases especially worth indexing. Once it was possible to search Twitter, finding a particular set of topics should have meant simply searching for particular keywords, making hashtags redundant. However, by using the # character to mark particular keywords, Twitter users communicate a desire to share particular keywords folksonomically. The approach also provided an opportunity for third-party providers to track hashtag use, and aggregate tweets with the same tag.

Unlike the use of the @ symbol, which was likely borrowed from other conversational media and appeared at many points at once, some have suggested that the hashtag does have an originator: Chris Messina (2007), who tweeted on 23 August 2007, "how do you feel about using # (pound) for groups. As in #barcamp [msg]?" Messina originally referred to these as "channel tags", an idea he had borrowed from the annotation used to reference IRC channels (Gannes, 2010). He championed its use, and it came to provide a way of tying together groups that were not engaged in conversation, creating new shared publics (Bruns & Burgess, 2011).

As Susan Orlean (2010) suggested, hashtags "have also undergone mission creep, and now do all sorts of interesting things", no doubt with some help from those who employ them. They have been used as prompts for conversation, to crowdsource ideas or resources, and often to express sarcasm or parenthetical commentary on a tweet. The original use may have been to help form topical groups, but because the hashtags were reused by so many people in so many tweets, they showed up more frequently in the Trending Tweets listed on Twitter. In 2009, for example, #iranelection topped the charts of trending news tweets, along with #musicmonday.

The focus on hashtags came with new attention to these trending topics. There has been a trending topics widget of one form or another on the Twitter home page for several years. Twitter was becoming less of a sociable medium, and more of a distributed, mass medium, at least in the eyes of the designers: "all of our recent changes embrace the notion that Twitter is not just for status updates anymore. It's a network where information is exchanged and consumed at a rapid clip every second of the day" (Twitter, 2010a).

When compared with other innovations, hashtags have largely been left out of the process of becoming integrated with the Twitter platform. There was an alternative proposal presented by Twitter in 2007 called 'tracking' that allowed for SMS tracking of keywords, but this made little impact. Third-party applications integrated hashtags early on, allowing for easy grouping of tweets by topic, and hashtags were eventually made clickable on the Twitter website as well.

CONCLUSION: PRACTICE AND PLATFORM

The pattern observed in these three cases can be found in many of the features that make up Twitter today. Twitter users find that they want to use tweets to accomplish something. They create a social and technological practice (whether that is the use of a single marker character, a set of common practices, or an application that accesses Twitter programmatically) that makes it possible to use Twitter to their intended ends. Eventually, the developers of the Twitter platform appropriate or incorporate some version of those practices, stabilising them into what every new user thinks of as 'Twitter'.

This makes up a large part of the creation myth of Twitter: it is a platform co-created by its users. But this is also a story that would be familiar to anyone who studies the evolution of socio-technical systems (Lamb & Kling, 2003). A range of relevant groups influence the shape of the technology, and it changes over time to meet the needs of the relevant groups. Not infrequently, those needs and applications are not the ones expected or intended by the developers. This applies not just to Web 2.0, but to earlier networked technologies, like the telephone and radio.

It is also important to note that with every appropriation of the practices of Twitter users into the platform itself, there is a loss of flexibility and diversity. The variety of different ways to signal a retweet has not been entirely subsumed by the internalised, hidden structure of the new-style retweet. But little by little, the alternatives are pushed to the margins by the default.

The problem with this process is that it is not always clear that the interests of the developers or owners of Twitter as a platform coincide with those of the users. More to the point, since the users' agendas do not and cannot coincide with one another, it would be impossible for the platform's standardisation of these practices to meet the needs of everyone. Just as blogging platforms restricted the universe of blogging practice over time (Siles, 2011), the process of taming Twitter behaviours ultimately reduces the possibilities and potential of the technology. When the practices are merely widespread—say, with the adoption of hashtags—likeminded users can find spaces of resisting this process, promising (as one user of Twitter did) to unfollow those who polluted his stream of tweets with hashtags. This becomes far more difficult when it becomes a part of the platform itself.

A post to the official Twitter blog in May of 2010 (Twitter, 2010b) attempted to more clearly delineate what 'The Twitter Platform' laid claim to, and where it provided space for open engagement. Even here, the language concealed as much as it revealed, calling Twitter variously a "platform", an "ecosystem", and an "investment". The former choice of identification, as Tarleton Gillespie (2010) suggested, is not an accident; platforms have politics. Like the creation myth of Twitter as a whole, which posits the community as a co-creator, this statement is marked by what remains unclear: how a platform can also be an investment.

Twitter has enjoyed a relatively glowing lack of criticism largely by comparison with Facebook, which has more users, tends to be more all-consuming, and whose changes therefore lead to more pointed rancour. Nonetheless, the processes are similar, as they are for many of the widespread platforms of interaction. Many of these hold themselves up as a blank slate, defined largely by the interactions of the user community. Even without ascribing Twitter's motivations to, for example, encouraging uses by celebrities and others who might provide a better opportunity for monetisation, the very process of appropriating user behaviours tends to foreshorten possibilities, leading to standardisation that excludes communities.

In another context, John Fiske (1986) argued that opening a broadcast text to multiple interpretations provided for a larger audience than a closed, stable narrative. Something similar might be said of social media platforms like Twitter: simple and open provides the greatest opportunity for interaction. Whether simple and open can lead to brand and profitability remains very much an open question, and perhaps the greatest tension in Twitter's future.

REFERENCES

Arceneaux, N., & Weiss, A. S. (2010). Seems stupid until you try it: Press coverage of Twitter, 2006–9. *New Media & Society, 12*(8), 1262–1279.

Barber, G. (2007, 7 Feb.). WTF Twitter: Interface and iterative development. *Man with no blog.* Retrieved from http://manwithnoblog.com/2007/02/07/wtf-twitter-interface-and-iterative-development/

Bijker, W. E. (1992). The social construction of fluorescent lighting, or how an artifact was invented in its diffusion stage. In W. E. Bijker & J. Law (Eds.), *Shaping technology/building society: Studies in sociotechnical change* (pp. 75–103). Cambridge, MA: MIT Press.

Bijker, W. E., & Law, J. (Eds.). (1992). *Shaping technology/building society: Studies in sociotechnical change.* Cambridge, MA: MIT Press.

boyd, d., Golder, S., & Lotan, G. (2010). Tweet, tweet, retweet: Conversational aspects of retweeting on Twitter. *Proceedings of the 43rd Hawaii International Conference on System Sciences (HICSS-43).* 16 Jan. 2010. Kauai, HI: IEEE. Retrieved from http://www.danah.org/papers/TweetTweetRetweet.pdf

Bruns, A., & Burgess, J. E. (2011). The use of Twitter hashtags in the formation of ad hoc publics. In *Sixth European Consortium for Political Research General Conference.* 25–27 Aug. 2011, University of Iceland, Reykjavik, Iceland. Retrieved from http://eprints.qut.edu.au/46515/

Calore, M. (2009). Much ado about @reply. *Webmonkey.* Retrieved from http://www.webmonkey.com/2009/05/much_ado_about_reply/

Cederholm, D. (2006, 20 Dec.). Reasons I like Twitter. *Simple Bits.* Retrieved from http://simplebits.com/notebook/2006/12/20/twitter-2/

Dyson, G. (1997). *Darwin among the machines: The evolution of global intelligence.* Cambridge, MA: Perseus Books.

Efimova, L., & de Moor, A. (2005). Beyond personal webpublishing: An exploratory study of conversational blogging practices. *Proceedings of the 38th Hawaii International Conference on System Sciences (HICSS-38).* 3–6 Jan. 2005. Big Island, HI. Retrieved from https://doc.novay.nl/dsweb/Get/Version-22432/HICSS05_Efimova_deMoor.pdf

Fiske, J. (1986). Television: Polysemy and popularity. *Critical Studies in Mass Communication, 3*(4), 391–408.

Fragoso, S., Rebs, R. R., & Barth, D. L. (2012). Interface affordances and social practices in online communication systems. In *Proceedings of the International Working Conference on Advanced Visual Interfaces* (pp. 50–57). 22–25 May 2012. Capri Island (Naples), Italy.

Gannes, L. (2010, 30 Apr.). The short and illustrious history of Twitter #hashtags. *Gigaom*. Retrieved from http://gigaom.com/2010/04/30/the-short-and-illustrious-history-of-twitter-hashtags/

Gillespie, T. (2010). The politics of 'platforms.' *New Media & Society, 12*(3), 347–364.

Gough, J. (2009, 28 Oct.). Twitter & tweets: Who can read what (and how and why to use the dot). *Julian Gough*. Retrieved from http://www.juliangough.com/journal/2009/10/28/twitter-tweets-who-can-read-what-and-how-and-why-to-use-the.html

Grifantini, K. (2009, 26 Aug.). The evolution of retweeting: Formalizing the retweet may change people's behavior. *MIT Technology Review*. Retrieved from http://www.technologyreview.com/news/415043/the-evolution-of-retweeting/

Halavais, A., & Martin-Elmer, H. (2009, October). *Back@you: Tracing the diffusion of a conversational convention*. Paper presented at the Association of Internet Researchers conference, Milwaukee, WI.

Honeycutt, C., & Herring, S. C. (2009). Beyond microblogging: Conversation and collaboration via Twitter. *Proceedings of the 42nd Hawaii International Conference on System Sciences (HICSS-42)* (pp. 1–10). 5-8 Jan. 2009. Big Island, HI.

Java, A., Song, X., Finin, T., & Tseng, B. (2007). Why we twitter: Understanding microblogging usage and communities. Joint 9th WEBKDD and 1st SNA-KDD Workshop, San Jose, CA. Retrieved from http://aisl.umbc.edu/resources/369.pdf

Johansmeyer, T. (2009, 21 Nov.). Twitter finally reveals revenue ambition, wants to be Google—but not yet. *BloggingStocks*. Retrieved from http://www.bloggingstocks.com/2009/11/21/twitter-finally-reveals-revenue-ambition-wants-to-be-google/

Kim, R. (2011, 3 Oct.). We are all human microphones now. *The Nation*. Retrieved from http://www.thenation.com/blog/163767/we-are-all-human-microphones-now

Kincaid, J. (2010, 19 Apr.). Facebook borrows another feature from Twitter (or was it FriendFeed?): The hovercard. *TechCrunch*. Retrieved from http://techcrunch.com/2010/04/19/facebook-borrows-another-feature-from-twitter-or-was-it-friendfeed-the-hovercard/

Kline, R., & Pinch, T. (1996). Users as agents of technological change: The social construction of the automobile in the rural United States. *Technology and Culture, 37*(4), 763–795.

Lamb, R., & Kling, R. (2003). Reconceptualizing users as social actors in information systems research. *MIS Quarterly, 27*(2), 197–235.

Messina, C. (2007, 23 Aug.). how do you feel about using # (pound) for groups. As in #barcamp [msg]? *@chrismessina* (Twitter). Retrieved from https://twitter.com/chrismessina/status/223115412

Meyer, E. (2007, 21 Jan.). The Twitters. *meyerweb.com*. Retrieved from http://meyerweb.com/eric/thoughts/2007/01/21/the-twitters/

Orlean, S. (2010, 29 June). Hash. *Free Range (New Yorker Blogs)*. Retrieved from http://www.newyorker.com/online/blogs/susanorlean/2010/06/hash.html

Pinch, T. J., & Bijker, W. E. (1984). The social construction of facts and artefacts: Or how the sociology of science and the sociology of technology might benefit each other. *Social Studies of Science, 14*(3), 399–441.

Porter, J. (2009). Relationship symmetry in social networks: Why Facebook will go fully asymmetric. *Bokardo*. Retrieved from http://bokardo.com/archives/relationship-symmetry-in-social-networks-why-facebook-will-go-fully-asymmetric/

Riemer, K., Diederich, S., Richter, A., & Scifleet, P. (2011). Tweet talking: Exploring the nature of microblogging at Capgemeni Yammer. (University of Sydney Business Information Systems Working Paper Series WP2011-02). Retrieved from http://hdl.handle.net/2123/7226

Siles, I. (2011). From online filter to Web format: Articulating materiality and meaning in the early history of blogs. *Social Studies of Science*, 41(5), 737–758.

Stutzman, F. (2007). The 12-minute definitive guide to Twitter. *Dev.aol.com*. Retrieved from http://dev.aol.com/article/2007/04/definitive-guide-to-twitter

Tropea, J. (2009). Twitter 3 years on, and why it's the killer app! *Library Clips*. Retrieved from http://libraryclips.blogsome.com/2009/03/04/twitter-3-years-on-and-why-its-the-killer-app/

Twitter. (2007, 29 May). Are you twittering @ me? *Twitter Blog*. Retrieved from http://blog.twitter.com/2007/05/are-you-twittering-me.html

Twitter. (2008, 12 May). How @replies work on Twitter (and how they might). *Twitter Blog*. Retrieved from http://blog.twitter.com/2008/05/how-replies-work-on-twitter-and-how.html

Twitter. (2009a, 14 May). The replies kerfuffle. *Twitter Blog*. Retrieved from http://blog.twitter.com/2009/05/replies-kerfuffle.html

Twitter. (2009b, 13 Aug.). Project retweet: Phase one. *Twitter Blog*. Retrieved from http://blog.twitter.com/2009/08/project-retweet-phase-one.html

Twitter. (2010a, 30 Mar.). Tweaking the Twitter homepage. *Twitter Blog*. Retrieved from http://blog.twitter.com/2010/03/tweaking-twitter-homepage.html

Twitter. (2010b, 24 May). The Twitter platform. *Twitter Blog*. Retrieved from http://blog.twitter.com/2010/05/twitter-platform.html

van Dijck, J. (2011). Tracing Twitter: The rise of a microblogging platform. *International Journal of Media & Cultural Politics*, 7(3), 333–348.

Werry, C. C. (1996). Linguistic and interactional features of Internet Relay Chat. In S. C. Herring (Ed.), *Computer-mediated communication: Linguistic, social, and cross-cultural perspectives* (pp. 47–64). Amsterdam, The Netherlands: John Benjamins.

The Politics of Twitter Data

CHAPTER **4**   Cornelius Puschmann and Jean Burgess

 there's #bigdata on Twitter, but the politics of working with it are highly complicated

THE BIG DATA MOMENT

> Data is not free, and there's always someone out there that wants to buy it. As an end-user, educate yourself with how the content you create using someone else's service could ultimately be used by the service-provider. (Jud Valeski, CEO of Gnip, as quoted in Steele, 2011, para. 19)

> There are significant questions of truth, control, and power in Big Data studies: researchers have the tools and the access, while social media users as a whole do not. Their data were created in highly context-sensitive spaces, and it is entirely possible that some users would not give permission for their data to be used elsewhere. (boyd & Crawford, 2012, p. 673)

Talk of Big Data seems to be everywhere. Indeed, the apparently value-free concept of 'data' has seen a spectacular broadening of popular interest, shifting from the dry terminology of labcoat-wearing scientists to the buzzword *du jour* of marketers. In the business world, data is increasingly framed as an economic asset of critical importance, a commodity on a par with scarce natural resources (Backaitis, 2012; Rotella, 2012).

It is social media that has most visibly brought the Big Data moment to media and communication studies, and beyond it, to the social sciences and humanities. Social media data is one of the most important areas of the rapidly growing data market (Manovich, 2012; Steele, 2011). Massive valuations are attached to companies that directly collect and profit from social media data, such as Facebook and Twitter, as well as to resellers and analytics companies like Gnip and DataSift. The expectation attached to the business models of these companies is that their privileged access to data and the resulting valuable insights into the minds of consumers and voters will make them irreplaceable in the future. Analysts and consultants argue that advanced statistical techniques will allow the detection of ongoing communicative events (natural disasters, political uprisings) and the reliable prediction of future ones (electoral choices, consumption).

These predictions are made possible through cheap, networked access to cloud-based storage space and processing power, paired with advanced computational techniques to investigate complex phenomena such as language sentiment (Thelwall, Chapter 7 in this volume; Thelwall, Buckley, & Paltoglou, 2011), communication during natural disasters (Sakaki, Okazaki, & Matsuo, 2010), and information diffusion in large networks (Bakshy, Rosenn, Marlow, & Adamic, 2012). Such methods are hailed as superior tools for the accurate modelling of social processes and have a growing base of followers among the proponents of "digital methods" (Rogers, 2009) and "computational social science" (Lazer et al., 2009). While companies, governments, and other stakeholders previously had to rely on vague forecasts, the promise of these new approaches is ultimately to curb human unpredictability through information. The traces created by the users of social media platforms are harvested, bought, and sold; an entire commercial ecosystem is forming around social data, with analytics companies and services at the helm (Burgess & Bruns, 2012; Gaffney & Puschmann, 2012, and Chapter 5 in this volume).

Yet, while the data in social media platforms is sought after by companies, governments, and scientists, the users who produce it have the least degree of control over "their" data. Platform providers and users are in a constant state of negotiation regarding access to and control over information. Both on Twitter and on other platforms, this negotiation is conducted with contractual and technical instruments by the provider, and with ad hoc activism by some users.

The complex relationships among platform providers, end users, and a variety of third parties (e.g., marketers, governments, researchers) further complicate the picture. These nascent conflicts are likely to deepen in the coming years, as the value of data increases while privacy concerns mount and those without access feel increasingly marginalised.

Our chapter approaches Twitter through the lens of "platform politics" (Gillespie, 2010), focussing in particular on controversies around user data access, ownership, and control. We characterise different actors in the Twitter ecosystem: private and institutional end users of Twitter, commercial data resellers such as Gnip and DataSift, data scientists, and finally Twitter, Inc. itself; and describe their conflicting interests. We furthermore study Twitter's Terms of Service and application programming interface (API) as material instantiations of regulatory instruments used by the platform provider, and argue for more promotion of data rights and literacy to strengthen the position of end users.

TWITTER AND THE POLITICS OF PLATFORMS

The creation of social media data is governed by an intricate set of dynamically shifting and often competing rules and norms. As business models change, the emphasis on different affordances of the platform changes, as do the characteristics of the assumed end user under the aspects of value-creation for the company. Twitter has been subject to such shifts throughout its brief history, as the service adapts to a growing user community with a dynamic set of needs.

In this context, there has been a recent critique of a perceived shift from an 'open' Internet (where open denotes a lack of centralised control and a divergent, rather than convergent, software ecosystem), towards a more 'closed' model with fewer, more powerful corporate players (Zittrain, 2008). Common targets of this critique include Google, Facebook, and Apple, who are accused of monopolising specific services, and of placing controls on third-party developers who wish to exploit the platforms or contribute applications which are not in accordance with the strategic aims of the platform providers. In Twitter's case, the end of the Web 2.0 era, supposedly transferring power to the user (O'Reilly, 2005), is marked by the company's shift to a more media-centric business model relying firstly on advertising and corporate partnerships and, crucially for this chapter, on reselling the data produced collectively by the platform's millions of users (Burgess & Bruns, 2012; van Dijck, 2011). This shift has been realised materially in the architecture of the platform—including not only its user interface, but also the affordances of its API and associated policies, affecting the ability of third-party developers, users, and researchers to exploit or innovate upon the platform.

There have been several recent controversies specifically around Twitter data access and control:

- The increasing contractual limitations placed on content through instruments such as the Developer Display Requirements (Twitter, 2012c), that govern how tweets can be presented in third-party utilities, or the Developer Rules of the Road (Twitter, 2012b), that forbid sharing large volumes of data;
- The requirement for new services built on Twitter to provide benefits beyond the service's core functionality;
- Actions against platforms which are perceived by Twitter to be in violation of these rules, e.g. Twitter archiving services such as 140Kit and Twapperkeeper.com, business analytics services such as PeopleBrowsr, and aggregators like IFTTT.com;
- The introduction of the Streaming API as the primary gateway to Twitter data, and increasing limitations placed on the REST API as a reaction to growing volumes of data generated by the service;
- The content licensing arrangements made between Twitter and commercial data providers Gnip and DataSift (charging significant rates for access to tweets and other social media content); and
- The increasing media integration of the service, emphasising the role of Twitter as "an information utility" (Twitter co-founder Jack Dorsey, as quoted in Arthur, 2012).

In the following, we relate these aspects to different actors with a stake in the Twitter ecosystem.

CONFLICTING INTERESTS IN THE TWITTER ECOSYSTEM

Lessig (1999) named four factors shaping digital sociotechnical systems: the market, the law, social norms, and architecture (code and data). The regulation of data handling by the service provider through the Terms of Service and the API is of particular interest in this context. As outlined above, Twitter seeks to regulate use of data by third parties through the Terms and the API, assigning secondary roles to the law (which the Terms frequently seek to extend) and social norms (which are inscribed and institutionalised in various ways through both the interface and widespread usage conventions).

TWITTER, INC.

Platform providers like Twitter, Inc. have a vested interest in the information that flows through their service, and as outlined above, these interests have become more pronounced over time, as the need for a plausible business model has grown more urgent. The users' investment of time and energy is the foundation of the platform's value, and therefore growing and improving the service is of vital importance. In the case of Twitter, this strategy is exemplified by the changes made to the main page over the years. Whereas initially Twitter asked playfully "What are you doing?", this invitation has long since been replaced by a more utilitarian and consumer-oriented exhortation to "Find out what's happening, right now, with the people and organizations you care about," stressing Twitter's relevance as a real-time information hub for business and the mainstream media.

Twitter's business strategy clearly hinges strongly on establishing itself as an irreplaceable, real-time information source, and on playing a vital part in the corporate media ecosystem of news propagation. Under its current CEO Dick Costolo, Twitter has moved firmly towards an ad-supported model of "promoted tweets" similar to Google's AdWord model. Exercising tighter control over how users experience and interact with the service than in the service's fledgling days is a vital component of this strategy.

Data is a central interest of Twitter's in its role as a platform provider, not solely because it aims to monetise information directly, but because the value of the data determines the value of the company to potential advertisers. Increasing the relevance of Twitter as a news source is crucial, while maintaining a degree of control over the data market that is evolving under the auspices of the company.

END USERS

Twitter's end users are private citizens, celebrities, journalists, businesses, and organisations; in other words, they can be both individuals and collectives, with aims that are strategic, casual, or a dynamic combination of both. What unites these different stakeholders is that they have an interest in being able to use Twitter free of charge, and that data is merely a by-product of their activity, but not their reason for using the platform. They do, however, have an interest in controlling their privacy and in being able to do the same things with their information that both Twitter and third-party services are able to do. While the Terms spell out certain rights that users have and constraints that they are under, the rights can only be exercised through the API, while the constraints are enforced by legal means (Beurskens, Chapter 10 in this volume).

End users have diverse reasons for wanting to control their data, including privacy concerns, impression management, fear of repressive governments, the desire to switch from one social media service to another, and curiosity about one's own usage patterns and behaviour. Giving users the ability to exercise these rights not only benefits users, but also platform providers, because it fosters trust in the service. The perception that platform providers are acting against users' interests behind their backs can be successfully countered by implementing tools that allow end users greater control of "their" information.

DATA TRADERS AND ANALYSTS

Both companies re-selling data under license from Twitter and their clients have interests which are markedly different from those of the company and platform end users. While Twitter seeks long-term profits guaranteed by controlled access to the platform and growing relevance, and end users may want to guard their privacy and control their information while being able to use a free service, data traders want access to vast quantities of data that allow them to model and predict user behaviour on an unprecedented scale. Access to unfiltered, real-time information (provided to them in the form of the Streaming API) is vital, while to their clients the predictive power of the analytics is important. Neither is very concerned with the interests of end users, who are treated similarly to subjects in an experiment of gigantic proportions. Privacy concerns are relegated to the background, as they would reduce the quality of the analytics, and they are effectively traded for free access to the platform. What is also neglected is the ability to access historical Twitter data, as businesses by and large want to monitor their current performance, with only limited need to peer into the past.

A key aim of data traders is to commodify data and to guard it carefully against infringers operating outside the data market. In an interview, data wholesaler Gnip's CEO Jud Valeski assigned the responsibility to end users, recommending that they educate themselves about the public and commodified status of the data generated by their personal media use:

> Read the terms of service for social media services you're using before you complain about privacy policies or how and where your data is being used. Unless you are on a private network, your data is treated as public for all to use, see, sell, or buy. Don't kid yourself. (Valeski, as quoted in Steele, 2011, para. 27)

Two things stand out in this statement: the claim that data on Twitter is public, and the inference that because it is public, it should be treated as "for all to use, see, sell, or buy." The public-private dichotomy applies to Twitter data only in the sense that what is posted there is accessible to anyone accessing the Twitter website or using a third-party client (with the exception of direct messages and protected accounts). But the question of access is legally unrelated to the issue of ownership—rights to data cannot be inferred from technical availability alone, otherwise online content piracy would be legal. In the same interview, Valeski also consistently referred to platform providers such as Twitter as "publishers", and warned of "black data markets".

TERMS OF SERVICE AND API AS INSTRUMENTS OF REGULATION

Since its launch in March 2006, Twitter has steadily added documents that regulate how users can interact with its service. In addition to the Terms (Twitter, 2012a), two items stand out: the Developer Rules of the Road (Twitter, 2012b) and the Developer Display Requirements (Twitter, 2012c), which were added to the canon in September 2012. Twitter's Terms have changed considerably since Version 1, published when the platform was still in its infancy. In relation to data access, they lay out how users can access information, what rights Twitter reserves to the data that users generate, and what restrictions apply. Initially, the Terms spelled out the users' rights with respect to their data, i.e., each user's own personal content on the platform:

> By submitting, posting or displaying Content on or through the Services, you grant us a worldwide, non-exclusive, royalty-free license (with the right to sublicense) to use, copy, reproduce, process, adapt, modify, publish, transmit, display and distribute such Content in any and all media or distribution methods (now known or later developed). (Twitter, 2012a, para. 5-1)

This permission to use the data is supplemented with the permission to pass it on to sanctioned partners of Twitter:

> You agree that this license includes the right for Twitter to make such Content available to other companies, organizations or individuals who partner with Twitter for the syndication, broadcast, distribution or publication of such Content on other media and services, subject to our terms and conditions for such Content use. (Twitter,2012a, para. 5-2)

Third parties are also addressed in the Terms and encouraged to access and use data from Twitter: "We encourage and permit broad re-use of Content. The

Twitter API exists to enable this" (Twitter, 2012a, para. 8-2). However, the exact meaning of *re-use* in this context remains unclear, and reading the other above-mentioned documents, the impression is that data analysis is not the kind of re-use intended by the Terms. Neither is it made explicit whether the content referred to is still the users' own content or all data on the platform (i.e., the data of other users). Furthermore, it seems that it is no longer Twitter's users who are addressed, but third parties, as no referent is given. Reference to the API also suggests that a technologically savvy audience is addressed, rather than any typical user of Twitter.

The claim of encouraging broad re-use is further modified by the Developer Rules of the Road, the second document governing how Twitter handles data:

> You will not attempt or encourage others to: sell, rent, lease, sublicense, redistribute, or syndicate access to the Twitter API or Twitter Content to any third party without prior written approval from Twitter. If you provide an API that returns Twitter data, you may only return IDs (including tweet IDs and user IDs). You may export or extract non-programmatic, GUI-driven Twitter Content as a PDF or spreadsheet by using 'save as' or similar functionality. Exporting Twitter Content to a datastore as a service or other cloud based service, however, is not permitted. (Twitter, 2012b, para. 8)

Here, too, developers, rather than end-users, are the implicit audience. Not only is the expression "non-programmatic, GUI-driven Twitter Content" fairly vague, the restrictions with regard to means of exporting and saving the data make the "broad re-use" that Twitter encourages in the Terms difficult to achieve in practice. They also stand in contradiction to the Terms, which state that

> Except as permitted through the Services (or these Terms), you have to use the Twitter API if you want to reproduce, modify, create derivative works, distribute, sell, transfer, publicly display, publicly perform, transmit, or otherwise use the Content or Services. (Twitter, 2012a, para. 8-2)

Thus, only by using the API and obtaining written consent from Twitter is it possible to redistribute information to others. This raises two barriers—requiring permission, and having the technical capabilities needed to interact with the data—that must both be overcome, narrowing the range of actors able to do so to a small elite. In relation to this form of exclusion, boyd and Crawford (2012) spoke of data "haves" and "have-nots", noting that only large institutions with the necessary computational resources will be able to compete. Studies such as those by Kwak, Lee, Park, and Moon (2010) and Romero, Meeder, and Kleinberg (2011) are only possible through large-scale institutional or corporate involvement, as both technical and contractual challenges must be met. While vast quantities of data are theoretically available via Twitter, the process of obtaining it is in practice complicated, and requires a sophisticated infrastructure to capture information (beyond one's personal archive) at scale.

Actions such as the one against PeopleBrowsr, an analytics company that was temporarily cut off from access to the API, support the impression that Twitter is exercising increasingly tight control over the data it delivers through its infrastructure (PeopleBrowsr, 2012). PeopleBrowsr partnered with Twitter for over four years, paying for privileged access to large volumes of data, but as a result of its exclusive partnerships with specific data resellers, Twitter unilaterally terminated the agreement, citing PeopleBrowsr's services as incompatible with its new business model.

CONCLUSION: DATA RIGHTS AND DATA LITERACY

Contemporary discussions of end user data rights have focussed mainly on technology's disruptive influence on established copyright regimes, and industry's attempts to counter this disruption. Vocal participants in the digital rights movement are primarily concerned with copyright enforcement and Digital Rights Management (DRM), which, so the argument goes, hinder democratic, cultural participation by preventing the free use, embellishment, and re-use of cultural resources (Postigo, 2012a, 2012b). The lack of control that most users can exercise over data they have themselves created in platforms such as Twitter seems, in some respects, a much more pronounced issue.

Gnip's CEO Jud Valeski framed the "owners" of social media data to be the platform providers, rather than end users, a significant conceptual step forward from Twitter's own characterisation, which endows the platform with the licence to reuse information, but frames end users as its owners (as cited in Steele, 2011). Valeski's logic is based on the need to legitimise the data trade—only if data is a commodity, and if it is owned by the platform provider rather than the individual users producing the content, can it be traded. It furthermore privileges the party controlling the platform technology as morally entitled to ownership of the data flowing through it.

Driscoll (2012) noted the ethical uncertainties surrounding the issues of data ownership, access, and control, and pointed to the promotion of literacy as the only plausible solution:

> Resolving the conflict between users and institutions like Twitter is difficult because the ethical stakes remain unclear. Is Twitter ethically bound to explain its internal algorithms and data structures in a language that its users can understand? Conversely, are users ethically bound to learn to speak the language of algorithms and data struc-

tures already at work within Twitter? Although social network sites seem unlikely to reveal the details of their internal mechanics, recent 'code literacy' projects indicate that some otherwise non-technical users are pursuing the core competencies necessary to critically engage with systems like Twitter at the level of algorithm and database. (p. 4)

In the current state of play, the ability of individual users to effectively interact with "their" Twitter data hinges on their ability to use the API, and on their understanding of its technical constraints. Beyond the technical know-how that is required to interact with the API, issues of scale arise: the Streaming API's approach to broadcasting data as it is posted to Twitter requires a very robust infrastructure as an endpoint for capturing information (see Gaffney & Puschmann, Chapter 5 in this volume). It follows that only corporate and government actors—who possess both the intellectual and financial resources to succeed in this race—can afford to participate, and that the emerging data market will be shaped according to their interests. End users (both private individuals and non-profit institutions) are without a place in it, except in the role of passive producers of data. The situation is likely to stay in flux, as Twitter must at once satisfy the interests of data traders and end users, especially with regard to privacy regulation. However, as neither the contractual nor the technical regulatory instruments used by Twitter currently work in favour of end users, it is likely that they will continue to be confined to a passive role.

REFERENCES

Arthur, C. (2012, 23 Jan). Twitter too busy growing to worry about Google+, says Dorsey. *Guardian.co.uk*. Retrieved from http://www.guardian.co.uk/technology/2012/jan/23/twitter-dorsey

Backaitis, V. (2012). Data is the new oil. *CMS Wire*. Retrieved from http://www.cmswire.com/cms/information-management/data-is-the-new-oil-014966.php

Bakshy, E., Rosenn, I., Marlow, C., & Adamic, L. (2012). The role of social networks in information diffusion. *Proceedings of the 21st International Conference on the World Wide Web (WWW '12)* (pp. 1–10). April, 2012. Lyon, France. New York, New York: ACM.

boyd, d., & Crawford, K. (2012). Critical questions for Big Data: Provocations for a cultural, technological, and scholarly phenomenon. *Information, Communication and Society 15*(5), 662–679.

Burgess, J., & Bruns, A. (2012). Twitter archives and the challenges of 'Big Social Data' for media and communication research. *M/C Journal, 15*(5). Retrieved from http://journal.media-culture.org.au/index.php/mcjournal/article/viewArticle/561

Driscoll, K. (2012). From punched cards to 'Big Data': A social history of database populism. *communication +1, 1*, article 4. Retrieved from http://scholarworks.umass.edu/cpo/vol1/iss1/4

Gaffney, D., & Puschmann, C. (2012). Game or measurement? Algorithmic transparency and the Klout score. *Proceedings of #influence12: Symposium & Workshop on Measuring Influence on Social Media* (pp. 1–2). 28–29 Sep. 2012. Halifax, Nova Scotia, Canada.

Gillespie, T. (2010). The politics of 'platforms.' *New Media & Society, 12*(3), 347–364.

Kwak, H., Lee, C., Park, H., & Moon, S. (2010). What is Twitter, a social network or a news media? *Proceedings of the 19th International Conference on the World Wide Web (WWW '10)* (pp. 591–600). 26–30 Apr. 2010. Raleigh, NC.

Lazer, D., Pentland, A., Adamic, L., Aral, S., Barabasi, A-L., Brewer, D., . . . Van Alstyne, M. (2009). Computational social science. *Science, 323* (5915), 721–723. doi:10.1126/science.1167742

Lessig, L. (1999). *Code and other laws of cyberspace.* New York, NY: Basic Books.

Manovich, L. (2012). Trending: The promises and the challenges of Big Social Data. In M. K. Gold (Ed.), *Debates in the digital humanities* (pp. 460–475). Minneapolis, MN: University of Minnesota Press.

O'Reilly, T. (2005). What is Web 2.0? Design patterns and business models for the next generation of software. *O'Reilly Network.* Retrieved from http://www.oreillynet.com/pub/a/oreilly/tim/news/2005/09/30/what-is-web-20.html

PeopleBrowsr. (2012). PeopleBrowsr wins temporary restraining order compelling Twitter to provide firehose access. Retrieved from http://blog.peoplebrowsr.com/2012/11/people-browsr-wins-temporary-restraining-order-compelling-twitter-to-provide-firehose-access/

Postigo, H. (2012a). Cultural production and the digital rights movement. *Information, Communication and Society, 15*(8), 1165–1185.

Postigo, H. (2012b). *The digital rights movement.* Cambridge, MA: MIT Press.

Rogers, R. A. (2009). *The end of the virtual: Digital methods.* Amsterdam, The Netherlands: Amsterdam University Press.

Romero, D. M., Meeder, B., & Kleinberg, J. (2011). Differences in the mechanics of information diffusion across topics: Idioms, political hashtags, and complex contagion on Twitter. *Proceedings of the 20th International Conference on World Wide Web (WWW'11)* (pp. 695–704). New York, NY: ACM.

Rotella, P. (2012). Is data the new oil? *Forbes.* Retrieved from http://www.forbes.com/sites/perryrotella/2012/04/02/is-data-the-new-oil/

Sakaki, T., Okazaki, M., & Matsuo, Y. (2010). Earthquake shakes Twitter users. *Proceedings of the 19th International Conference on the World Wide Web (WWW '10)* (pp. 1–10). New York, NY: ACM. doi:10.1145/1772690.1772777

Steele, J. (2011). Data markets aren't coming, they're already here. *O'Reilly Radar.* Retrieved from http://radar.oreilly.com/2011/01/data-markets-resellers-gnip.html

Thelwall, M., Buckley, K., & Paltoglou, G. (2011). Sentiment in Twitter events. *Journal of the American Society for Information Science, 62*(2), 406–418. doi: 10.1002/asi.21462

Twitter. (2012a, 25 June). Terms of service. Retrieved from http://twitter.com/tos

Twitter. (2012b, 5 Sep.). Developer rules of the road. Retrieved from https://dev.twitter.com/terms/api-terms

Twitter. (2012c, November 20). Developer display requirements (updated version). Retrieved from https://dev.twitter.com/terms/display-requirements

van Dijck, J. (2011). Tracing Twitter: The rise of a microblogging platform. *International Journal of Media and Cultural Politics, 7*(3), 333–348.

Zittrain, J. (2008). *The future of the Internet and how to stop it.* New Haven, CT: Yale University Press.

Data Collection on Twitter

5

CHAPTER   Devin Gaffney and Cornelius Puschmann

where does the data come from?
describing the #streaming, #REST, and
#search APIs to social scientists

In addition to being a versatile communications platform to users around the globe, Twitter is also an excellent source of current information. Data extracted from Twitter is used by researchers with different backgrounds (pollsters, marketers, academics from different disciplines) to answer a variety of questions, ranging from simple information about particular users or events (How many followers does a given user have? Who is the most active user tweeting under a certain hashtag?) to complex queries (Which users are central in a large network? How does information propagate among groups of users?). Some studies examine select individuals or small communities, while others require large volumes of information collected over long periods. Depending on the aims, different tools can be used to collect data—from Web-based analytics services that combine collection, analysis, and visualisation, to directly mining the Twitter API and interpreting the data using a dedicated statistics package.[1] Collecting data as part of a project, whether directly through the API or by using a dedicated software package, remains one of the most challenging aspects of Twitter-based research. While the technical and methodological requirements may seem daunting at first glance, an in-depth knowledge of the tools and the kind of data available through them can address many common concerns. In this chapter, we provide an overview of different techniques and their respective advantages and limitations. First, we discuss collecting data via Twitter's API, both directly and using a set of software packages, and then we turn to the question of how to integrate Twitter data into common social scientific study designs.

THE TWITTER APIS

Rather than offering a single API, three different Twitter data interfaces are available to researchers wanting to query the service: the Streaming API, the REST API, and the Search API. With few exceptions, the corpus of research generated to this point has relied on data collection through one of these three sources.

THE STREAMING API

The Streaming API is likely the most widely used data source for Twitter research. Typically, large-scale quantitative analyses of Twitter data are based on raw data collected through this source (Hong, Convertino, & Chi, 2011; Wu, Hofman, Mason, & Watts, 2011). It is worth pointing out that the Streaming API is a highly unorthodox kind of resource compared to how most other APIs function. In more traditional configurations, an API is "pull" based—the researcher requests a page of data from the server by requesting a URL, at which point the server returns the requested page. The Streaming API, however, is "push" based—that is, data is constantly flowing from the requested URL (the endpoint), and it is up to the researcher to develop or employ tools that maintain a persistent connection to this stream of data while simultaneously processing it. This stream of data is provided exclusively as a live poll, meaning that the moment a tweet is posted on Twitter, it becomes available. Because streaming data is supplied in the fashion of a live polling system not designed for historical analysis, research that takes a diachronic perspective is much more difficult than it would be via a traditional pull system, as the researcher must essentially operate on Twitter's schedule. When studying forms of relatively spontaneous organisation, such as the 'Arab Spring', 'Occupy', and 'Indignados' movements, data collection is especially difficult, as it may only be in hindsight that the event is recognised as such and its beginnings become significant (cf. Juris, 2012; Lotan et al., 2011; Vallina-Rodriguez et al., 2012 for such research). Studies of scheduled events, such as elections, require the researcher to be conscientious in establishing a stream for collecting data, ideally long before the event, in order to compile an analytically useful corpus. A central future challenge to the academic community will be to conduct complex and multifaceted analyses despite such restrictions, rather than tailoring research questions to data availability.

The Streaming API: Representative sampling

Fortunately the need to capture live data does not apply in the same way to all research contexts, and much of the research on Twitter to date asks more general questions, for example, by focussing on the platform's macroscopic structural properties (Kwak, Lee, Park, & Moon, 2010), or by describing how users conceptualise their communicative practices (Baym, Chapter 17 in this volume; Marwick & boyd, 2011). While Twitter is used in many countries and languages, user communities differ significantly in relation to their size, composition, and usage habits. Researchers should be keenly aware of seemingly small details that may be reflected in the data, for example, usage spikes over the course of a day, or fluctuations in activity during the weekend in comparison to workdays. Collecting data for prolonged periods of time is always preferable when possible, even if not all the data collected is used in the analysis. Many active users do not tweet daily, or perhaps even weekly, while others are highly active and skew the representativeness of a sample accordingly. Finally, not all quantitative research of Twitter is based on the contents of tweets: works such as Cha & Haddadi (2010) employ the social graph data available and focus entirely on follower-followee relations, rather than message content.

The Streaming API: Bandwidth limitations

The Streaming API is delivered in three bandwidths: "spritzer", "gardenhose", and "firehose", which deliver up to 1%, 10%, and 100% of all tweets posted on the system, respectively. By default, any regular user account on Twitter is granted spritzer access to the system, which is frequently sufficient for research purposes. The gardenhose is granted occasionally to users with defensible and compelling reasons for increased access, and the firehose is only available as a component of "a business relationship" with Twitter directly or through authorised re-sellers (Singletary, 2012). For the spritzer and the gardenhose, the percentage cap comes into effect only when more than the respective percentage of all tweets match the conditions placed on the stream. If, for example, a researcher is collecting data for a small conference, spritzer access will be sufficient to capture every tweet posted under the conference hashtag, since in only the most extreme cases will tweets about such an event exceed 1% of all traffic on the platform. It should be pointed out that according to the Twitter documentation, the respective sample sizes are based on the entirety of all information posted to Twitter as it is streamed, rather than on a subset of tweets to which a certain filtering criterion (e.g., a hashtag or keyword) applies. When in doubt about whether all desired content has been captured, researchers should check if their query returns a number of results close to 1% or 10% of the APIs' current total throughput. The streaming service also returns status messages indicating how many tweets have been missed if a cap has come into effect, notifying the researcher of the total number of tweets missed since the poll began.

The Streaming API: Endpoints and parameters

Beyond the three bandwidth options, the Streaming API offers two different methods, *sample* and *filter*, as points of access to data. *Sample* simply provides up to 1% or 10% of all tweets, selected at random. While the data has never been independently verified as random, it is generally assumed that it is of an acceptable degree of randomness. Crucially, two samples taken at the same point in time are identical, making reproducibility of results possible (Bruns & Liang, 2012; Hecht, Hong, Suh, & Chi, 2011). Inside the *filter* method, the *track*, *follow*, and *locations* parameters can be used to select specific results from the stream.

Track allows for researchers to search for multiple comma-delimited terms to be sent into Twitter's streaming request as an option. When Twitter receives the request, it only returns tweets that include those words, separated by non-word characters, to the researcher. In all but very few cases, the result will not exceed 1% of the total traffic on the platform, making this method combined with the *track* parameter a convenient way of compiling a keyword- or hashtag-based corpus.

Follow returns only tweets from a set of users represented by their collective comma-delimited user IDs. Currently, this parameter allows for collection from up to 5,000 accounts. Researchers intent on studying specific communities of users may find this method particularly useful for researching known groups of people for extended periods of time.

Locations provides an ideal access point for researchers interested in geographically bounded research. As of this writing, approximately 1% of all traffic on Twitter is "geotagged"—that is, an additional metadata object is appended to a tweet indicating its geographic origin. These geotagged tweets are represented as points, or latitude/longitude pairs that indicate a precise location, and polygons, or rectangles drawn by four pairs of points that can be as small as a city park or as large as a province. While the data is only 1% of all traffic, it is likely that this proportion will increase in the future, making it a highly attractive instrument for various kinds of geographically bounded research.

THE REST API

The REST (REpresentational State Transfer) API provides a set of methods for data interaction that is fundamentally different from the Streaming API, using the more traditional pull model. In total, over a hundred active methods are available in the REST API, few of which have been explored for research purposes. Using a combination of methods, the social graph data of a group of users can be assembled through this system, i.e. information on who is following whom and other data beyond the immediate content of individual tweets. Specifically, given a user of interest, two methods (*followers/ids* and *friends/ids*) can return listings of other users that follow or are followed by the user of interest at up to 5,000 user IDs per request. Further useful methods in the REST API provide access to trending topics, allow batch user lookups with groups of user IDs, and generally perform functions which are interesting in concert with the information that can be collected through the Streaming API.

REST API: Rate limiting

The REST API carries one heavy restriction: it is a rate-limited resource. Just as the Streaming API's spritzer and gardenhose levels of access are artificially limited to only a portion of traffic, the rate limit is in place largely to ensure reasonable traffic expectations for Twitter's infrastructure. This makes it very difficult for researchers to collect the data they desire in a timely manner, particularly in the case of REST requests. In the past, an un-authenticated computer could make 150 requests per hour. When any account logged into Twitter via Open Authentication (OAuth), this rate limit increased to 350 requests per hour. For using the followers/ids or friends/ids methods, this meant that at best, only 150 (or 350, when logged in) users could be processed for each method per hour. As of March, 2013, however, these limitations will be further reduced to approximately 60 requests per hour, and only OAuth requests to the API will be honoured (Sippey, 2012). If, for example, the researcher collects information about a user who has 563 friends and 178 followers, it would be possible to collect all friends and followers with one request to each of these methods, for a total cost of two requests in a one-hour window.

If that user has 5,630 friends and 1,780 followers, the amount would increase to three requests (as friends/ids can, at most, return 5,000 accounts per page of data). Previously, researchers could apply for IP-based and account-based white-listings, which, if granted, increased this limitation to 20,000 requests per hour. Twitter has since ended this practice, and does not hand out new white-listings. boyd and Crawford (2012) have put forward a compelling argument around the artificial class division that is created with this distinction of high-throughput accounts and everyone else (see also Puschmann & Burgess, Chapter 4 in this volume). While there are ways to circumvent limitations—for example, by setting up large networks of computers that collect data in tandem—such practices are actively monitored by Twitter, and violators are punished by blacklisting their accounts.

THE SEARCH API

Early analyses of Twitter, such as Gaffney's (2010) work on the Iran election of 2009, were largely based on the Search API. Originally, this was the only point of access for searching for tweets that mentioned hashtags, and was therefore widely used for event-based research. Similar to the REST API, it is a pull-based resource, and essentially replicates the functionality of Twitter's search function. While, in theory, some historical collection of data is still possible through the Search API, in practice its utility is severely limited. Data loosely falls off of the search system within a week of being posted, and no reliable information is available on its completeness. Twitter actively discourages use of the Search API and plans to discontinue it in the near future, as it is costly to maintain and was never intended for high-throughput real-time data dissemination.

TOOLS

While virtually all access to Twitter data takes place through one of the APIs (and often via a combination of several API methods), a number of tools exist to simplify this process. Rather than having to make API calls directly, researchers can use them to specify what data they want to collect. Client-based programs such as The Archivist or TAGS come with the constraint that they must be run on a regular basis from the user's computer to collect data. By contrast, server-based collection methods such as yourTwapperKeeper and Twitter Database Server run around the clock, collecting data whenever it is made available. This process is further simplified by Web-based services such as 140kit, that provide more in-depth analytical capabilities than services not designed for research, but at the same time restrict download access to tweets to conform with Twitter's Terms of Service, which bar the republication of full tweets without the company's consent (see Beurskens, Chapter 10 in this volume). Finally, data resellers such as Gnip and DataSift provide extensive historical data without the challenges of collection, but at a premium.

Table 5.1: Software Packages for Twitter Data Collection

Tool	Requires Hosting?	Requires Programming?	Provides Raw Data?	Provides Analytics?	Paid Service?
140kit	No	No	No	Yes	No
140kit Source Code	Yes	Yes	Yes	Yes	No
yourTwapperKeeper	Yes	Yes	Yes	No	No
The Archivist	No	No	No	Yes	No
TAGS	No	No	Yes	Yes	No
Twitter Database Server	Yes	Yes	Yes	No	No
Gnip	No	No	Yes	Yes	Yes
DataSift	No	No	Yes	Yes	Yes

THE ARCHIVIST

The Archivist (TA) is a free and open-source desktop application that runs on Windows based on Microsoft's .NET framework, and is among the simplest tools for saving and analysing tweets. In contrast to most other available tools, TA does not require a Web server. Each instance can collect tweets that include a certain keyword or hashtag, retrieved through the Search API. Use of the Search API makes TA subject to its limitations, a problem likely to become more severe in the future.

Retrieving large volumes of information or historical data is not generally possible, and for continuous retrieval, the collecting machine must run constantly. TA is recommendable only for small collections of tweets that can be manually verified for consistency, such as small hashtag archives and individual user streams. For any research requiring a reliable sample that cannot be manually verified, using TA or any other desktop software cannot be advised, as issues of latency, bandwidth, and stability are likely to impact the quality of the sample.

TAGS

The Twitter Archiving Google Spreadsheet (TAGS) is a Web-based script that can be used for the cloud-based collection of tweets. Running under Google Spreadsheets, TAGS is able to retrieve Twitter data through the REST API and consequently is subject to rate limiting, especially when used without authenticating with Twitter. It sidesteps some of the limitations of The Archivist by being hosted, while at the same time not requiring users to run their own server. Like The Archivist, TAGS performs a number of statistical operations on the extracted data, facilitating analysis. While it is not necessary to install any software, TAGS has an interface that is slightly less intuitive than that of The Archivist and requires a Google account.

YOURTWAPPERKEEPER

YourTwapperKeeper (YTK) is one of the most popular tools available to researchers wanting to simplify the process of extracting data. Written in PHP, YTK is among the most accessible projects currently available. It leverages both Twitter's Streaming API and the Search API to collect tweets that match a given term. In order to run the code, a researcher must employ an active PHP connection and be able to run a pair of scripts—the stream (Streaming API requests) and crawl (Search API requests) scripts. Additionally, it requires a MySQL database in order to store the information collected.

YTK (like its precursor TwapperKeeper) has been used in a number of research projects (Papacharissi & de Fatima Oliveira, 2012; Wilson & Dunn, 2011), and is frequently cited as an appropriate route for data collection (Bruns & Liang, 2012). There are some drawbacks to employing YTK, however. Most importantly, it only captures a small portion of the range of metadata currently available through the Streaming API with each tweet. Of the data that Twitter provides, YTK only collects the tweet's text and a few basic metadata attributes, which are then stored in a single table, rather than saving all available information and performing preprocessing to facilitate analysis. As a result, questions related to links, geographic places, and less orthodox questions focussed on particular metadata attributes cannot be easily investigated unless the researcher post-processes the data or alters their installation's source code.

TWITTER DATABASE SERVER

Twitter Data Server (TDS) is another server-based solution for collecting hashtag data. Like YTK, it is based on PHP and the MySQL database server. It provides only an absolutely minimal interface for browsing the captured data, instead relying on the user's ability to interact with the MySQL tables that contain the captured tweets through the command line or via a third-party utility such as the popular phpMyAdmin. While running and interacting with TDS is somewhat more complicated than YTK, TDS appears to consume less computational resources and capture additional fields, such as resolved URLs, not provided by YTK.

140KIT

140kit[2] is a Web-based tool for the analysis of Twitter data. Unlike the other services mentioned, 140kit ensures that all metadata fields are collected. In its online version, no raw data is allowed to be downloaded due to Terms of Service limitations imposed by Twitter on public datasets (Twitter, 2012). The software is available as a hosted platform and as a stand-alone package similar to yourTwapperKeeper (140kit Source Code), though the language that the program is written in, Ruby and Ruby on Rails, is less prevalent than PHP and may be more difficult to implement, depending on the kind of Web server available. Like YTK and TDS, it also requires a computer to run the software constantly in order to stream data. Additionally, it only supports the Streaming API, though researchers can extend functionality if required. In contrast to most other analytics tools, 140kit is built specifically with researchers in mind as its target audience, and may thus be more suitable for answering questions outside of commercial contexts.

GNIP AND DATASIFT

Gnip is one of the best tools available to researchers in terms of data quality, but it comes at a premium. The company collects data through firehose access to Twitter, and re-sells this data to both the research community and private businesses. Access for 10% and 50% of the Streaming API, alongside several other datapoints of note (such as integration with Klout scores), costs US$5,000 and US$30,000 per month, respectively (Small, Kasianovitz, Blanford, & Celaya, 2012). While the availability of historical data makes the service potentially relevant, the considerable costs of the service and a lack of complete metadata (similar to YTK) may deter researchers from using it.

Similar to Gnip, DataSift provides much of the same functionality, though its pricing scheme differs, with the lowest level of access costing US$3,000 per month at the time of writing. Also similar to Gnip, some metadata fields that may be of interest to the researcher are omitted, though it also adds other data-points of note (again, integration with Klout scores). A downside of both services is that by targeting businesses, not all of the information that may be relevant to researchers is made available through them.

CONCLUSION: LIMITATIONS

The limitations of social scientific research based on Twitter data stem from constraints which impact research projects on different levels. As with any other methodology, not all types of data and forms of analysis align themselves equally well with all kinds of research questions. Because of its tendency to be data- rather than question-driven, much of the current quantitative research on Twitter focusses on measuring and comparing specific structural parameters in very large data samples, sometimes with little regard for the theoretical salience of these parameters. This is understandable before the backdrop of Big Data research as a fundamentally new approach to finding patterns, relationships, and links between elements, rather than paradigmatically theorising the meaning of said elements beforehand (Lazer et al., 2009). An ideal study should be well grounded in a specific set of research questions and query the data in accordance with them. In contrast to traditional instruments such as surveys and conventional content analysis, it is important to note that even the exploratory phase of research is markedly quantitative when exploring social media. Since searching, filtering, and ranking are the only feasible way to make masses of content readable to the human researcher, they form a logical first step in any analysis, even in qualitative studies. At the same time, quantitative research should present data as it pertains to the questions asked, rather than simply because it is possible and large volumes of data have been collected. Furthermore, there is the question of how representative Twitter users are of the overall population—both on Twitter and beyond it. Adoption rates and usage strategies differ greatly, casting doubt on claims of representativeness. When making judgments about populations of Twitter users based on tweets, those users who mainly read but hardly post may be overlooked, while the significance of highly vocal users may be given too much weight. Inferences about the population at large based on Twitter are difficult as a result of this inherent skew, yet without generalisation the potential for sociological research is limited, in spite of much enthusiasm for Twitter as a data source (e.g., Golder & Macy, 2012).

A third and distinct set of limitations is technology-based. As has been pointed out, there is no way of checking how completely a given data set captures what flowed through Twitter at the time that it was compiled. Without firehose access, researchers rely entirely on Twitter to provide a representative sample of what is there. At the same time, it is important to note that this is not solely because of Twitter's need to monetise its data, but a result of the unique challenge of building an infrastructure powerful enough to store such vast quantities of information in real time. Incomplete data sets can hamper an analysis, yet asking in hindsight for a complete archive of tweets related to a past event is impossible. Not only does this make new research difficult, it also makes absolute reproducibility extremely hard to achieve, with obvious implications for the validity of research results.

Why quantify to begin with? Qualitative sociological research on Twitter comes with its own unique potentials, but also with its own set of constraints, for example, with regard to privacy. Arguably, Twitter's strength lies in the ability to gain interesting insights from short and often highly context-bound messages, yet these are also difficult to interpret and carry a range of meanings for different stakeholders. While a "deep", qualitative approach is more nuanced than computational procedures, it is also severely limited in its scale. By choosing an object that can be studied in detail, the researcher makes specific choices about her object of study. On the other hand, the "shallow" aggregation of data always risks arriving at judgments that are ill-supported because they are based on incorrect or overreaching implicit assumptions. Relying on one's informed experience from other contexts when considering Twitter as a source of data, and pragmatically deciding how research objectives can be aligned with the technical and methodological challenges at hand will usually produce the best result.

NOTES

1   The acronym API stands for application programming interface. APIs are data interfaces offered by many Web platforms. Their main purpose is to provide software developers an unambiguous, data-only version of a site's content for use in their own software.

2   One of the authors of this chapter is the principal developer of 140kit.

REFERENCES

boyd, d., & Crawford, K. (2012). Critical questions for Big Data: Provocations for a cultural, technological, and scholarly phenomenon. *Information, Communication & Society, 15*(5), 662–679.

Bruns, A., & Liang, Y. E. (2012). Tools and methods for capturing Twitter data during natural disasters. *First Monday, 17*(4), 1–8.

Cha, M., & Haddadi, H. (2010). Measuring user influence in Twitter: The million follower fallacy. *Proceedings of the Fourth International AAAI Conference on Weblogs and Social Media (ICWSM '10)* (pp. 10–17). Menlo Park, CA: AAAI Press.

Gaffney, D. (2010). #iranElection: Quantifying online activism. *Proceedings of the WebSci10: Extending the Frontiers of Society On-Line* (pp. 1–8). 26–27 Apr. 2010. Raleigh, NC.

Golder, S., & Macy, M. (2012). Social science with social media. *ASA Footnotes, 40*(1), 7.

Hecht, B., Hong, L., Suh, B., & Chi, E. H. (2011). Tweets from Justin Bieber's heart: The dynamics of the 'location' field in user profiles. *Proceedings of the International Conference on Human Factors in Computing Systems (CHI '11)* (pp. 1–10). Vancouver, British Columbia, Canada: ACM Press.

Hong, L., Convertino, G., & Chi, E. H. (2011). Language matters in Twitter: A large scale study characterizing the top languages in Twitter characterizing differences across languages including URLs and hashtags. *Proceedings of the Fifth International AAAI Conference on Weblogs and Social Media (ICWSM '11)* (pp. 518–521). Menlo Park, CA: AAAI Press.

Juris, J. S. (2012). Reflections on #Occupy Everywhere: Social media, public space, and emerging logics of aggregation. *American Ethnologist, 39*(2), 259–279.

Kwak, H., Lee, C., Park, H., & Moon, S. (2010). What is Twitter, a social network or a news media? *Proceedings of the 19th International Conference on the World Wide Web (WWW'10)* (pp. 1–10). Raleigh, NC.

Lazer, D., Pentland, A., Adamic, L., Aral, S., Barabasi, A-L., Brewer, D., . . . Van Alstyne, M. (2009). Computational social science. *Science, 323*(5915), 721–723.

Lotan, G., Graeff, E., Ananny, M., Gaffney, D., Pearce, I., & boyd, d. (2011). The revolutions were tweeted: Information flows during the 2011 Tunisian and Egyptian revolutions. *International Journal of Communication, 5*, 1375–1405.

Marwick, A., & boyd, d. (2011). To see and be seen: Celebrity practice on Twitter. *Convergence: The International Journal of Research Into New Media Technologies, 17*(2), 139–158.

Papacharissi, Z., & de Fatima Oliveira, M. (2012). Affective news and networked publics: The rhythms of news storytelling on #Egypt. *Journal of Communication, 62*(2), 266–282.

Singletary, T. (2012). How do I get firehose access? *Twitter Developer Forums*. Retrieved from https://dev.twitter.com/discussions/2752

Sippey, M. (2012). Changes coming in Version 1.1 of the Twitter API. *Twitter Developer Blog*. Retrieved from https://dev.twitter.com/blog/changes-coming-to-twitter-api

Small, H., Kasianovitz, K., Blanford, R., & Celaya, I. (2012). What your tweets tell us about you: Identity, ownership and privacy of Twitter data. *International Journal of Digital Curation, 7*(1), 174–197.

Twitter. (2012, 25 June). Terms of service. Retrieved from http://twitter.com/tos

Vallina-Rodriguez, N., Scellato, S., Haddadi, H., Forsell, C., Crowcroft, J., & Mascolo, C. (2012). Los Twindignados: The rise of the Indignados movement on Twitter. *Proceedings of the ASE/IEEE International Conference on Social Computing (SocialCom'12)* (pp. 1–6). Amsterdam, The Netherlands.

Wilson, C., & Dunn, A. (2011). Digital media in the Egyptian revolution: Descriptive analysis from the Tahrir datasets. *International Journal of Communication, 5*, 1248–1272.

Wu, S., Hofman, J. M., Mason, W. A., & Watts, D. J. (2011). Who says what to whom on Twitter. *Proceedings of the 20th International Conference on the World Wide Web (WWW '11)* (pp. 705–714). New York, NY: ACM Press.

Metrics for Understanding Communication on Twitter

6

CHAPTER   Axel Bruns and Stefan Stieglitz

 .@sender, @receiver, timestamp, http://url.org/, #hashtags—a tweet consists of much more than just 140 characters

As the systematic investigation of Twitter as a communications platform continues, the question of developing reliable comparative metrics for the evaluation of public, communicative phenomena on Twitter becomes paramount. What is necessary here is the establishment of an accepted standard for the quantitative description of user activities on Twitter. This needs to be flexible enough in order to be applied to a wide range of communicative situations, such as the evaluation of individual users' and groups of users' Twitter communication strategies, the examination of communicative patterns within hashtags and other identifiable *ad hoc* publics on Twitter (Bruns & Burgess, 2011), and even the analysis of very large datasets of everyday interactions on the platform. By providing a framework for quantitative analysis on Twitter communication, researchers in different areas (e.g., communication studies, sociology, information systems) are enabled to adapt methodological approaches and to conduct analyses on their own. Besides general findings about communication structure on Twitter, large amounts of data might be used to better understand issues or events retrospectively, detect issues or events in an early stage, or even to predict certain real-world developments (e.g., election results; cf. Tumasjan, Sprenger, Sandner, & Welpe, 2010, for an early attempt to do so).

In principle, the exploration of such universal metrics for the analysis of Twitter communication is straightforward, and builds immediately on the communications data and metadata which is available through the Twitter Application Programming Interface (API; see Gaffney & Puschmann, Chapter 5 in this volume). Given the range of metadata which is associated with each tweet retrieved through the API, and the additional data points which may be extracted from the tweet text itself, a series of key metrics emerge; we outline these in the first part of this chapter. However, the effective use of such metrics also depends on a deeper understanding of the communicative phenomena which they describe; as with any quantitative approach, a focus merely on the raw figures themselves is likely to obscure more important patterns within the data. These can only be uncovered by the careful consideration of the provenance of the overall data set, as well as through the sensible selection of specific subsets of the overall dataset for further analysis. We point to such considerations by providing a discussion across specific data sets of Twitter communication in the second part of the chapter. Finally, we provide a short overview about how the metrics we identified can be beneficially combined with other well-established methods.

The concepts we introduce here provide a fundamental set of analytical tools for the study of public communication on Twitter, but they do not purport to represent an exhaustive list of possible metrics for the description of Twitter-based user activities. Additional, more specific metrics which relate to particular communicative contexts on Twitter may also be developed; we encourage researchers to document their analytical choices in such specific cases in similar detail, so that these metrics can also become part of the wider toolkit of conceptual models and practical methods which is available to social media researchers.

BASIC METRICS

Centrally, the Twitter API provides the tweet text itself, the username and numerical ID of the sender, and a timestamp which is accurate to the second; further metadata (which are likely to be of use only in more specific cases) include fields providing the—at present, rarely used—geolocation of the sender at the time of tweeting, the client used to send the tweet (e.g., Web, *Tweetdeck*, Blackberry), and a reference to the user's Twitter profile picture. Additionally, structural analysis of the tweet text itself will be able to reveal whether the tweet contains one or multiple hashtags, one or multiple @mentions of other users, and/or references to any URLs outside of Twitter. Finally, it may also be possible to identify whether @mentions of other users represent (manual) retweets in one of a number of the widely used syntactical formats (e.g., RT *@user*, MT *@user*, via *@user*, or "*@user*) which indicate retweets. Outside of retweets, a distinction between mere @mentions—that is, references to another user which are not inherently intended to strike up a conversation—and intentional @replies is likely to be much more difficult to establish, not least also because the transition between both is gradual: the first @mention in what eventually becomes a multi-turn @reply chain is always both @mention *and* @reply.

For each message, then, the following key data points can be established by analysing the tweet itself and its associated metadata:

- sender: Twitter username and numerical ID
- recipient(s):   @mentioned usernames in the tweet (if any)
- timestamp: accurate to the second
- tweet type: retweet, genuine @reply (non-retweet), or original tweet (no @mentions)
- hashtag(s):   hashtags referenced in the tweet (if any)
- URLs: URLs included in the tweet (if any)

As noted above, further metrics may also be developed—for example by examining whether the tweet contains mentions of specific keywords or named entities which are of interest in the particular research context, or whether the tweet is composed in a specific language and/or character set. As these metrics are case-specific, however, they are unlikely to be generalisable for comparative Twitter research beyond such individual cases, and do not concern us here.

Automated parsing of all tweets within a given dataset (see Bruns, 2012, for an implementation in the pattern-matching language Awk), then, is able to determine these data points on a tweet-by-tweet basis. This information may then be aggregated into a detailed set of metrics which describe the communicative patterns captured in the dataset; such aggregation can be performed, *inter alia*, for each specific timeframe within the dataset (minutes, hours, days, . . .); for each individual user participating (as active sender of tweets, or as recipient of @mentions) in the dataset; and for larger groups of users which have been identified on the basis of specific criteria. A further combination of these approaches (to develop diachronic metrics for specific users, for instance) is also possible, of course. (Several of the chapters in the "Practices" section of Part II of this volume pursue such approaches.)

TEMPORAL METRICS

Metrics which describe the communicative patterns captured in a given Twitter dataset over time are a crucial tool for the identification of important phenomena for further—not least also qualitative (cf. Einspänner, Dang-Anh, & Thimm, Chapter 8 in this volume)—investigation. At their simplest, such metrics may simply outline the overall volume of tweets within a dataset (which may comprise messages containing a given hashtag or keyword, for example) statically or dynamically (e.g., to show particular spikes or lulls in user activity; see, for example, Stieglitz & Krüger, 2011). Following the syntactical parsing of tweets, however, it also becomes possible to track such volumes separately for original tweets, @replies, and retweets, or for tweets containing URLs or hashtags; this can trace, for example, the dissemination of key information (URLs) or the emergence of new concepts and memes (hashtags) on Twitter.

Additionally, it may also be important to examine the number of unique users participating in the communicative process at any one time, and compare this with the volume of tweets; this may help to distinguish moments of especially heated discussion (marked by an increase in tweets per user) from spikes in activity that are caused by an influx of active users (marked by an increase in tweet volume, but not in tweets per user). Similarly, researchers may wish to track the activities of specific users or groups of users over time, to examine how these users respond (differently) to particularly communicative events, and even to explore the types of tweets such users send at different points in time; we return to these questions below.

USER METRICS

Such user-based metrics may also be calculated independently of the temporal dimension, of course. In this case, what emerges is a more comprehensive picture of the respective communicative strategies employed by different users on Twitter: most importantly, this approach can determine the overall balance between original tweets, @replies, and retweets for each user, and thereby draw distinctions between users who take a largely annunciative approach (mainly original tweets), conversational approach (mainly @replies), or dis-seminative approach (mainly retweets). Various combinations between such approaches—potentially shifting over time—are also possible. Further, the extent to which users include URLs in their tweets may also be included in this analysis. In addition to examining user activity, similar metrics are also available for the recipients of @mentions within the dataset. Here, it is possible to examine the balance between @replies and retweets received by each user referenced in tweets contained in the dataset. Such metrics can be understood, in the first place, to pro-vide an evaluation of the visibility and importance of each user to those of their peers who actively sent tweets in the dataset, and a further distinction between @replies and retweets may also point to whether these recipients are mainly posi-tioned as partners in conversation (@replies received) or sources of information (retweets received; also cf. Weller, Dröge, & Puschmann, 2011, on this point). Indeed, a further comparison between the metrics for incoming and outgo-ing tweets for each user provides additional detail on their specific placement within the communicative context contained in the dataset. Users who receive many @mentions, but rarely @reply in return, must be seen mainly as subjects *of* conversation; users who both receive and send @replies frequently, by contrast, are active subjects *within* conversation. Similarly, users who receive substantial retweets without having sent a substantial number of tweets themselves may be seen as having provided more important impulses to the dataset than users who tweet frequently, but receive a relatively low number of retweets from others.

GROUP METRICS

While such per-user metrics are useful for an identification of the most active and most visible users within a dataset, and for a detailed evaluation of their specific types of communicative activity on Twitter, it will often also be use-ful to aggregate these metrics both for known, pre-existing groups of Twitter accounts (as determined by the specific research agenda), or for groups of users which emerge from the quantitative analysis of the dataset itself. As the first of these possibilities is necessarily case-specific, we discuss only the second here, focussing on a grouping of users by their level of contribution to the dataset itself. From the per-user metrics, the total number of tweets sent by each contributor to the dataset is already known; on this basis, users can be ranked and distinguished into a number of specific, more and less active groups. While such distinctions may in principle be made along any line, user activity in most communicative situations on Twitter and other platforms will be distributed in keeping with a power law: a comparatively small number of highly active users are likely to dominate the dataset, while a much larger "long tail" (Anderson, 2006) of far less active users will be responsible for a smaller volume of tweets. Therefore, a distinction of users using the 10/90 or 1/9/90 rule (Tedjamulia, Dean, Olsen, & Albrecht, 2005) is sensible here: a 1/9/90 division, for example, groups the one per cent of lead users (as measured by the number of tweets they contributed to the dataset) separately from the next nine per cent of still highly active users, and separately in turn from the remaining 90% of least active users in the long tail of participants. Using such distinctions, it is then again possible to determine the tweeting patterns for these three groups: the number of original tweets, @replies, and retweets they have sent, as well as the number of tweets containing URLs (or other, specific communicative markers as relevant to the research project). Additionally, it may also be important to examine the number of users from each of the three groups (as defined over the *entirety* of the dataset) who are active during any individual temporal period covered by the dataset: this indicates, for example, whether established lead users were highly active throughout the time frame under examination, or whether there were times when normally less active users gained a greater share of the overall discussion.

INTERPRETING TWITTER METRICS

Such standard metrics represent a powerful tool for the analysis of communicative activities and interactions on Twitter; however, they must also be employed correctly in order to generate a reliable (and ultimately, comparable) picture of communicative processes on Twitter. Here, it becomes crucial to consider the provenance of the dataset under examination, in order to determine the limits of what forms of communicative activity it may or may not contain. Most commonly, at present, the metrics which we have described here are extracted from Twitter datasets which have been raised on the basis of keyword or hashtag filters. This means that they necessarily contain only a selection of all communication taking place on Twitter, and indeed, even represent only a subset of all communicative activity which may be relevant to the themes described by the keywords or hashtags themselves. Hashtags, for example, are used to explicitly mark tweets as relevant to a specific theme, but this also means that hashtag datasets do not contain *all* relevant tweets, but only those whose authors knew of and felt motivated enough to include the hashtag in the tweet. Furthermore, hashtags may be misused (accidentally or on purpose, e.g. by spammers; cf. Mowbray, Chapter 14 in this volume). In this case, some tweets will be included in the dataset which are not actually related to the intended topic. Most importantly, what is missing from such datasets are the messages which engage in follow-up conversation to a hashtagged tweet, but were not deemed important enough by their authors to receive a hashtag themselves. When our standard metrics are applied to such hashtag datasets, therefore, it is likely that the metrics which describe communicative interaction through @replying—though correct for the hashtag dataset itself—may significantly underestimate the full volume of @replies which was prompted by hashtagged tweets. Conversely, since hashtags most centrally represent a convention designed to make tweets more easily discoverable, it is also likely that metrics for hashtag datasets overestimate the extent to which retweeting of messages relating to the hashtag topic takes place on Twitter: hashtagged tweets may be retweeted disproportionately much, by the very virtue of *being* hashtagged.

For keyword datasets, on the other hand, the situation is different again. While hashtags can (but not always do) serve as a means to enable the coming-together of *ad hoc* publics which interact with one another, the same is not usually true of mere keywords; a keyword dataset, therefore, constitutes a cross-section through the Twitter activities of users who are largely unlikely to be aware of one another, while hashtags inherently provide at least the potential for such awareness. Although hashtag datasets themselves already miss much of the @replying which may take place *around* the hashtag (but without using it in tweets), keyword datasets may well be likely to further underestimate @replying activity, as they will contain @replies only if they contain the selected keyword, but will rarely pick up full threads of communication. Keyword datasets necessarily contain fragments of wider conversation, therefore, and their metrics must be understood from that perspective. Such critiques are not meant to fundamentally dismiss the value and validity of research which utilises such datasets; rather, they seek to highlight the advantages and disadvantages of specific sampling approaches for Twitter data in the context of the metrics which may be established for such datasets.

As long as hashtag or keyword datasets remain an important tool for Twitter research, at any rate (and there is no reason why they should not), what is important is simply to recognise these inherent distortions in observable communication patterns which are caused by the approaches chosen to observe them, and to ensure that in broader, comparative investigations across individual cases, like is compared with like. Where these limitations are understood, then, the standardised metrics which we have outlined here can generate important new insight into the divergence or systematicity of communicative patterns on Twitter, as we demonstrate in the following.

METRICS COMPARISONS ACROSS SPECIFIC CASES

The metrics we have outlined here are valuable for an examination of individual communicative phenomena as described by specific datasets; however, by providing a standard approach to quantifying communicative activity on Twitter, they also especially lend themselves to cross-comparisons. Such comparisons are able to uncover significant differences in how the same communicative affordances (Twitter itself, as well as specific mechanisms such as hashtags, @replies, or retweets) are used in different contexts and by different groups of users, hinting at a range of more fundamental patterns which may well reflect deep-seated principles in human communication well beyond the Twitter platform itself. We illustrate this through two comparative analyses. Figure 6.1 shows the relative contributions of more and less active user groups (determined according to the 1/9/90 rule outlined above) to a range of hashtag datasets:

- #auspol: Australian political discussion, 8 February to 8 Dec. 2011.
- #occupy: political discussion about the Occupy movement, 19 Dec. 2011 to 19 Apr. 2012.
- #masterchef: backchannel for a popular Australian television show, 1 May to 8 Aug. 2011.
- #royalwedding: backchannel for the wedding of Prince William and Catherine Middleton, 29 Apr. 2011.
- #stopkony: viral campaign to arrest Ugandan warlord, Joseph Kony, 8 to 21 Mar. 2012.

Our analysis of the relative contributions made by the three groups of users in each case reveals some stark differences between these cases. The #auspol hashtag, containing some 850,000 tweets during the period analysed, is clearly dominated by a small group of some 260 lead users, who posted well over 60% of all tweets; indeed, in combination, the two most active groups of users, representing ten per cent of the total number of unique users participating in the hashtag, posted more than 90% of all tweets captured in the dataset. #occupy and #masterchef display similar patterns: in each case, these two groups are responsible for more than 60% of all tweets. This can be seen as evidence of a well-established elite of Twitter users which dominates these hashtags, and

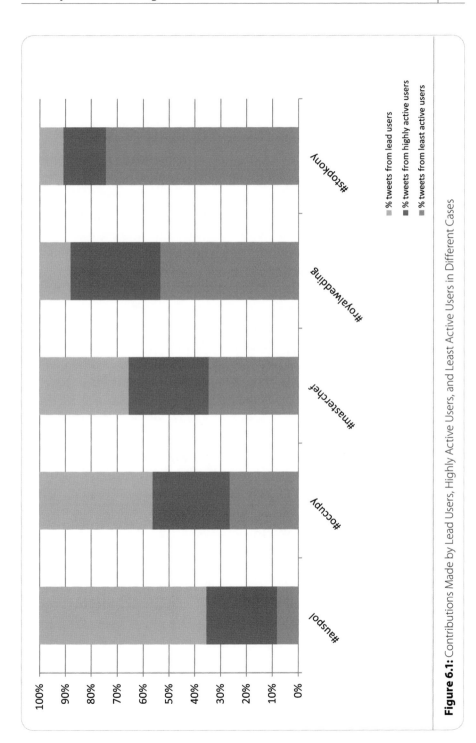

Figure 6.1: Contributions Made by Lead Users, Highly Active Users, and Least Active Users in Different Cases

may point to the presence of genuine community structures, centred around the leading users.

Hashtags such as #stopkony and #royalwedding show a considerably less pronounced domination by leading users; here, the most active one per cent of users accounts for just over ten per cent of all tweets, and the least active 90% of the user base comes much closer to contributing to the hashtag to an extent that reflects their numerical advantage. For #stopkony, this underlines the viral nature of this campaign: although made visible by the public endorsements from a handful of Twitter celebrities which were deliberately targeted by the Kony 2012 campaign (cf. Paßmann, Boeschoten, & Schäfer, Chapter 25 in this volume), the campaign itself (and its associated hashtags) gained and maintained momentum because many of the millions of followers of these celebrities in turn retweeted their #stopkony tweets. The bulk of hashtag activity, therefore, results from individual users whose involvement may remain marginal (at its most basic, in the form of single retweets); only a few users participated in more comprehensive ways.

The #royalwedding hashtag, finally, represents a far more time-limited event, unfolding on a single day. Here, although there is substantial activity in the hashtag itself (with over 920,000 tweets from close to half a million unique users), there may not have been enough time for community structures and a recognised group of leading users to emerge; it is the intermediate group of highly active (but not leading) users which is especially prominent in this case, therefore. Given the necessarily singular nature of the event, we can only speculate that, had the hashtag continued for a longer period of time, the balance between lead and highly active users may gradually have shifted, finally resulting in a more dominant group of lead users, recruited from this pool of already highly active participants.

If such comparisons of the relative structures of different hashtag user communities (to the extent that they indeed act *as* communities) can reveal important differences in how hashtag publics operate, further comparison of actual communicative patterns within hashtags is also valuable. Figure 6.2 presents a comparison of two key metrics for a wide selection of hashtags (also cf. Bruns & Stieglitz, 2012, for a more wide-ranging comparison): for each hashtag, it plots the percentage of tweets containing URLs against the percentage of tweets which are retweets. Two broad clusters of hashtags are immediately obvious. One set of hashtags is marked by a low percentage both of URLs and of retweets; these hashtags represent foreseen, well-publicised, television events, and include (in addition to #royalwedding and #masterchef) popular shows such as the Eurovision and

Oscars awards, the Australian Football League and Australian National Rugby League grand finals, the Tour de France, and the Australian political talk show *Q&A*. In such cases, Twitter serves as a backchannel to television, and enables its users to participate in a mediated, communal form of audiencing (Fiske, 1992) which—because of the shared television text upon which it is based—requires neither the exchange of additional information in the form of URLs nor substantial retweeting of messages to raise awareness of an issue or topic. The long-term discussion of Australian politics in #auspol behaves in a similar fashion; we might speculate that #auspol participants are similarly engaging in a form of audiencing, if in reaction to the overall media coverage of political matters rather than in relation to one unified televisual text. They are, in essence, *fans* of politics who use Twitter as a backchannel for the play-by-play discussion of plot developments in the Australian political narrative.

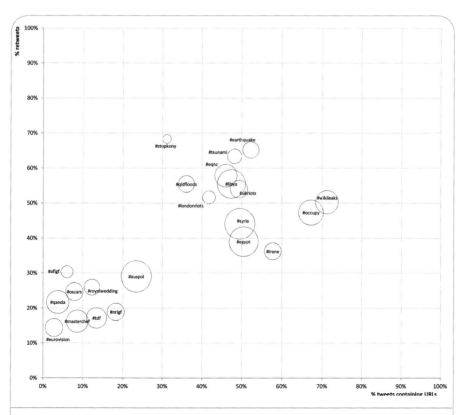

Figure 6.2: Share of Retweets and Tweets Containing URLs in Specific Data Sets

The second cluster, whose hashtags contain substantial percentages of both URLs and retweets, comprises events such as the popular revolts in Libya, Egypt, and Syria in the course of the 2011–2012 Arab Spring; natural disasters such as the earthquake and tsunami on the Sendai coast in Japan, the 2010–2011 earthquakes in Christchurch, New Zealand, the 2011 floods in Queensland, Australia, and the 2011 Hurricane Irene which affected the US; and the riots in London and the wider UK. Hashtags such as #stopkony, #occupy, and #wikileaks are also associated with the cluster, though showing some divergence from common communicative patterns.

For hashtags within this cluster, finding and sharing information by posting and retweeting tweets which contain URLs is a core practice; this is in keeping with a process of collaborative curation of information on the hashtag topic through gatewatching (Bruns, 2005). Such activities are commensurate with breaking news: at times when there is an information deficit about the exact situation on the ground, Twitter users seem to come together to pool resources, and thereby curate what information is coming to hand. This may also explain the differences between individual hashtags within the cluster itself: as later and comparatively well-reported stages of the Arab Spring, the uprisings in Egypt and Syria were able to tap into already relatively well-established networks of Twitter users, requiring comparatively less retweeting to disseminate information; similarly, as an anticipated weather event, tweets about Hurricane Irene did not need to be retweeted widely in order to become widely visible. By contrast, the earthquakes as well as the Queensland floods or UK riots could not be foreseen, and therefore represent potentially more shocking breaking news events; widespread retweeting to raise awareness is to be expected in such situations.

A viral campaign such as Kony 2012 is comparable to such crisis events; indeed, the very principle of such viral campaigning is to achieve widespread visibility within a very short space of time, and thereby to generate further follow-on media coverage. The Kony 2012 campaign effectively managed to instil this sense of crisis in its supporters. By contrast, however, movements such as Occupy and platforms such as WikiLeaks are responses to a sense of 'permanent crisis' in conventional democratic systems; additionally, they are marked by a deep distrust of the mainstream media's ability to provide balanced information. Therefore, the extensive presence of URLs in #occupy and #wikileaks tweets is unsurprising.

This necessarily brief discussion points to an underlying systematicity in how Twitter users utilise the platform to communicate. The patterns which we have outlined here are by no means exhaustive, of course; other common uses of Twitter may be uncovered by examining a wider range of hashtag datasets, by exploring communicative patterns in Twitter datasets which are based on principles of selection other than the presence of hashtags in tweets, or by exploring the correlations between other elements of the standard metrics we have outlined above. What even these brief examples highlight, however, is the inherent value of such systematic approaches to generating standardised metrics for the description of communicative processes on Twitter.

CONCLUSION: COMBINING METRICS AND METHODS

As we have demonstrated, the investigation of communicative metrics on Twitter provides relevant findings to better understand the overall patterns within this communication. Combining these different metrics with other well-established methods such as manual content analysis, sentiment analysis, or social network analysis allows researchers to derive further, in-depth results. Of course, the appropriateness of such combinations depends strongly on the specific research question. For instance, sentiment analysis combined with temporal metrics might deliver more information about changes in sentiment among Twitter users in a specific time frame and in relation to certain issues (see, for example, Papacharissi & de Fatima Oliveira, 2012). Manual content analysis combined with user metrics, by contrast, might enable a detailed analysis of the communicative efforts of specific actors.

Obviously, there are several more ways to combine the metrics outlined in this chapter with well-established methodologies. However, to date, such mixed-method approaches are used only very rarely (e.g., Stieglitz & Dang-Xuan, 2012). Researchers in this field must continue to work on identifying and documenting metrics, as well as on developing more comprehensive frameworks to combine metrics and methods.

REFERENCES

Anderson, C. (2006). *The long tail: Why the future of business is selling less of more.* New York, NY: Hyperion.

Bruns, A. (2005). *Gatewatching: Collaborative online news production.* New York, NY: Peter Lang.

Bruns, A. (2012, 31 Jan). More Twitter metrics: Metrify revisited. *Mapping Online Publics.* Retrieved from http://mappingonlinepublics.net/2012/01/31/more-twitter-metrics-metrify-revisited/

Bruns, A., & Burgess, J. (2011). The use of Twitter hashtags in the formation of *ad hoc* publics. In *Proceedings of the European Consortium for Political Research Conference, Reykjavik,*

Iceland. Retrieved from http://snurb.info/files/2011/The%20Use%20of%20Twitter%20 Hashtags%20in%20the%20Formation%20of%20Ad%20Hoc%20Publics%20(final).pdf

Bruns, A., & Stieglitz, S. (2012). Quantitative approaches to comparing communication patterns on Twitter. *Journal of Technology in Human Services, 30*(3–4), 160–185. Retrieved from http://dx.doi.org/10.1080/15228835.2012.744249

Fiske, J. (1992). Audiencing: A cultural studies approach to watching television. *Poetics, 21*(4), 345–359.

Papacharissi, Z., & de Fatima Oliveira, M. (2012). Affective news and networked publics: The rhythms of news storytelling on #egypt. *Journal of Communication, 62*(2), 266–282.

Stieglitz, S., & Dang-Xuan, L. (2012). Social media and political communication: A social media analytics framework. *Social Network Analysis and Mining*. Retrieved from http://link. springer.com/article/10.1007%2Fs13278-012-0079-3#page-1. doi: 10.1007/s13278-012-0079-3

Stieglitz, S., & Krüger, N. (2011). Analysis of sentiments in corporate Twitter communication: A case study on an issue of Toyota. In *Proceedings of the 22nd Australasian Conference on Information Systems (ACIS)*, Paper 29. Retrieved from http://aisel.aisnet.org/acis2011/29

Tedjamulia, S. J. J., Dean, D. L., Olsen, D. R., & Albrecht, C. C. (2005). Motivating content contributions to online communities: Toward a more comprehensive theory. In *Proceedings of the 38th Annual Hawaii International Conference on System Sciences (HICSS '05)* (p. 193b). Washington, DC: IEEE.

Tumasjan, A., Sprenger, T. O., Sandner, P. G., & Welpe, I. M. (2010). Predicting elections with Twitter: What 140 characters reveal about political sentiment. In *Proceedings of the Fourth International AAAI Conference on Weblogs and Social Media* (pp. 178–185). Retrieved from http://www.aaai.org/ocs/index.php/ICWSM/ICWSM10/paper/viewFile/1441/1852

Weller, K., Dröge, E., & Puschmann, C. (2011). Citation analysis in Twitter: Approaches for defining and measuring information flows within tweets during scientific conferences. In *Proceedings of #MSM2011: 1st Workshop on Making Sense of Microposts*. Retrieved from http://ceur-ws.org/Vol-718/paper_04.pdf

Sentiment Analysis and Time Series with Twitter

7
CHAPTER   Mike Thelwall

 time series and #sentiment analysis as #quantitative methods reveal trends, points of interest and changes in mood

The contributions of ordinary members of the public to microblogging services like Twitter and Weibo can give social scientists unique insights into public reactions to specific events and to changes in public opinion over time. For example, the changing volume of tweeting around an event, such as the U.K. riots of 2011 or a United Nations Conference on Sustainable Development, can reveal when the public first became interested in it and when this interest started to fade. Peaks in the volume of tweeting can also point to which instances within a broad event generated the most interest. Both blogs and microblogs are particularly useful for monitoring public opinion in this way because (a) they are reliably time-stamped, unlike most of the rest of the Web, so that they can be analysed from a temporal perspective, (b) they are relatively easy to create, so that a wide segment of the population with Internet access could, in theory, create them, and (c) they are public and hence accessible to researchers, unlike most social network sites. Microblogs are probably created by a wider segment of the population than blogs in some countries, and are naturally updated more frequently (for USA Twitter usage information, see Smith & Brenner, 2012), making them a better source of public opinion information than blogs for many purposes. A number of previous studies have analysed Twitter or blogs over time. An early study of blogging analysed the Danish Cartoons affair, which centred around the publication of a number of cartoons depicting the prophet Muhammad in the Danish newspaper *Jyllands-Posten*, and which became an international incident. The study showed that there was almost no blogging about it when the cartoons were initially published on 30 Sep. 2005 and that it became a major news story in Feb. 2006 because of two related events (a boycott of Danish products in Saudi Arabia and the withdrawal of the Saudi ambassador from Denmark) that occurred five months after the cartoons were published, rather than directly because of the publication of the cartoons themselves (Thelwall, 2007). The unique advantage of blogs for this study, in comparison to other sources of information at the time (Twitter launched in July 2006), was to show that there was little interest in the topic, at least in the blogosphere, when the cartoons were published. To reiterate this point, no other source of information could reveal whether there had been any public interest in the issue between Sep. 2005 and Jan. 2006: the closest approximation would be to measure the extent of press or media coverage of the topic during this period.

Another study compared the events attracting the most Twitter attention in six different English-speaking countries, finding significant overlaps in interest that could be explained by geopolitical factors (Wilkinson & Thelwall, 2012). On a smaller scale, an investigation of 7,184 tweets relating to three different campus shootings was used to show how the medium was used to help make sense of the situation at different points in time (Heverin & Zach, 2012). Trends in tweeting have even been used to automatically identify emerging news stories (Becker, Naaman, & Gravano, 2011).

The time series approach can also be used to investigate changes in sentiment over time, either to understand the role of sentiment in an event or changes in popularity over time. One large-scale study compared overall changes in sentiment in tweets over time with external social, political, cultural, and economic phenomena, finding a connection between offline events and online sentiment (Bollen, Pepe, & Mao, 2011). A number of studies have focussed on sentiment in relation to elections to assess whether it is possible to predict outcomes (Chung & Mustafaraj, 2011). Whilst it seems logical that sentiment expressed in Twitter would reflect the public mood, there is a problem with spam, attempts to manipulate Twitter for political goals, and different levels of Internet use for people of differing political persuasions that makes this difficult in practice (cf. Mowbray, Chapter 14 in this volume). The remainder of this chapter describes how to conduct a time series and sentiment analysis of Twitter. It gives an overview of methodological issues, and describes in broad detail how to use a specific set of software tools developed at the University of Wolverhampton, UK, in order to gather and analyse tweets for a specific topic.

CREATING A CORPUS OF TWEETS

At the time of writing, the official Twitter search engine search.twitter.com seemed to only return tweets that were up to two weeks old (cf. Gaffney & Puschmann, Chapter 5 in this volume, for details on the functionality of the Twitter API). In addition, there are currently no free archives of tweets, and hence no convenient source of tweets for a Twitter time series analysis. A partial exception is Topsy Analytics (http://analytics.topsy.com/), which provides free graphs of the level of tweeting of any user-entered keyword for the previous month. Researchers interested in analysing large sets of tweets therefore need to either buy a relevant collection of tweets or monitor Twitter over a period of time in order to build their own corpus of tweets (see Gaffney & Puschmann, Chapter 5 in this volume). This chapter describes the latter approach.

A simple way to create a corpus of tweets would be to search Twitter periodically, such as daily, recording and saving the results. For instance, the search might be a hashtag or a more complex query or set of queries designed to match topic-relevant tweets. This would be time-consuming and would also be ineffective if there were too many matching queries to save. A practical alternative is to use a computer program to automatically submit the queries periodically and save the results. This approach is described below, following a brief discussion of ethical considerations and limitations.

Creating a corpus of tweets does not have the same ethical and privacy implications as interview transcripts or questionnaire data, because tweets are inherently public and readable, when posted to a public account, by anyone with an Internet connection (but see Beurskens, Chapter 10 in this volume, for legal considerations, and Zimmer & Proferes, Chapter 13 in this volume, for issues related to privacy). Hence, it can be argued that they should be regarded as documents rather than as human-related data (Wilkinson & Thelwall, 2011). Researchers should nevertheless avoid republishing individual tweets or their corpus of tweets, as this could have privacy implications caused by drawing attention to the individuals concerned (and this would also avoid violating Twitter's conditions of use). For example, if investigating suicide-related tweets, it could have negative consequences if a suicidal tweet was quoted in an article and the Twitter user or their acquaintances found it. In practical terms, the software Webometric Analyst (http://lexiurl.wlv.ac.uk) can be used to gather tweets. This program can automatically submit a pre-defined set of queries to Twitter every hour and record the results for later analysis. Webometric Analyst works by submitting the queries to Twitter via the officially permitted route, the Twitter API (Applications Programming Interface). At the time of writing, the Twitter API permitted keyword queries and allowed the query to specify the language of the tweets and the approximate geographic location of the Twitter users. These features are useful to ensure that only the most relevant tweets are gathered. To use this facility, the following steps are recommended:

1. Construct and test a set of queries for the topic researched. As far as possible, these queries should collectively give good coverage of the topic so that most tweets relevant to the topic would match at least one of the queries. As far as possible, the queries should not match tweets that are spam or otherwise irrelevant to the topic. Normally, the second consideration is most important for time series analyses, because it is difficult to filter out the irrelevant tweets and they can result in irrelevant analysis results. The queries should be tested in search.twitter.com before being used, and the first results from Webometric Analyst should also be checked for accuracy and appropriateness.

2. Run the tested queries in Webometric Analyst on a computer that is permanently switched on. This will collect tweets in real time so the queries should be set up before, or shortly after the start of, the event to be monitored. Webometric Analyst should be left going for the duration of the event to be monitored, or as long as makes sense for the analysis.

3. Process the tweets gathered using the methods below.

Gathering tweets using the method above has a sampling limitation. Twitter may not return all tweets that match a query and may impose arbitrary restrictions to conserve resources. Hence the results should be treated as a sample rather than as a comprehensive collection. Twitter also returns a maximum number of results per query depending on the API used, so a query may give especially incomplete results under certain circumstances (see Gaffney & Puschmann, Chapter 5 in this volume, for details).

A SIMPLE TIME SERIES ANALYSIS OF A TWITTER CORPUS

A simple way to analyse temporal trends in a Twitter corpus would be to sample a specified number of tweets at different time periods, such as at the beginning, middle, and end, and then use a content analysis (Neuendorf, 2002) to classify the samples. A comparison of the results at different time periods could then be used to identify changes over time. In contrast, the methods described in the remainder of this section are more quantitative, and use a graphical approach to identify trends in volume of tweeting over time. Nevertheless, it is a good idea to use content analysis in conjunction with the graphical approach, in order to get deeper and more qualitative insights into the data.

The graphical time series approach is essentially to construct a graph of the volume of topic-relevant tweeting over time, and to use the shape of the graph to identify trends in interest in the topic as well as individual events of interest during the time monitored. The graph used is the number of tweets in the corpus (i.e., matching the set of queries used to generate the corpus) plotted against time. The time interval used is normally hours (days are used instead if there are too few relevant posts per hour), so that each point on the graph represents the number of topic-relevant tweets gathered within a single hour. Figure 7.1 is an example of such a graph created from a corpus of tweets about the U.K. riots

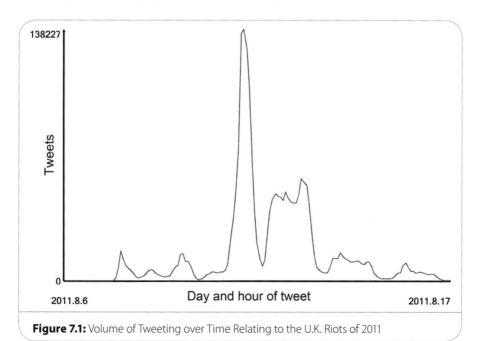

Figure 7.1: Volume of Tweeting over Time Relating to the U.K. Riots of 2011

of 2011. This event was a wave of rioting, looting, and arson in some English cities in early August 2011 initially triggered by the police shooting of Mark Duggan in Tottenham, London (see also Vis, Faulkner, Parry, Manyukhina, & Evans, Chapter 29 in this volume).

At the peak, at least 138,227 relevant tweets were sent in a single hour. It is clear from the graph that the peak in interest was considerably after the start of the riots. Note that some of the fluctuations are caused by time of day, with troughs during each night.

Visual inspection of the graph can be used to identify the overall trend in interest in the topic as well as key points of interest. If the graph is for a specific, time-limited topic (e.g., a conference), then the following should be observed:

- The initial increase in volume of tweeting points to the time at which the topic started to gain interest. A content analysis of tweets at the initial stages can point to people's initial reactions to the topic, which may be different from their later thoughts, after being exposed to more mass-media coverage of it.
- The graph is likely to show a decrease in volume at some stage, pointing to the time at which interest peaked, with people starting to lose interest in the topic afterwards. A content analysis of tweets at the peak will suggest the major issues about the topic that were discussed at the time that it was most significant in the public consciousness.
- The graph may reach a point at which the level of tweeting about the topic is almost the same as before the event. This suggests that people have either forgotten about the issue or have ceased to find new information about it to post. A content analysis of tweets after this time can point to the legacy of the event—such as whether it became a meme, referred to in other contexts, or if discussions about the key issues resurfaced periodically.

If the topic is a long-running or permanent issue, such as interest in a political party, then the following should be observed. For this issue, a comparative content analysis of tweets at the start of the period investigated and tweets at the end is recommended. This can point to changes in interest that have taken place.

- Is the volume of interest increasing or decreasing over time?
- Is any increase or decrease approximately constant, or are there changes in the broad pattern?

For either type of graph the following should be searched for:

- Spikes indicating specific events of interest for the topic. To detect what the event of interest is, a sample of tweets from the spike should be read. This can be conducted as a formal content analysis, if necessary. Spikes can sometimes be caused by spam, and so it is important to check that each spike analysed is genuine by checking the sample for suspicious content.
- Any other strange behaviour in the graph. These may be caused by external events, but care should be taken to check that they are not anomalies due to technical reasons, such as temporary breaks in the Twitter service.

The identification of spikes can be difficult, because Twitter time series graphs are likely to be quite spiky due to natural variation in the data rather than due to external events. Hence, when detecting spikes, only the largest spikes should be investigated. Moreover, if a graph is very jagged, so that it appears to be spiky without any trend, then it should be redrawn with a longer time period (e.g., with each point corresponding to the number of tweets in a whole day rather than a whole hour) in order to get a less jagged line.

TIME SERIES ANALYSIS OF QUERIES WITHIN A TWITTER CORPUS

The above analyses can be repeated for a subset of a corpus by constructing a graph for the tweets matching a given query. This can reveal patterns of interest for the subtopic represented by the query within the overall topic. The methods described above can be repeated, except based on a graph of the *percentage* of tweets matching the query out of all tweets in the corpus. This is useful for tracking an issue or other theme throughout the corpus. Figure 7.2 shows how this works for the issue of the police. As is clear from the graph, the police were central to tweeting about the riots, with 25% of tweets explicitly mentioning them. They were subsequently less frequently tweeted about, but remained a significant topic of interest throughout. Graphs can also be constructed with more complex queries to capture issues that may be expressed with different words. For instance, to track gender issues in any topic, the query *male female man woman men women* could be submitted to capture and graph several different ways in which gender may be mentioned. It is primarily up to the researcher to produce a list of topics to analyse in this way, and to convert them into effective queries to produce a graph.

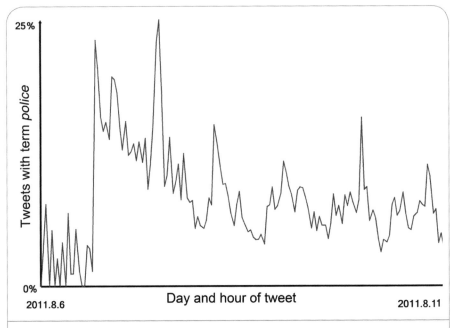

Figure 7.2: Percentage of Tweets from a U.K. Riots 2011 Corpus that Mentioned the Term *Police*

The graph shows a clear decrease in interest over time, although even at the lowest periods, 5–10% of tweets mentioned the police and there are periodic spikes in interest in the police. The corpus was collected via a set of hashtags relating to the riots and the cities involved.

A Twitter corpus can be converted into a time series graph using software. This can be much more efficient than using a manual approach—perhaps aided by a spreadsheet. The time series graphing and analysis program Mozdeh is described here. Mozdeh is a Windows-based program that is designed to run on most Windows-based systems without the need to be formally installed. A Twitter corpus gathered can be converted into a graph by Mozdeh in a number of steps, as described fully online (http://mozdeh.wlv.ac.uk). First, the corpus must be converted into the file format used by Mozdeh, and then Mozdeh has to index the tweets in order to create a graph. Once the indexing is complete, Mozdeh will create a time series graph of the data. This can then be used to identify the spikes or trends discussed above. Mozdeh can also produce the graphs for subsets of the tweets by entering a query with the graph so that it produces a graph of how frequently the tweets within the corpus match the query. The same steps described above can be used to analyse this graph. It makes sense to analyse keywords relevant to the broad topic to identify how they relate to the overall topic.

Mozdeh also has a more complex function that automatically identifies the 1,000 individual words within the corpus that exhibit the biggest increase in frequency over time. This option can be used to automatically identify important events within the topic as an alternative to the spike method. This is because words associated with an increase in frequency within the corpus are normally associated with spikes of increases in interest caused by a particular topic.

SENTIMENT ANALYSIS FOR TWITTER

Automatic sentiment analysis has become popular over the past decade, especially for web data. A sentiment analysis program predicts the sentiment content of texts based upon features it identifies, such as the words used and the presence of emoticons. This section describes a sentiment analysis program designed for social Web data, SentiStrength, and explains how it can contribute to analyses of a Twitter corpus. The availability of software to conduct sentiment analysis makes it possible to run large-scale investigations into sentiment on Twitter. Whilst SentiStrength is not the only sentiment analysis program that has been designed for the social Web and applied to Twitter data (e.g., Kouloumpis, Wilson, & Moore, 2011), it is one of the few designed to produce results that are valid for social science research purposes (Thelwall & Buckley, 2013). Moreover, it gives an explicit rationale for each sentiment classification made, which most sentiment analysis programs do not, and this can help with follow-up qualitative analyses. For more information about the many alternative sentiment analysis methods there are books that give an overview (e.g., Liu, 2012).

SentiStrength is primarily based upon a list of words known to normally be used in a positive or negative context. Each word in SentiStrength's lexicon is associated with a positive or negative score for the polarity and strength of the sentiment term. The score is on a scale of 1 (neutral) to 5 (very positive), or -1 (neutral) to -5 (very negative). For example, *love* scores +3 and *hate* scores -4 in this scale. When fed with a new text, SentiStrength checks it for the presence of sentiment terms from its lexicon, and predicts the sentiment of the text based upon the scores of the words found, subject to about 12 additional rules (Thelwall, Buckley, & Paltoglou, 2012). Each text is given two scores, one for the strength of positivity contained within it (on the 1 to 5 scale), and one for the strength of negativity (on the -1 to -5 scale). Hence, the sentence "I love to hate you" would score -4 and +3. The results of a sentiment analysis program applied to a Twitter corpus can be used to identify trends in sentiment. This is normally achieved by plotting the average sentiment over time in a time series graph. For example, one study investigated whether peaks of interest in a major media event were associated with increases in positive or negative sentiment (Thelwall, Buckley, & Paltoglou, 2011). Alternatively, the average sentiment before a particular event could be compared to the average sentiment after it, to assess the impact of the event on tweeting; or the average sentiment of tweets mentioning one keyword could be compared to the average sentiment of tweets mentioning a different one. As illustrated by these examples, the goals of a sentiment analysis typically need to involve comparing the average sentiment of multiple sets of tweets or over time. As described on the Mozdeh website, SentiStrength can be used to conduct a sentiment analysis of a Twitter corpus in two ways. It can be applied directly to the corpus, as saved by Webometric Analyst, to record positive and negative sentiment strengths for each tweet. This would help to compare the average sentiment of one or more sets of tweets—for example, the average sentiment strengths could be worked out via a spreadsheet after loading the corpus and sentiments scores into it. Alternatively, SentiStrength can be applied to the tweets indexed by Mozdeh, and then Mozdeh will produce a graph of the average positive and negative sentiment strengths of all tweets over time, or of all tweets matching a particular query over time, as shown in Figure 7.3 for the 2011 U.K. riots.

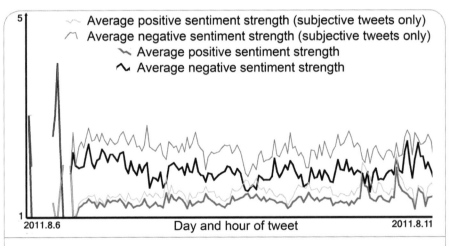

Figure 7.3: Average Positive and Negative Sentiment Strength for U.K. Riots Tweets Mentioning the Police

The tweets are consistently more negative than positive (the negative sentiment strength lines are higher than the positive sentiment strength lines). Negativity seems to decrease over time, and there seem to be two important spikes in negativity and two different spikes in positivity towards the end.

Any type of automatic sentiment analysis, including SentiStrength, uses heuristics to identify sentiment in text, and will therefore make mistakes by misinterpreting the sentiment content of a text. This can occur, for example, if sentiment is expressed using unusual words, using complex linguistic formulations, such as sarcasm or irony, or is implicit rather than explicit (e.g., "The rioters bypassed my shop", or "The politician avoided the question"). A program like SentiStrength has, on average, human-level accuracy for sentiment-strength detection, and so incorrect sentiment readings should normally not cause problems when averaging sentiment over many texts, including when sentiment is incorporated in a diagram like Figure 7.3. Nevertheless, if unusual sentiment expressions are repeated and become a meme or common expression in the context of a particular topic, then this can systematically distort the results of a sentiment analysis. Some methods have been suggested to partly remedy this issue (Thelwall & Buckley, in press) by modifying the parameters of a sentiment analysis program for a particular topic, but this is a time-consuming task. The practical implication of this is that sentiment analysis results should not be taken at face value, but should be checked for anomalies, and if such anomalies are found, then steps should be taken to improve the algorithm.

CONCLUSION

The above description covers how to collect tweets relevant to a topic, and how to conduct a sentiment analysis or time series analysis of tweets for that topic. It is designed to give practical advice about what kinds of problems can be investigated and how to conduct analyses using University of Wolverhampton software. Whilst alternative approaches are available, and some are discussed in this book, the methods in this chapter provide an integrated and coherent set that have been used in previously published research. The methods here are quantitative, and work best when a reasonably large collection of tweets can be collected for a topic, such as over 10,000. This restriction means that the methods are suitable for topics that are not of purely niche interest and hence generate a reasonable volume of tweeting. When such a corpus can be collected, then the methods described here can give insights into how public reactions to the topic change over time, including the identification of significant events, overall trends in topic interest, and changes in average sentiment strength over time. Since Twitter can contain spam, and a corpus may accidentally contain irrelevant tweets, it is important to check the validity of the results at each stage—for example, by examining a proportion of tweets—to check for anomalies and spam. With such safeguards, these studies can either give insights into how Twitter is used to discuss an event, or insights into the offline event, as mediated through tweeting. In the future, the methods described here may become more widely available and perhaps even commonly used. Topsy Analytics, with its one-month Twitter time series graphs, has already made Twitter time series analysis practical if a long time period and sentiment analysis are not necessary. This site may disappear at any time, but similar sites may be created in the future. A similar site is Google Trends (http://www.google.com/trends), which gives time series graphs for Google users' keyword searches, and can give graphs for specific locations. The main limitation of this tool is that it is impossible to carry out a content analysis of the results or to use any other method to be sure of the cause of any spikes identified. More generally, future research is needed to fully evaluate the methods described here and to identify the situations in which they are most useful, as well as the kinds of information that they are likely to give in particular contexts.

REFERENCES

Becker, H., Naaman, M., & Gravano, L. (2011). Beyond trending topics: Real-world event identification on Twitter. In N. Nicolov & J. G. Shanahan (Eds.), *Proceedings of the 5th International Conference on Weblogs and Social Media (ICWSM 2011)*. Menlo Park, CA: AAAI Press. Retrieved from http://sm.rutgers.edu/pubs/becker35-icwsm2011.pdf

Bollen, J., Pepe, A., & Mao, H. (2011). Modeling public mood and emotion: Twitter sentiment and socioeconomic phenomena. *Proceedings of the Fifth International AAAI Conference on Weblogs and Social Media (ICWSM 2011)*. 17–21 July 2011, Barcelona, Spain. Retrieved from http://arxiv.org/abs/0911.1583

Chung, J. E., & Mustafaraj, E. (2011). Can collective sentiment expressed on Twitter predict political elections? In W. Burgard & D. Roth (Eds.), *Proceedings of the Twenty-Fifth AAAI Conference on Artificial Intelligence (AAAI 2011)* (pp. 1768–1769). Menlo Park, CA: AAAI Press.

Heverin, T., & Zach, L. (2012). Use of microblogging for collective sense-making during violent crises: A study of three campus shootings. *Journal of the American Society for Information Science and Technology, 63*(1), 34–47.

Kouloumpis, E., Wilson, T., & Moore, J. (2011). Twitter sentiment analysis: The good the bad and the OMG! In *Proceedings of the Fifth International AAAI Conference on Weblogs and Social Media (ICWSM 2011)*. 17–21 July 2011. Barcelona, Spain. Palo Alto, CA: AAAI Press.

Liu, B. (2012). *Sentiment analysis and opinion mining.* San Rafael, CA: Morgan & Claypool.

Neuendorf, K. (2002). *The content analysis guidebook.* London, UK: Sage.

Smith, A., & Brenner, J. (2012). Twitter use 2012. *Pew Internet.* Retrieved from http://www.pewinternet.org/Reports/2012/Twitter-Use-2012.aspx

Thelwall, M. (2007). Blog searching: The first general-purpose source of retrospective public opinion in the social sciences? *Online Information Review, 31*(3), 277–289.

Thelwall, M., & Buckley, K. (2013). Topic-based sentiment analysis for the Social Web: The role of mood and issue-related words. *Journal of the American Society for Information Science and Technology, 64*(8), 1608–1617.

Thelwall, M., Buckley, K., & Paltoglou, G. (2011). Sentiment in Twitter events. *Journal of the American Society for Information Science and Technology, 62*(2), 406–418.

Thelwall, M., Buckley, K., & Paltoglou, G. (2012). Sentiment strength detection for the social Web. *Journal of the American Society for Information Science and Technology, 63*(1), 163–173.

Wilkinson, D., & Thelwall, M. (2011). Researching personal information on the public web: Methods and ethics. *Social Science Computer Review, 29*(4), 387–401.

Wilkinson, D., & Thelwall, M. (2012). Trending Twitter topics in English: An international comparison. *Journal of the American Society for Information Science and Technology, 63*(8), 1631–1646.

Computer-Assisted Content Analysis of Twitter Data

8
CHAPTER

Jessica Einspänner, Mark Dang-Anh, and Caja Thimm

to understand what people are saying, special tools and methods are needed #CAQDAS

CONCEPTUAL OVERVIEW: STATE OF THE ART OF ONLINE CONTENT ANALYSIS

Content analysis can be understood as a methodological framework within which various approaches of textual and non-textual analyses can be applied. The research technique of content analysis facilitates the systematic coding and analysing of the content of spoken, written, or audio-visual communication (Berelson, 1952; Krippendorff, 2004). It is used in order to identify and classify words, phrases, or other meaningful matter, such as images, sounds, or even numerical records in terms of their structure and semantics. By interpreting frequency distributions and co-occurrence patterns of the single analytical units, this methodological approach allows for systematically drawing valid conclusions from data "to the context of their use" (Krippendorff, 2004, p. 18). Early content analyses trace back to the 17th century, when the Church started to examine the content of the first newspapers systematically (Krippendorff, 2004, p. 3). As a fully developed scientific method, however, content analysis was not employed until the 1940s, when it was used for analysing mass-media content (Herring, 2010). In the Information Age, the Internet has become an important means for interpersonal communication and social interaction. In order to assess the relevance of online communication, the "careful and systematic observation of its contents seems inevitable" (Rössler, 2002, p. 301). Chat protocols, weblog content, social network communication, or other multimedia content is especially of interest to researchers, as this kind of online communication is supposed to be "the bearer of human existence" (Capurro & Pingel, 2002, p. 192). Almost instant access to people's utterances, uploaded pictures, or videos that could give information about certain characteristics and preferences of their behaviour (e.g., consumption, political opinion, manners of interaction), make the online environment an attractive research area for politics, economy, and science. Following a broad interpretation such as proposed here, researchers often draw on content analysis as an established methodological framework, and extend its traditional concepts while applying them to the online world (Herring, 2010).

The objectives of a content analysis of Twitter data can be as diverse as the possible methodological procedures. For example, the metrics of tweets can be analysed, i.e. how many @replies did two particular users exchange within a certain hashtag-based discourse? Which were the most common phrases used by a certain group of users in the data set? It might also be interesting to go into a detailed qualitative analysis of the tweets and find out about, for example, the linguistic characteristics of Twitter language and its speech acts, argumentative schemas, or semantic co-occurrences. One might also want to compare the topics that emerge on Twitter and the types of users who talk about similar or diverging topics, for example, politicians versus citizens. The examination of conversational structures through Social Network Analysis (Magnani, Montesi, Nunziante, & Rossi, 2011)—which can be regarded as one form of content analysis (Herring, 2010)—is just as interesting as doing opinion mining through Sentiment Analysis (Kumar & Sebastian, 2012; Nielsen, 2011), or using a mixed-method approach—for instance, a combined statistical and hermeneutical analysis—in order to assess the diffusion of information on Twitter (Huang, Thornton, & Efthimiadis, 2010; Jansen, Zhang, Sobel, & Chowdury, 2009). Content analysis is an approach to empirical research based on pre-existing material. On Twitter, we deal with high amounts of naturally occurring data, i.e. data that is usually produced without being motivated by any research intent, unlike elicited data from interviews, surveys, etc. Traditionally, content analysis does not necessarily require special software, and might as well be carried out manually or with common spreadsheet software. However, due to the large sample sizes that can be collected for the analysis of Twitter data, we recommend using data analysis software to support the research process along its different stages. Especially when it comes to more sophisticated research questions that demand statistical analysis; large, automated coding processes; or coding procedures that involve several coders, it might be useful to choose specific software to process the digital data at hand.

There is a wide range of Computer-Assisted Qualitative Data AnalysiS (CAQDAS) software that can be used for different types of digital content analyses. Whereas most of the common tools incorporate instruments to analyse quantitative (numeric) data as well as qualitative data (e.g., MAXQDA, QDAMiner, ATLAS.ti, Qualrus, NVivo), the range of the analytical features varies. Some of the programmes offer basic dictionary-based text analysis (that enables adding codes and hierarchies to text segments); others also allow for analysing audio, video, and other non-textual data. Although using CAQDAS software for Twitter research is not the most widely used approach, it can in fact make a content analysis more efficient, and thus provide alternatives to using automated approaches when dealing with larger datasets.[1] A well-organised coding scheme can handle extensive lists of codes and categories to be applied to the material, as well as a large number of statistical procedures. If multiple coders analyse the same data, simultaneously or at different times, CAQDAS software can be used to determine intercoder or intracoder agreement.

In this chapter, we will discuss speech act analysis of tweets as an example of software-assisted content analysis. We start with some elementary thoughts on the challenges of the collection and evaluation of Twitter data before we give a brief description of the potentials and limitations of using the software QDA Miner (as one typical example for possible analysis programmes). Our focus will lie on analytical features that can be particularly helpful in speech act analysis of tweets.

SAMPLING DATA IN TWITTER

One of the great challenges in analysing Twitter data—not only in content analysis—is to choose a sample that is appropriate to answer a research question. Collecting an exhaustive sample or a true random sample is hardly, if ever, possible in terms of scraping the required data in a consistent manner (Bruns & Liang,

2012). Limited access to the Twitter API, as well as specific hardware require-ments, often prevent researchers from collecting a representative sample of all Twitter users, let alone identifying and collecting an entire population of post-ings or users (see Gaffney & Puschmann, Chapter 5 in this volume). As long as researchers are not granted direct access by Twitter, the data-scraping process is restricted. Nonetheless, open-source tools such as yourTwapperKeeper allow col-lecting tweets from the search API and the streaming API (Bruns & Liang, 2012). To decide which sample should be collected for an analysis, the researchers should familiarise themselves with possible collection criteria. Content-based samples can, for example, be selected by collecting tweets that contain certain hashtags, words, or phrases. When it comes to event-related discourses, hashtags can be used for both labelling and identifying relevant postings. Tracking tweets that contain certain hashtags is a way "to establish a dataset of the most vis-ible tweets relating to the event in question" (Bruns & Liang, 2012). The same applies for hashtags as topical markers. However, not every posting contains a hashtag, and researchers should always be aware of the incompleteness of a sample based on hashtags, words, or phrases.

Alternatively, a sample can be created by collecting tweets from a specific account. However, Twitter limits the number of postings one can scrape from a users' account. Only if the total number of sent messages is below the cur-rent API limit, which is changed off and on, is it possible to collect all tweets sent by a user. In order to track account-related conversations, it is necessary to additionally collect tweets that are addressed to an account by using @replies. A third dimension that has to be considered in sampling Twitter data is that of time. Collecting a consistent random sample within a specific time frame is virtually impossible because of the API restrictions. Nevertheless, an appro-priate scraping period must be chosen to build up a data set, besides applying word-based or account-based criteria. Again, depending on the research ques-tion, one might, for example, collect a few hashtag-based postings over a lon-ger period of time, a large number of word-based postings over a short period of time, or postings from a specified account over a long period of time. Bruns and Liang (2012) provide deeper insights into ways of collecting Twitter data.

When performing content analysis on Twitter data, tweets can be regarded as single sampling units (cf. Krippendorff, 2004, pp. 98–99). In principle, defining a tweet as the sampling unit follows clear-cut formal means (syntax): a posting, restricted to 140 characters, sent by a unique user at a particular moment; but, except for a few cases, tweets can usually also be regarded as units of meaning (semantics). Considering tweets as sampling units allows for a metadata-per-tweet approach by which metadata like account name, timestamp, geo coordinates (if provided), etc. are distinctly assigned to each tweet.

SPEECH ACT ANALYSIS OF TWEETS WITH CAQDAS SOFTWARE

CAQDAS tools allow for combining automated (quantitative) with manual (quantitative or qualitative) content analysis. It is often appropriate to identify noticeable patterns and structures of the metrics of the data. This can be achieved by measuring the number of tweets from a particular user or group of users (*metrics per user*), analysing the Twitter communication over a certain period of time (*metrics per time frame*), or the development of a given topic (*metrics per hashtag*; see Bruns & Stieglitz, Chapter 6 in this volume; Bruns & Burgess, 2012). Peaks in patterns of communication (e.g., significantly more or less tweets containing a certain hashtag in a given time frame) or distinctive features within a user's tweeting style (e.g., changing retweeting or linking habits) can be the (exploratory) basis for formulating specific research questions and hypotheses, and give the researcher an idea of where to start with a qualitative, more in-depth analysis.

In the following, we give a short outline of some of the possible (first) steps of a tweet analysis carried out with the help of a CAQDAS tool, QDA Miner. By describing some of the possible analytical processes with this particular tool, we do not necessarily consider these options to be the best way of using it. Usually, there are several ways of approaching one task within this software—or there may be better ones with another programme. However, we think that QDA Miner, as rather typical CAQDAS software, is not only a fairly comprehensible, but also a suitable tool to analyse the content of tweets. In our discussion, we thus refer to QDA Miner as a token of content analytical software.

We start with some basic settings, and end with a more detailed description of speech act analysis within the methodological framework of content analysis.

BASIC CONTENT ANALYSIS (FIRST-LEVEL ANALYSIS)

The computer-assisted content analysis of tweets can be organised on two levels. The first level allows for a basic content analysis suitable for big and small data. Basic analytical functions are word- or phrase-frequency analyses, keyword-in-context lists (KWIC), and some basic data visualisations, such as hierarchical word tree diagrams. Word-frequency lists help provide a quick overview of the words or phrases that occur in the analysed text a certain number of times. Such frequency lists can also be customised by excluding inappropriate terms (e.g., common strings like "www", "http", "RT", etc., or the (key)word that occurs in every tweet because it was the criterion for selecting the data). QDA Miner also facilitates first-level computational coding of the imported tweets. Here, character strings are lemmatised, i.e. shortened to their word stem, in order to assign inflected word forms to dictionary entries. Such automated content analysis is limited to a dictionary with fixed thesauri implying a complex, but rather static and thus superficial relation between words and meanings, as illustrated by the following example:

> As an example, the Linguistic Inquiry and Word Count (LIWC) dictionary maps the word set {ashes, burial*, buried, bury, casket*, cemet*, coffin*, cremat*, dead, death*, decay*, decease*, deteriorat*, die, died, dies, drown*, dying, fatal, funeral*, grave*, grief, griev*, kill*, mortal*, mourn*, murder*, suicid*, terminat*} to LIWC category 59, death. The asterisks are 'wild-card' characters telling the program to treat 'cremating', 'cremated' and 'cremate', as all matching cremat*, and thus all mapping to category 59. (Lowe, 2003, p. 2)

One problem with the automatic categorising is that misspelled words or chat language (e.g., "rotfl", "lol", etc.) are usually not classified in standard dictionaries. However, applying a user-defined dictionary where new words and expressions can be entered may solve this problem.

Another problem is the correct allocation of identified words for one category and their contextual meaning. Both can differ: whereas the software may categorise the word *play* under HUMOR, it actually does mean something else in the context of the tweet, "there is a video link on the page, play it." Another example is the ambiguity of the word *beat* that may be automatically classified as AGGRESSION (e.g., by the RID.CAT-dictionary), but can have another connotation in the context of "Obama beat Romney in the general election." These examples illustrate limitations of automated content analysis. Researchers should not solely rely on existing dictionaries and mere statistical frequencies, but need to carefully scrutinise these first-level findings and consider manual coding.

SPEECH ACT ANALYSIS (SECOND-LEVEL ANALYSIS)

The bigger the data set, the more difficult it gets to analyse it in-depth. After frequency counts on the first level, coding of the tweets takes place on the second. Coding means categorising text fragments or multimedia content. Categories are defined in a coding scheme. They can be generated deductively from an existing theory or inductively "as near as possible to the material" (Mayring, 2000, p. 2). However, most coding schemes are being developed in a more iterative and cyclic process (Teddlie & Tashakkori, 2010), constantly refining categories considering the pertinent, theoretical literature and the material coded so far. Annotating text segments with codes means interpreting and quantifying these segments in order to make them computable. As our focus in this chapter lies on coding speech acts in tweets, we will briefly introduce speech-act theory before giving some examples of how CAQDAS software can help with the manual coding process.

The linguistic evaluation of tweets can be quite challenging due to possible grammatical inconsistencies of computer-mediated language. As Twitter is widely used for conversation (Bruns, 2012; Magnani et al., 2011), an analysis of speech acts is highly interesting, as it can give information about the types of actions that people want to accomplish through communication (Nastri, Peña, & Hancock, 2006). The objective of a speech act analysis is to identify different types of purposeful utterances, such as command, complain, compliment, etc. There are several taxonomies categorising speech acts with regard to their intention (illocutionary acts). Often, Searle's (1976) basic classification of illocutionary acts, which again is based on Austin's (1962) work, is adopted for analysing computer-mediated language (e.g., Nastri et al., 2006). Searle (1976) categorised purposeful utterances as assertives or representatives (commiting the producer of an utterance to the truth of the proposition), directives (attempting to get the receiver to do something), commissives (commiting the producer to some future course of action), expressives (expressing the psychological state of a situation), and declarations (bringing about a change in a state of affairs). Table 8.1 gives some examples of possible verb groups for each category.

Table 8.1: Basic Classification of Illocutionary Acts (Purposeful Speech Acts) by Searle (1976)

Speech Act	Paradigms of Verbs (Examples)
Assertive / representative	Describe, call, conclude, deduce
Directive	Ask, order, command, request, beg, invite, permit
Commissive	Promise, swear
Expressive	Thank, congratulate, apologise, condole, welcome
Declaration	Declare, nominate

While discussing Searle's theory in more depth is beyond the scope of this chapter, it should have become clear that analysing speech acts in Twitter communication demands a lot of interpretative effort, and may not be possible without some theoretical considerations. One difficulty lies in the linguistic specifics of tweets. For example, the researcher needs to specify if a hyperlink can be identified as a speech act. One could regard a hyperlink as an implied request to click on it and code it as directive. However, if codes are supposed to give information about the meaning of the material, this would probably not be really helpful. It could instead be reasonable to explore the content behind the hyperlink and code it in a way that appropriately determines the underlying speech act. If one is instead merely interested in the structure of a tweet, the hyperlink could simply be coded as such (the same procedure can be applied to the other Twitter-specific signifiers, i.e. the @-symbol, RT, or #, in order to quantify these functional operators, cf. Thimm, Einspänner, & Dang-Anh, 2012).

A similar decision must be made in the case of chat language (or rather, Internet slang), especially emoticons. Sometimes one tweet only consists of a slang utterance, e.g. "lol", or just a smiley. This could point to some form of humour or self-expression (Nastri et al., 2006). Here, the traditional speech-act classification may not be sufficient. It could therefore be reasonable to consider creating a new category (and a new code) for these or similar cases of Twitter language. Sometimes speech-act categories can also overlap, i.e. directives and commissives. This makes determining the "right" speech act even more difficult, especially if several coders work on the same material and individual intuitions have be harmonised to assure consistent coding decisions.

Most of these difficulties cannot be resolved by computer software, as they are inherent to the data or require theoretical evaluation. However, using content-analysis software has the advantage that codes and labels can be modified or merged at any time. It can be helpful to use a "work in progress" category in the beginning of the coding process, for example, if the rules for distinguishing speech acts are not yet defined conclusively. However, any final decision on the definition of the categories must be explicated in the coding scheme as clearly as possible. Based on the coding of speech acts, CAQDAS software can run correlations on different speech-act codes in order to identify argumentative patterns in Twitter communication. One result may be, for example, that in a high number of cases assertives co-occur with commissives, or that expressives contain a high number of emoticons (if coded respectively). Such results can then again be statistically correlated with different variables—for example, a groups of users—in order to find out how certain social groups use which kind of linguistic strategies or argumentation patterns on Twitter. This way of analysing the language of Twitter is one of the most useful features of content analysis software. At the same time, however, statistical parameters such as correlations may be difficult to interpret, and researchers need to decide which analytical procedures can be meaningfully applied in light of their hypotheses or research questions, to avoid drawing artificial, data-centric conclusions.

CONCLUSION

Content analysis provides a useful and multifaceted, methodological framework for Twitter analysis. CAQDAS tools support the structuring of textual data by enabling categorising and coding. Depending on the research objective, it may be appropriate to choose a mixed-methods approach that combines quantitative and qualitative elements of analysis and plays out their respective advantages to the greatest possible extent while minimising their shortcomings. Big data (from several thousand up to millions of tweets) should rather be considered for a quantitative assessment of, for instance, communication patterns within the data set. It can subsequently be reasonable to extract a subsample (= small data) and analyse it qualitatively with the help of CAQDAS software. Basic functions such as word, phrase, or category count analyses as well as features like co-occurrence or KWIC-analyses can be useful additions for a systematic interpretation of the data. The process of coding speech acts within tweets as a form of qualitative content analysis can be very demanding, as (re-)contextualising tweets, differentiating similar speech acts (or topics, arguments, etc.), categorising Twitter-specific symbols, and finally, interpreting the co-occurrences can be quite challenging. Table 8.2 summarises the main advantages and limitations of CAQDAS in Twitter analysis.

Conducting content analysis with the use of CAQDAS software can expand the researcher's capability to interpret Twitter data. However, due to various limitations, qualitative data analysis software should rather be used as a supportive tool than a product that drives the whole research process. In the end, the interpretation of the findings still has to be done by the researcher.

Table 8.2: Overview of the Advantages and Disadvantages of Using CAQDAS Software for Analysing Twitter Messages

CAQDAS and Twitter Analysis: Advantages	CAQDAS and Twitter Analysis: Disadvantages
Allows for mixed-methods approaches.	CAQDAS packages are very complex; need a lot of time and effort to get to know the particular features and functions.
Metrical analyses as well as frequency analyses can be carried out quickly; give a good first impression on the data.	Dictionary entries/categories not sufficient for language-in-context.
Basic analysis (word/phrase/category count) and visualisation possible with small and big data.	Limited automated coding processes; manual coding required.
Codes can be arranged hierarchically and be modified during coding and analysis; overlapping of codes possible.	In-depth content analysis (semantic analysis) hardly possible with big data.
Inter- and intracoder reliability tests can be performed.	Most software is proprietary and costly.

NOTE

1 More information on CAQDAS can be found, for example, on the website of the Surrey CAQDAS networking project (http://caqdas.soc.surrey.ac.uk).

ACKNOWLEDGMENTS

This chapter originates from the context of the research project "Political Deliberation on the Internet" (as part of the DFG SPP 1505 "Mediatized Worlds"), headed by Caja Thimm, University of Bonn, Germany. We would like to thank the German Research Foundation (DFG) for funding our work.

REFERENCES

Austin, J. L. (1962). *How to do things with words*. Oxford, UK: Oxford University Press.

Berelson, B. (1952). *Content analysis in communication research*. New York, NY: Free Press.

Bruns, A. (2012). How long is a tweet? Mapping dynamic conversation networks on Twitter using Gawk and Gephi. *Information, Communication & Society, 15*(9), 1323–1351. doi:10. 1080/1369118X.2011.635214

Bruns, A., & Burgess, J. (2012). *Notes towards the scientific study of public communication on Twitter.* In A. Tokar, M. Beurskens, C. Puschmann, S. Keuneke, M. Mahrt, I. Peters, . . . Weller, K. (Eds.), *Science and the Internet* (pp. 159–169). Düsseldorf, Germany: Düsseldorf University Press.

Bruns, A., & Liang, Y. E. (2012). Tools and methods for capturing Twitter data during natural disasters. *First Monday, 17*(4). Retrieved from http://firstmonday.org/htbin/cgiwrap/bin/ ojs/index.php/fm/article/view/3937/3193

Capurro, R., & Pingel, C. (2002). Ethical issues of online communication research. *Ethics and Information Technology, 4*(3), 189–194.

Herring, S. C. (2010). Web content analysis: Expanding the paradigm. In J. Hunsinger, L. Klastrup, & M. Allen (Eds.), *International handbook of Internet research* (pp. 233–249). Dordrecht, The Netherlands: Springer.

Huang, J., Thornton, K. M., & Efthimiadis, E. N. (2010, June). *Conversational tagging in Twitter.* Paper presented at the 21st ACM Conference on Hypertext and Hypermedia (HT '10). Retrieved from http://jeffhuang.com/Final_TwitterTagging_HT10.pdf

Jansen, B. J., Zhang, M., Sobel, K., & Chowdury, A. (2009). Twitter power: Tweets as electronic word of mouth. *Journal of the American Society for Information Science and Technology, 60*(11), 2169–2188.

Krippendorff, K. (2004). *Content analysis: An introduction to its methodology.* Thousand Oaks, CA: Sage.

Kumar, A., & Sebastian, T. M. (2012). Sentiment analysis on Twitter. *International Journal of Computer Science Issues, 9*(4), 372–378.

Lowe, W. (2003). Software for content analysis—A review. Retrieved from http://www.ou.edu/ cls/online/lstd5913/pdf/rev.pdf

Magnani, M., Montesi, D., Nunziante, G., & Rossi, L. (2011). Conversation retrieval from Twitter. In P. Clough, C. Foley, C. Gurrin, G. J. F. Jones, W. Kraaij, H. Lee, & V. Murdoch (Eds.), *Advances in information retrieval* (pp. 780–783). Berlin, Germany: Springer.

Mayring, P. (2000). Qualitative content analysis. *Forum: Qualitative Sozialforschung / Forum: Qualitative Social Research, 1*(2), Article 20. Retrieved from http://nbn-resolving.de/ urn:nbn:de:0114-fqs0002204

Nastri, J., Peña, J., & Hancock, J. T. (2006). The construction of away messages: A speech act analysis. *Journal of Computer-Mediated Communication, 11*(4), 1025–1045.

Nielsen, A. (2011). *A new ANEW: Evaluation of a word list for sentiment analysis in microblogs.* Paper presented at the Extended Semantic Web Conference, Workshop Making Sense of Microposts (#MSM2011), Heraklion, Crete. Retrieved from http://sunsite.informatik.rwth-aachen.de/Publications/CEUR-WS/Vol-718/paper_16.pdf

Rössler, P. (2002). Content analysis in online communication: A challenge for traditional methodology. In B. Batinic, U. D. Reips, & M. Bosnjak (Eds.), *Online social sciences* (pp. 301–317). Seattle, WA: Hogrefe & Huber.

Searle, J. R. (1976). A classification of illocutionary acts. *Language in Society, 5*(1), 1–23.

Teddlie, C., & Tashakkori, A. (2010). Overview of contemporary issues in mixed methods research. In A. Tashakkori & C. Teddlie (Eds.), *Sage handbook of mixed methods in social & behavioural research* (pp. 1–41). Thousand Oaks, CA: Sage.

Thimm, C., Einspänner, J., & Dang-Anh, M. (2012). Twitter als Wahlkampfmedium: Modellierung und Analyse politischer Social-Media-Nutzung [Twitter as a medium in election campaigns: Model and analysis of the political use of social media]. *Publizistik, 57*(3), 293–313.

Ethnographic and Qualitative Research on Twitter

9

CHAPTER   Alice E. Marwick

needed: qualitative methods like interviews and #ethnographic fieldwork to understand the diverse practices and perspectives of Twitter users

Twitter's success has made it a rich research site for scholars interested in online interaction, information dissemination, activism, and a plethora of other subjects. The sheer volume of users, tweets, and hashtags has made the site a favourite for quantitative data analysis and "big data" number-crunching. For instance, in an early study of Twitter, Krishnamurthy, Gill, and Arlitt (2008) collected information about nearly 100,000 users, including number of accounts followed, number of accounts following them, and frequency of status updates. The authors created a taxonomy of Twitter users, grouping them into *broadcasters*, *acquaintances*, *miscreants*, and *evangelists* based on the ratio of following-to-follower. Similarly, Java, Song, Finin, and Tseng (2007) used a sample of 1.3 million tweets from 76,177 users to describe why people use Twitter, which they summarised as "information sharing, information seeking, and friendship-wise relationship [sic]" (p. 60). While such studies are valuable, inferences made on the basis of the properties of a large data set are limited in what they can explain. In the latter study, asking people about their motivations for using Twitter would probably reveal an array of interesting motivations that do not neatly map on to these three groups. Because Twitter is such a vast network with so many user groups, simply collecting a great deal of data may not be adequate for describing use beyond simple queries. Qualitative methods, such as interviews, ethnographic observations, and content analysis, provide a rich source of data that allow us to go beyond description. For instance, qualitative methods can help unpack user presumptions about individual technologies, distinguishing general communicative or social media behaviour from behaviour that is specific to a platform.

Qualitative methods can also reveal much about social norms, appropriateness, or larger social concerns about technology. Twitter's breadth and diversity requires recognising that different user groups have different social norms and idioms of practice (Gershon, 2010). Generalisations made about one hashtag, meme, or network of users may not apply to another, providing only a small portion of the picture. Qualitative research allows scholars to investigate the practices of a particular user group, as it can go beyond tracking follower counts or hashtag use to include many more sources of input about a specific community or user segment. Moreover, qualitative data can often be useful for triangulating and augmenting quantitative results (see, for example, Honeycutt & Herring, 2009; Naaman, Becker, & Gravano, 2011). This chapter discusses a variety of qualitative research methods, including interviews, ethnographic fieldwork, and textual analysis.

INTERVIEWS

Interviews are a basic tool of qualitative methods in a range of disciplines, including sociology, media studies, anthropology, and human-computer interaction (Spradley, 1979; Wengraf, 2001). The content and protocol of the interview will depend on the research questions being asked and type of interview method (semi-structured, ethnographic, narrative, and so forth). While interviews can be conducted via direct conversations on Twitter, this approach produces a very particular and constrained style of interview, due to the 140-character limit. More common is interviewing Twitter users in person, or using a medium like the telephone or Skype to conduct long-distance interviews.

INTERVIEWS ON TWITTER

The simplest way to interview Twitter users is to ask one's own Twitter follow-ers, or to @reply individual users and ask them quick questions. This approach has several advantages. It is quick and easy, does not cost anything, and allows the researcher to target broad populations in relatively small amounts of time. On the other hand, it is hardly representative (although one could argue that virtually nothing on Twitter would represent "society as a whole"). Besides the obvious bias of using a convenience sample made up of one's own followers, many Twitter users will not reply to @replies from people they do not know, and get-ting the attention of specific accounts is easier said than done. The researcher's earnest question may look like intrusive marketing spam, or simply get lost in the rapidly changing stream of tweets. And, obviously, it is difficult to conduct interviews of any depth using the service. Question-and-answer tweets might more properly be referred to as a very short survey.

Even with these limitations, I found this method quite useful as part of a larger project. I worked on one study that examined how highly followed indi-viduals conceptualised their audience (Marwick & boyd, 2011a). My co-author and I were interested in "context collapse", the phenomenon where large social network sites like Facebook and Twitter "collapse" acquaintances from different social contexts into the single word *friend* or *follower*. We wondered if Twitter users recognised the coexistence of these multiple audiences, or had only a subgroup of followers in mind when they tweeted. Using the site Twitterholic. com, which ranks Twitter accounts by number of users, we generated a list of the top 300 most-followed individual users on Twitter, removing media and business accounts. I created a research Twitter account separate from my per-sonal account, which clearly identified my affiliation and purpose. I then sent individual tweets to each person, asking them who they thought of when they tweeted. My response rate was very low, but a number of people did respond. I then created a similar list of 300 accounts with 10,000–100,000 followers and repeated the process. The response rate was higher, and I followed up with each responder via Twitter. Two agreed to be interviewed, one via email and one over the phone. I then tweeted my own followers and asked for responses. The response rate was still higher. My co-author danah boyd had approximately 15,000 followers at the time (a very high number for 2009), and she retweeted my inquiry, garnering still more responses.

At this point we still did not have anything resembling a 'representative' sample, but we had several hundred responses and could group them into a rough taxonomy of "how people thought about their audiences." We noticed that these categories remained constant regardless of the number of followers; in other words, many of the accounts with only a few hundred followers carefully curated their tweets in the same way that people with hundreds of thousands of followers did, and several of the most highly followed accounts claimed that they tweeted only for themselves. We also found several categories we had not considered while formulating our research questions. We could use this information, combined with what we had gleaned from our literature review of previous studies of the audience, to draw some rough conclusions about conceptions of the audience on Twitter. We used full-length interviews to test these assumptions (Marwick & boyd, 2011a).

The goal of the second study was to understand how teenagers use Twitter, and whether there are significant differences between teenage and adult Twitter use. We collected a large sample of tweets (400,000) that contained the hashtag, "#IGoToASchoolWhere". This topic involved young people complaining or making funny observations about their high schools (the most popular tweet was "#IGoToASchoolWhere the kids are higher than the grades!"). An intern and I spent many hours going through the corpus, determining the most frequently retweeted tweets, the most prolific authors, and the highest-followed accounts that participated. We used quantitative methods to determine these three factors, but I also spent a lot of time reading through the tweets to get a "feel" for the sample. I searched for various college-related terms, and randomly sampled accounts to feel confident enough to make the assumption that most of the participants were teenagers, not adults.

However, in order to test this assumption, we needed to talk to the people participating in the #IGoToASchool hashtag. I again used my research account to send inquiries to the 300 most frequent tweeters in our #IGoToASchool sample. I created a webpage with a URL-shortened link (e.g., bit.ly/teentwitter) which I included in the tweets, so users could verify that the study was legitimate. I got a single response, and it was of the "Uh, what?" variety. Unfortunately, the methods I had used in the audience study did not work. Teenagers are less likely than highly followed adult accounts to @reply strangers, and they change their usernames more frequently than the average Twitter user. I had waited too long after data collection to talk to the participants; I should have tweeted participants while the hashtag was trending. In general, when studying a particular hashtag or event, it is best to act quickly and try to get requests out while the topic is still trending or current. We had to abandon Twitter interviews and rely primarily on quantitative data and content analysis of the sample, along with a close reading of the tweets themselves.

After this experience, I think it is best to use the Twitter interviews as a supplement to triangulate results gleaned through other methods such as in-person interviews, content analysis, or quantitative analysis. Designing a research project so that it required interviews with specific Twitter users (as opposed to "Twitter users") was a mistake, given the low response rate.

INTERVIEWS ABOUT TWITTER

A preferred interview method is to conduct long-form in-person, email, phone, or Skype interviews with Twitter users. These have the advantage of providing more information and background than can be garnered in 140 characters. In their study of "unfollowing," Kwak, Chun, and Moon (2011) interviewed 22 Korean users about why they unfollowed people on Twitter, both in person and on Skype. The researchers compared this interview data with quantitative analysis of the following behaviour of 1.2 million Korean Twitter users. While some of the interview data confirmed their quantitative findings, other findings were surprising—such as people following others reciprocally, even if they did not know the person who had followed them (Kwak et al., 2011). Thus, on the one hand, as in this case, even a small number of interviews may help to augment the quantitative findings.

On the other hand, long-form interviews require more time and dedication, which may be difficult, depending on the population under investigation. The logistics of interview coordination are often difficult. Participants can be recruited over Twitter, but many researchers find that, out of necessity, they must use email or Facebook to reach out to a broader group of individuals, as the response rate on Twitter may be low. For instance, in a study of fans of the Brazilian band Restart, who use Twitter, Recuero, Amaral, and Monteiro (2012) recruited 43 fans at Restart concerts and another 23 through social media. However, it may be difficult to recruit a very specific sample (e.g., "#IGoToASchoolWhere hashtag users") or a representative sample, as the only people who will respond are those willing to talk to researchers. In this case, interviews may be used as part of a multi-methodological study to confirm or complicate previous findings. For example, Letierce, Passant, Breslin, and Decker (2010), in their study of how Twitter is used to spread scientific methods, surveyed scientists, collected tweets, and interviewed 10 researchers to clarify points in the data analysis.

As stated in the introduction, interviews can be an effective way to investigate normative assumptions about technology. When I interview people about individual social media technologies (like Twitter or Facebook), I ask a lot of basic questions (e.g., "What is a hashtag?"), and pay attention to how people explain their actions. When I first began interviewing technologists about Twitter (Marwick, 2010), I was sometimes tempted to show off my technical knowledge, but I found it more effective to feign ignorance and ask users to explain principles of the technology to me, which can reveal a lot about implicit norms and social practice. Depending on the study, I have also found it useful to ask interview participants to show me their Twitter accounts and walk me through individual tweets. This can reveal a lot of rich information about content strategies and presumptions that the user makes (as well as a gap between self-reported data and practice!). I have also found that it is necessary to understand Twitter as part of a multiplex of communication options (Haythornthwaite, 2001). Studies show that virtually all Twitter users use another social network, usually Facebook, in addition to Twitter (Brenner, 2012). Thus, it is important to distinguish social media behaviour *in general* from social media behaviour *on Twitter*.

ETHNOGRAPHIC RESEARCH ON TWITTER

For the purposes of this article, I will differentiate ethnographic interviews (which involve understanding participants' meaning-making processes) from ethnographic fieldwork, which involves in-person observation and participation, ideally over a lengthy period of time, either online or in a particular physical location (Fetterman, 2009; Madden, 2010).

"IN-PERSON" FIELDWORK

For my doctoral dissertation, I conducted more than a year's worth of ethnographic fieldwork in San Francisco among members of the "Web 2.0 scene" (Marwick, 2010). My participants were avid users of Twitter and were happy to discuss it in interviews, but I also observed their use of technology in the field. While I was not always able to see people tweeting in social situations, the technologies constantly came up in conversation. I tried to keep records of even small mentions of the technology. I paid close attention to discussions and conversations about the "right" or "wrong" ways to use technology, which revealed many normative assumptions about the "best" way to use Twitter. I tried to track when participants chose to use Twitter (e.g., "I have to 'overheard' that", or "that's going on Twitter"), when it was inappropriate, when people refused to tweet, and when people discussed Twitter in groups. When informants mentioned that they read something on Twitter, or explained how they learned to use Twitter, this information was quite useful.

Comparisons of a person's discussions of Twitter with their Twitter stream can reveal an added layer of useful information. For example, the information gathered by researchers in face-to-face settings may be consistent, or divergent, from the uses demonstrated by collected tweets or the type of information considered proper to share. Moreover, Twitter provides an articulated social graph in the form of the lists of following/followers that appears on every Twitter profile. Examining who chooses to follow—or not follow—whom can enable greater understanding of a particular social scene in which ethnographic fieldwork is being conducted. This also applies to tweets about events, such as parties or conferences. Reviewing tweets about an event where ethnographic data was gathered can help flesh out participants' meaning-making practices about their activities.

Twitter exists as part of an ecosystem of communicative options for users, and often what is posted on Twitter is not limited to that medium. Participants may discuss specific tweets or accounts on Tumblr or blogs; repost certain tweets to Facebook; use Twitter to post Instagram pictures or Foursquare check-ins; or take part in a variety of other social media interactions. Thus, Twitter must be understood as part of a mediascape which includes other forms of social media, as well as texting, phone calls, emails, and in-person discussions. Contextualising tweets within this rich social web is important.

While it is a cliché to affirm the importance of field notes, they are the most important source of information a researcher will have once fieldwork is complete. I carried a small notebook in my purse and frequently left events to scribble down notes about what was happening. I wish I had not assumed that I would remember certain things that happened. While I have found that only the most disciplined researchers write up their fieldnotes every night, there is a reason that this is consistently recommended (Emerson, Fretz, & Shaw, 1995).

DIGITAL OR VIRTUAL ETHNOGRAPHY

Digital, or virtual, ethnography refers to the practice of observing and/or participating in a particular online group or community over a period of time (Hine, 2000; Miller & Slater, 2000). Given the traditional definition of a field site as a *space*, "the stage on which the social processes under study take place" (Burrell, 2009), many such ethnographies have investigated bounded online "places" such as bulletin boards, forums, or multi-player games like World of Warcraft (Boellstorff, 2008; Kendall, 2002; Nardi, 2010). Twitter challenges this model because it is a large, public site, making it difficult to bound, or even determine, exactly who or what one is studying. Jenna Burrell's (2009) networked field site approach may be more appropriate, reframing Twitter as one part of a "network composed of fixed and moving points including spaces, people, and objects" (p. 189). In other words, Twitter may be one node on a network of field sites which include other social media sites, in-person locations, and material objects. (This was the case in my own dissertation project.) Twitter can be used as the primary place to observe interactions between people over a period of time, but these may be transient, ephemeral, and difficult to pin down.

Several approaches can be taken in determining the boundaries of Twitter as a field site. For example, a project could "follow" a set number of subjects who have been identified based on other research, such as "feminist bloggers" or members of a specific gaming guild. When tracking interactions between subjects, and indeed any Twitter users, conversations must be persistently rebuilt "by way of exploring several previous messages that form the conversation threads" (Bougie, Starke, Storey, & German, 2011, p. 5). This can be difficult, as Twitter's tools for such things are limited. Even when expanding an individual tweet to "conversation view", items are often missing, such as contributions by other users and messages sent as new tweets rather than as replies. The search function on Twitter is notoriously problematic. The only way to see all messages tweeted by a particular account is from the individual profile page, where all @replies are aggregated, or by collecting tweets through the API. While such tools can aid in tracking down components of conversations, they can also be painfully slow.

Another, albeit incomplete, way to bound a group is to track the use of hashtags. For example, I worked on a collaborative study in which the authors were interested in fan practices around the television show, *Glee* (Marwick, Gray, & Ananny, 2013). We collected tweets that contained one of three hashtags: #glee, #klaine, and #brittania (the last two are portmanteaus for names of queer couples on the show). However, it is difficult to call people who use a particular hashtag a "community" by any strict definition of the term. Some hashtags do function as spaces of expression with recurring actors (Bruns & Burgess, 2011), but in other hashtags the participants do not interact with each other. Moreover, hashtags can be used for a wide variety of purposes besides identification. And the majority of Twitter users do not use hashtags, as they only appear in between 5–11% of tweets (boyd, Golder, & Lotan, 2010; Suh, Hong, Pirolli, & Chi, 2010). While this can be a convenient method, it is also an inadequate one.

TEXTUAL INTERPRETATION

Because Twitter is partially a giant corpus of text, many textual analysis methods are appropriate for analysing Twitter interaction, from qualitative coding of individual tweets to close readings of particular accounts.

TEXTUAL AND DISCOURSE ANALYSIS

Qualitative research on Twitter also includes textual analysis and discourse analysis of individual tweets. Typically, these tweets are collected using an automated tool such as HootSuite Archives (formerly TwapperKeeper) or The Archivist, creating a fairly large corpus (discussed in detail in Gaffney & Puschmann, Chapter 5 in this volume). A subset is then selected for analysis and individually coded using textual analysis software such as Atlas.ti, NVivo, or Dedoose. For example, Zizi Papacharissi (2012) used textual analysis in her study of performative self-presentation in Twitter trending topics. Working with a sample of 1,798 tweets, the research team manually coded for descriptive features such as @replies and hashtags, as well as specific performative strategies which were operationalised based on concepts drawn from performance theory. Papacharissi also undertook discourse analysis on the same sample, identifying patterns and repetition in the text. She concluded that play is a primary performative strategy on Twitter, suggesting that "individuals confronted with a restricted stage for self-presentation seek to overcome expressive restrictions through imaginative strategies that include play" (Papacharissi, 2012, p. 1998). In other words, play provides a measure of deniability when voicing possibly controversial statements in a public forum rife with context collapse. In both these studies, qualitative textual analysis was used to unearth subtleties of interaction on Twitter which may have been missed using more quantitative methods.

Coding itself is a complex process which can be approached in a variety of ways. In Papacharissi's study, variables were strictly operationalised; for instance, a tweet was coded for "play" if it contained reordering, exaggeration, repetition, fragmentation, exaggeration and repetition, or (in)completion. Each of these strategies was carefully defined so as to make coding easier (for example, "reordering" was defined as "playing around with syntactical or grammatical rules, rearranging conventional sequencing of words to form sentences, and generally going against the norm of presenting thoughts into a written sentence" (Papacharissi, 2012, p. 1996). Other approaches include coding for the presence of a particular word (e.g., "drama" if the tweet contains the word "drama"), coding for particular names or hashtags, and so on; the right coding method will primarily depend on your research questions. For more on coding, see Charmaz, 2006; Corbin & Strauss, 2007; and Patton, 2002. When manually coding, I have found it easier to create a codebook based on pilot coding a subset of tweets, rather than rely entirely on grounded theory methods where categories come up during coding. This is primarily because I tend to create superfluous, repetitive codes without some sort of reference to draw from (for example, "celebrity", "celebrities", and "micro-celebrity" as three separate codes). Even though it is inevitable that the codebook will change throughout the coding process, having a fixed reference is invaluable and usually saves time in the end.

CLOSE READING AND CRITICAL DISCOURSE ANALYSIS

In addition to social science methodology, humanities scholarship has provided methods that can be useful when considering Twitter. Close reading is a primary method in literary criticism, in which texts are read paying rigorous attention to individual words, syntax, and diction. Critical discourse analysis is a similar close reading strategy in which the researcher focusses on power relationships and links between texts and ideology (Fairclough, 2003). In both instances, the researcher will need to choose a relatively small sample of tweets to analyse. This may be tweets from top users; all tweets from certain users; tweets containing a particular hashtag; tweets to a particular user, and so forth. In a study of celebrity interaction on Twitter, my co-author and I chose three case studies—Mariah Carey, Miley Cyrus, and Perez Hilton—to demonstrate particular aspects of power relationships inherent in fan-audience interactions. I conducted a close reading of three months of tweets from each celebrity, paying close attention to their interactions with other Twitter users, particularly @replies (Marwick & boyd, 2011b). In providing thick description of specific tweets and interactions, we were able to illuminate specific patterns of use that would have been difficult, if not impossible, to ascertain with a more automated method.

CONCLUSION

Twitter is an immensely rich site for analysis, with a diverse array of users, multiple language communities, and a variety of subcultures who have taken to it. While, as we have seen, virtually any qualitative method can be better used to understand Twitter, the majority of studies on Twitter to date have been quantitative. While the "big data" approach has advantages, it also has limitations (boyd & Crawford, 2011). Identifying large-scale patterns can be useful, but it can also overlook *how* people do things with Twitter, *why* they do them, and how they *understand* them. Quantitative studies often determine connections and networks, and interpret them "objectively" *ex post facto*, based on statistics and numbers. Instead, qualitative research seeks to understand meaning-making, placing technology use into specific social contexts, places, and times. Moreover, the claims to "truth" often made by "big data" methods frequently ignore the difficulty in finding any representative sample of Twitter, Twitter users, users, or people in general. Tweets gathered are often incomplete, even from APIs or the public "firehose". The search function is imprecise. Twitter is used by a relatively small number of people to begin with, and leaves out entirely those who do not use the Internet. Rather than taking statistics for granted, the methods outlined in this chapter, and demonstrated throughout this book, show alternate ways to make sense of user practices, social norms, and power relations as they play out on Twitter, and throughout the digital world.

REFERENCES

Boellstorff, T. (2008). *Coming of age in second life: An anthropologist explores the virtually human.* Princeton, NJ: Princeton University Press.

Bougie, G., Starke, J., Storey, M. A., & German, D. M. (2011). Towards understanding Twitter use in software engineering: Preliminary findings, ongoing challenges and future questions. In *Proceedings of the 2nd International Workshop on Web 2.0 for Software Engineering* (pp. 31–36). Retrieved from http://dl.acm.org/citation.cfm?id=1984707

boyd, d., & Crawford, K. (2011, Sep.). Six provocations for Big Data. Paper presented at A Decade in Internet Time: Symposium on the Dynamics of the Internet and Society, Oxford, UK. Retrieved from http://papers.ssrn.com/sol3/papers.cfm?abstract_id=1926431

boyd, d., Golder, S., & Lotan, G. (2010). Tweet, tweet, retweet: Conversational aspects of retweeting on Twitter. In *Proceedings of the Forty-Third Hawai'i International Conference on System Sciences (HICSS-43)* (pp. 1–10). Kauai, HI: IEEE Computer Society.

Brenner, J. (2012). Social networking. *Pew Internet & American Life Project.* Retrieved from http://pewinternet.org/Commentary/2012/March/Pew-Internet-Social-Networking-full-detail.aspx

Bruns, A., & Burgess, J. (2011, August). The use of Twitter hashtags in the formation of *ad hoc* publics. In *Proceedings of the European Consortium for Political Research Conference, Reykjavik, Iceland.* Retrieved from http://eprints.qut.edu.au/46515/

Burrell, J. (2009). The field site as a network: A strategy for locating ethnographic research. *Field Methods, 21*(2), 181–199.

Charmaz, K. (2006). *Constructing grounded theory: A practical guide through qualitative analysis* (1st ed.). Thousand Oaks, CA: Sage.

Corbin, J., & Strauss, A. (2007). *Basics of qualitative research: Techniques and procedures for developing grounded theory* (3rd ed.). Thousand Oaks, CA: Sage.

Emerson, R. M., Fretz, R. I., & Shaw, L. L. (1995). *Writing ethnographic fieldnotes* (1st ed.). Chicago, IL: University of Chicago Press.

Fairclough, N. (2003). *Analysing discourse: Textual analysis for social research.* New York, NY: Routledge.

Fetterman, D. M. (2009). *Ethnography: Step-by-step.* Newbury Park, CA: Sage.

Gershon, I. (2010.) *The breakup 2.0: Disconnecting over new media.* Ithaca, NY: Cornell University Press.

Haythornthwaite, C. (2001). Exploring multiplexity: Social network structures in a computer-supported distance learning class. *The Information Society, 17*(3), 211–226.

Hine, C. (2000). *Virtual ethnography.* Thousand Oaks, CA: Sage.

Honeycutt, C., & Herring, S. (2009). Beyond microblogging: Conversation and collaboration via Twitter. In *Proceedings of the Forty-Second Hawai'i International Conference on System Sciences (HICSS-42)* (pp. 1–10). Los Alamitos, CA: IEEE Computer Society.

Java, A., Song, X., Finin, T., & Tseng, B. (2007). Why we Twitter: Understanding microblogging usage and communities. In *Proceedings of the Joint 9th WEBKDD and 1st SNA-KDD Workshop* (pp. 56–65). San Jose, CA: ACM. Retrieved from http://portal.acm.org/citation.cfm?id=1348556. doi:10.1145/1348549.1348556

Kendall, L. (2002). *Hanging out in the virtual pub: Masculinities and relationships online.* Berkeley, CA: University of California Press.

Krishnamurthy, B., Gill, P., & Arlitt, M. (2008). A few chirps about Twitter. In *Proceedings of the First Workshop on Online Social Networks* (pp. 19–24). Seattle, WA: ACM. doi:10.1145/1397735.1397741. Retrieved from http://portal.acm.org/citation.cfm?id=1397741

Kwak, H., Chun, H., & Moon, S. (2011). Fragile online relationship: A first look at unfollow dynamics in Twitter. In *Proceedings of the 2011 Annual Conference on Human Factors in Computing Systems* (pp. 1091–1100). Vancouver, British Columbia, Canada. Retrieved from http://dl.acm.org/citation.cfm?id=1979104

Letierce, J., Passant, A., Breslin, J., & Decker, S. (2010). Understanding how Twitter is used to spread scientific messages. In *Proceedings of the WebSci10: Extending the Frontiers of Society On-Line.* Raleigh, NC. Retrieved from http://journal.webscience.org/314/

Madden, R. (2010). *Being ethnographic: A guide to the theory and practice of ethnography.* Thousand Oaks, CA: Sage.

Marwick, A. (2010). Status update: Celebrity, publicity and self-branding in Web 2.0. (Unpublished doctoral dissertation). New York University, New York, NY.

Marwick, A., & boyd, d. (2011a). I tweet honestly, I tweet passionately: Twitter users, context collapse, and the imagined audience. *New Media & Society, 13*(1), 114–133.

Marwick, A., & boyd, d. (2011b). To see and be seen: Celebrity practice on Twitter. *Convergence, 17*(2), 139–158.

Marwick, A., Gray, M. L., & Ananny, M. (2013). 'Dolphins are just gay sharks!' *Glee* and the queer case of transmedia as text and object. *Television & New Media*. Advance online publication. doi: 10.1177/1527476413478493

Miller, D., & Slater, D. (2000). *The Internet: An ethnographic approach*. New York, NY: Berg.

Naaman, M., Becker, H., & Gravano, L. (2011). Hip and trendy: Characterizing emerging trends on Twitter. *Journal of the American Society for Information Science and Technology (JASIST), 62*(5), 902–918.

Nardi, B. (2010). *My life as a night elf priest: An anthropological account of World of Warcraft*. Ann Arbor, MI: University of Michigan Press.

Papacharissi, Z. (2012). Without you, I'm nothing: Performances of the self on Twitter. *International Journal of Communication, 6*, 1989–2006.

Patton, M. Q. (2002). *Qualitative research and evaluation methods*. Thousand Oaks, CA: Sage.

Recuero, R., Amaral, A., & Monteiro, C. (2012). Fandoms, trending topics and social capital in Twitter. *Selected Papers of Internet Research, ir 13.0*. Retrieved from http://spir.aoir.org/index.php/spir/article/view/7.

Spradley, J. P. (1979). *The ethnographic interview*. New York, NY: Harcourt Brace Jovanovich.

Suh, B., Hong, L., Pirolli, P., & Chi, E. H. (2010). Want to be retweeted? Large scale analytics on factors impacting retweet in Twitter network. In *2010 IEEE Second International Conference on Social Computing (SocialCom)* (pp. 177–184). doi:10.1109/SocialCom.2010.33

Wengraf, T. (2001). *Qualitative research interviewing: Biographic narrative and semi-structured methods*. Thousand Oaks, CA: Sage.

Legal Questions of Twitter Research

10
CHAPTER   Michael Beurskens

 not all that's #legal is also #ethical, and legal
rules for #privacy, #dataaccess or #copyright
may vary across countries

To a large extent, Twitter research is subject to legal uncertainty. Yet, the (legally appropriate) answer, "It depends . . . ", is insufficient for research practice. Therefore, most researchers rely on Twitter's Terms of Service (Twitter, 2012b) or on their belief that their activities are considered "ethical". However, ethics and law do not necessarily go hand in hand (Beurskens, 2012). While not every researcher has to be a lawyer, a brief overview of the issues and policies involved should shed some light on future directions and roadblocks in Twitter research.

At first glance, use of Twitter seems to be governed by three core documents: their "Terms of Service" (Twitter, 2012b), governing their relationship to all users; their "Rules of the Road" (Twitter, 2012c), specifically tailored to developers; and their "Privacy Policy" (Twitter, 2012a), which states their intended use of data gathered by use of their service. However, several issues arise in this context—for instance, what about users who access Twitter's website without having an account, who therefore never had to read or accept any agreement?

Additionally, will agreement on the basis of these documents be held enforceable in every possible jurisdiction around the world? Furthermore, while the relationship between Twitter, Inc. and third parties is detailed in the aforementioned documents, use of tweets might well infringe on rights of other users or even third parties. It is therefore inevitable to examine the legal framework upon which Twitter, Inc. bases their agreements and to which any use of data derived from Twitter, Inc. must conform.

Specifically, this chapter takes a closer look at "ownership" in tweets and associated data as well as in data sets gathered by researchers or provided by Twitter (which might be subject to different rules). Another issue relates to demands to either anonymise data on the one hand or attribute tweets to their authors on the other. Finally, a discussion of further use of the collected data is needed, including (internal) archiving and making it available to third parties.

After giving some (minimal) background on the legal framework, the analysis will examine the legal position of the "creator", i.e. a Twitter user who writes or "creates" tweets. Subsequently, I will look at the legal protection of Twitter, Inc. as the data provider and finally discuss the rights researchers accessing such data acquire and may eventually grant to third parties.

BACKGROUND

The intuition that anything "ethical" will most likely also be "legal" is generally correct. Yet, occasionally, ethical requirements might be stricter than required by law, for example, requiring attribution of the source of datasets, even when they are not protected by copyright law. On the other hand, there are cases when a perfectly ethical practice might be a violation of legal rules (cf. Beurskens, 2012).

The core problem of legal rules is that they are highly divergent among different states. There is no global consensus on copyright, data protection, or privacy laws, and especially not on contract law or torts (i.e., intentional or negligent loss caused to third parties with whom there is no contractual agreement). This also means that every state is limited to enforcing its laws within its borders (but cf. Bradley, 1997, on the tendency of the United States to enforce IP rights on the Web as a whole). On the other hand, Twitter (like most Internet services) is accessible worldwide, and research may well be conducted cross-border. In general, this would imply that Twitter, Inc. must comply with the legal rules by any state where its website or data streams are accessible. Many legal systems provide safe harbour rules shielding providers of Internet services from liability as long as they do not create or influence content themselves. Yet, these rules apply to neither the authors of a tweet nor to researchers. Thus, they might be subject to infringement of torts law in all states where the content may be accessed. The activity of researchers is generally only governed by the laws and possibly additional contractual agreements or policies of the place where they conduct their research, i.e. the location of their research institution.

Due to the different legal systems involved, a complete legal analysis of Twitter research is impossible, as it would require a definite assessment of the rules of all countries where content may be created, or accessed, or where research may be conducted. Yet, some fundamental issues are common among most industrial states, allowing for generalisation and reduction of obscurity.

THE ROLE OF THE USERS

Clause 5 of the Twitter Terms of Service (Twitter, 2012b) acknowledges that users retain their rights to any content, making Twitter a mere licencee, who is allowed to (*inter alia*) make content available to others. This inherently leads to the question of what rights a Twitter user may claim.

Even though a layperson would probably consider a user to be the "author" of a tweet, this does not necessarily imply that the tweet is actually protected by copyright law. While the applicable standards in copyright law differ among states, protection generally requires either "originality" or "sweat of the brow", i.e. significant expenditure of labour.

In the majority of cases, a tweet will meet neither of these requirements. However, the amount of text is not a relevant criterion for determining protection. Things might be different in Chinese or Japanese, for example—where 140 characters may contain a lot of meaning—or with scientific formulae, which can also express complex ideas extremely briefly ("$E = mc^2$"). In addition, creativity may be found even in few words, for example in haiku.

It is noteworthy that protection is focussed on the means used to express an idea, not on the facts expressed. Thus, the number of retweets is not a relevant indicator with regard to a tweet's possible protection. Consequently, the general rule is that tweets will not be protected by copyright law, and such protected tweets are extremely rare (for a more detailed analysis, see Haas, 2010; Teebagy North, 2011).

In fact, overly broad protection would contravene copyright policy. Even though copyright only prohibits reproduction and distribution and not independent re-creation of a work, such protection would have a chilling effect on communication: Since it may be hard to determine whether content was independently created or merely copied, any tweet would run the risk of being considered an infringement. For example, the simple tweet "Rain again. I'll stay at home" might be independently used by multiple persons. If a legal system would assign copyright to the first author of such a tweet, anyone later using identical terms would run the risk of being considered a copyist (and thus be subject to injunctions or damages). With copyright running for life + 70 years in most states, use of Twitter would be impossible, or its use would be limited to sharing links, as the number of possible expressions for certain facts or situations is limited.

The lack of copyright protection is noteworthy, as it means that there is no legal requirement to attribute tweets, i.e. name the original author, and no need for a licence agreement (but see Nelson, 2012, who would grant protection to tweets, but allow third parties access to tweets without any licence agreement under the U.S. fair use doctrine, and thus ensure attribution). Information or electronic data, as such, is generally protected neither by copyright nor by any other intellectual property right or other statutes. Only certain conduct (deleting data from a storage space used by someone else, breaching protective measures, e.g. by hacking) is prohibited. Thus, lack of copyright protection actually allows for a great variety of reuse without having to worry about violation of intellectual property law.

While not protected by copyright law, news about certain events, e.g., sports competitions, may be covered by specific legal rules or subject to a general prohibition against misappropriation (McDonnell, 2012; see also Twitter & Google, 2010). Such rules are intended to protect news agencies, but generally only apply to commercial use, thus not to non-commercial tweets by private individuals. Still, there are some noteworthy attempts to prevent visitors to sporting events, for instance, from posting about the event in real time (Sheppard, 2010).

Even though a tweet may include protected trademarks ("I'm lovin' it") or trade names, this will generally not be considered a use in commerce, or at least not a use to name a competing good or service. Thus, trademark law is generally irrelevant to the legal protection of Twitter content. Just as a side note, while the users themselves might be subject to specific obligations—including not disclosing trade secrets, avoiding slandering of other users, or distributing links to illegal material (ranging from child pornography to copyright infringing downloads)—these obligations are not transferred to Twitter, Inc. or researchers, as long as they act in good faith (see clause 4, clause 9 of the Terms of Service, Twitter, 2012b).

Finally, reuse of a tweet in certain projects might infringe on personal rights of the authors—which are mainly protected by privacy, data protection, and anonymity laws (cf. Graham & Anderson, 2010). These are of fundamental relevance to any kind of use, and will be covered in detail below. They must be distinguished from cases of "twitterjacking"—where users pose as celebrities, and thus infringe a right of publicity (Jung, 2011)—which are irrelevant in the context of research discussed here.

A naive researcher might assume that making content available on Twitter implies a *prima facie* agreement to reuse such content. It seems contradictory to post something on the public Internet (Twitter is open to people who have not expressly agreed to any terms of use; even searching is possible on twitter.com) and simultaneously reserve rights to such content. Consequently, there is little a user can do to prevent others from reading their tweets. Still, possible use of tweets is not limited to mere passive access. By automatically consuming and analysing data, a tweet might find unexpected and undesired uses. Such use is not limited to market research, but also covers monitoring and profiling. Insofar, most legal systems try to protect users' rights, even though information is made available publicly and is thus unprotected by technical measures.

THE ROLE OF TWITTER, INC.

While Twitter, Inc. operates the necessary services and is thus indispensable in the distribution of tweets, its role in the legal framework is more questionable. On the one hand, they try to act as a clearinghouse, making licenced use of the content posted and granting sub-licences to third parties (usually through Gnip, Inc.) based on an agreement with them (clause 5 of the Terms of Service, Twitter, 2012b). Use of data is Twitter, Inc.'s only potential source of income. In selling "Promoted Tweets", "Promoted Trends", or "Promoted Accounts" they rely on data provided by their users, thus allowing for targeted advertisement. Of course, one might argue that the optical placement of promoted data in Twitter's website would already give sufficient incentive to pay. The amount of spam proves that direct advertisement without official approval seems to have at least some success. If everyone had full access to all data available to Twitter, the amount of advertisement would increase, while Twitter would have little reason to ask for payment to create targeted commercials. Monopolising and direct or indirect control over licencing of data is thus indispensable to any potential business model for social media services. As long as Twitter serves that purpose, direct negotiations between researchers and individual users become unnecessary. On the other hand, Twitter denies any responsibility regarding tweets (clause 4 of their Terms of Service), while reserving the right to remove content as well as to suspend or "terminate users" (verbatim from clause 8 of their Terms of Service), and providing for a notice-and-take-down-procedure (clause 9) for copyright violations (see also Puschmann & Burgess, Chapter 4 in this volume).

Twitter, Inc. is in a rather uncomfortable position regarding possible responsibility for misconduct using their service. Since their centralised server technology allows for full and unlimited control of each and every tweet, a legal regime might well impose responsibility and liability for such content. This in turn would require Twitter to censor undesired tweets. There are numerous cases where civil liability might arise. For example, someone might allege to be a certain public figure ("twitterjacking"; Jung, 2011) or use a trademarked hashtag for commercial purposes ("cybersquatting"; Curtin, 2010). As a consequence, Twitter allows for "Verified Accounts" which are distinguished from normal user accounts by a visible "badge", i.e., a symbol next to the user name (Twitter, 2012d). Nevertheless, this only applies to "business partners . . . and individuals at high risk of impersonation" (Twitter, 2012d). The resources necessary to eliminate any illegal, inappropriate, or unwanted tweets would eventually kill off the service itself. Yet, most jurisdictions provide for immunity of Internet service providers from liability (but see Monaghan, 2011, who argued that these rules should be abolished to create an incentive to develop appropriate filtering mechanisms; and Helman & Parchomovsky, 2011, who proposed a "best available technology standard" for automatic filtering).

Furthermore, Twitter's self-envisioned (and as shown above, commercially necessary) role as a clearinghouse for commercial or research use of tweets, or as a steward for its users, is hindered by the legal framework as well. Since national laws are highly fragmented and, as pointed out, users have little say in the use of their tweets, this approach will not work. The only way Twitter, Inc may acquire the necessary licences and the right to sub-licence, is on the basis of contract law. Yet, some jurisdictions might object to the contents of such contracts—they might be considered unconscionable and therefore invalid in relevant parts. Furthermore, the comparison of Twitter's Terms of Service to a state's constitution is hindered by the fact that a contract is based on agreement between the parties. Thus, amendments have to be expressly or impliedly agreed upon. Generally, standardised contracts agreed upon by a mere mouse click are subject to a strict scrutiny review by courts, especially if one party is a mere consumer, as the typical non-commercial Twitter user is.

Taken to the opposite extreme, Twitter, Inc. might be considered a "transparent" (i.e., irrelevant) entity. If it held no rights and had no basis to intervene against abuse of its data, it should be possible to simply ignore Twitter, Inc. in legal analysis. However, such an understanding would go too far as well. Twitter, Inc. holds no intellectual property rights or licences regarding tweets or collections thereof. The reason is simple—neither the selection nor the organisation of tweets (by hashtags) is imposed by Twitter as a central authority; instead they are collaboratively created by the platform's users. The total data set of tweets is more akin to a phonebook than a best-of-selection of poems or music, and therefore not considered worthy of copyright protection. The European Union *sui generis* right in databases also does not apply, as it is limited to companies with their seat in the European Union, whereas Twitter, Inc. is registered in the United States. With regard to commercial reuse, e.g. use for market research or selling advertisements by grabbing tweets and integrating them into a competing social network service, unfair competition law applies. The aforementioned misappropriation doctrine works in favour of Twitter, Inc. as it would apply to news agencies.

Of higher relevance is the factual power Twitter has over its users as well as anyone interested in making use of data stored in its system. Twitter, Inc. is neither legally required to make tweets available to anyone, nor is the company obliged to open accounts for whoever wants to use its service. Antitrust law does not apply, as it is unlikely that Twitter would be considered a dominant competitor (or even a monopolist) in the market for Internet communication (but see Weber Waller, 2012). Therefore, Twitter, Inc. may indeed lock out users at any time. In addition, the company has factual power of the data, which in turn forms the basis for the requirements imposed under its "Rules of the Road" for developers (Twitter, 2012c) or the more general Terms of Service (Twitter, 2012b).

In fact, these terms give Twitter, Inc. a rather strong position. They expressly allow for suspension of access to the API or even any content on Twitter at any time (clause V/1 of the Rules of the Road, Twitter, 2012c). Twitter may terminate any licences granted for any reason. Use of Twitter data in violation of the Terms of Service or the Rules of the Road would constitute a breach of contract and may give cause to a claim for damages. Again, enforceability of these agreements may be questionable in some jurisdictions (but see Taylor, 2011, on user influence); thus the factual power to exclude is Twitter's most significant power.

SCIENTIFIC USE OF TWITTER DATA

The aforementioned principles apply analogously to any researcher making use of Twitter data. Copyright in data sets would require some originality in their selection or structure, which may be found in, for example, creative filtering techniques. It does not depend on the protection of the tweets themselves (which would be owned by their respective authors if such protection existed). In the European Union, data sets are protected whenever a qualitatively and/or quantitatively substantial investment was required in obtaining, verifying, or presenting the data, even if the collection lacks any kind of originality or creativity. In addition, sophisticated graphs and charts may be protected under copyright law as long as they meet the originality standard.

Often, data sets will meet neither the originality standard of copyright law nor the substantial investment standard under the European Union's Database Directive (Database Directive, 1996). In such cases, there is little to no legal protection of data sets. Unfair competition law does not apply as long as only non-commercial use is involved. Torts law and criminal law mainly focus on hacking and modification of data stored on protected systems by third parties. There is not even a requirement to attribute such unprotected data to its source.

Most research is, however, conducted in universities and funded by institutions requiring certain ethical standards (see Zimmer & Proferes, Chapter 13 in this volume). These rules usually go beyond the legal standards and require attribution even of mere factual data. These ethical standards should not be taken lightly, as they may also incur legal consequences and are part of employment contracts or funding agreements. These may lead to damages or even termination of employment. Thus, stealing data from other researchers might not violate the law, but is still undesirable.

CONCLUSION: (RE-)USING "YOUR" DATA SETS

Transparency and reproducibility are fundamental principles of any scholarly research. When dealing with Twitter data, this seemingly implies that researchers must make their whole data set available. Two issues give cause for concern, though. First, Twitter disapproves of such shadow databases and has tried to ban their dissemination. Secondly, such datasets may be subject to strict review under laws protecting privacy.

Researchers accessing Twitter using the API are subject to Twitter's general Terms of Service (Twitter, 2012b) as well as the Rules of the Road (Twitter,

2012c). While research should not negatively affect Twitter's business model, the Rules allow Twitter to restrict access at any time. The Rules of the Road expressly prohibit "exporting Twitter content to a datastore as a service or other cloud based service" (Twitter, 2012c, clause I.4.A). However, eternally accessing historically archived data directly using the public Twitter website is not possible, nor may a user provide a filter providing others access to the exact data set used for research. Legally, Twitter, Inc. cannot do much to prevent archiving, sharing, or reuse of data transferred to researchers once it leaves its servers. Twitter can only rely on contractual claims subject to the aforementioned reservations regarding enforceability. Damage claims seem hopeless, as Twitter suffers no actual loss. Nevertheless, Twitter may well block out users or even employ blocks of IP addresses. They might also redesign their API at any time, thus eliminating essential features (see Gaffney and Puschmann, Chapter 5 in this volume). Furthermore, even though legal claims are unlikely, university administration tends to be careful with regard to approving research on Twitter data. Due to the lack of legal certainty regarding privacy issues and Twitter, Inc.'s restrictive terms of use, research might seem a liability risk.

As a guideline, storing data on an in-house workstation or server and providing it to third parties on demand should be possible under the current regime. Conversely, a public archive would violate the rules set up by Twitter, Inc.

On a policy basis, good arguments may be made that Twitter's rules are contradictory. Through donating archives to the U.S. Library of Congress, they emphasise the relevance of tweets as documentation of modern culture and history. Yet, their terms of use do not allow for use of active data sets in scholarly research, or mirroring of limited data sets needed to provide evidence necessary under good scholarly practice. Requiring non-profit research to use commercial services such as Gnip would impose high costs. Therefore, Twitter, Inc. evidently distinguishes research from archiving. Indeed, future legislation, especially in the United States, might resolve the issue (but see Graham, 2011, who agreed with current drafts which mainly rely on libraries as archiving institutions).

Privacy is an important issue at the core of social networking. While users voluntarily publish their everyday activities and opinions, they do not automatically also agree to the use of such data in any way imaginable. Indeed, Twitter's Terms of Service (Twitter, 2012b) provide for very broad licences (Terenzi, 2010, believed that such agreements should resolve any potential issues regarding privacy). However, these rules might be in violation of possible privacy laws, especially in the European Union, and might thus be held void. Thus, researchers who gather and store data sets are subject to privacy laws. While these laws generally allow for use in research, they require anonymisation of any data used and shared. It is certainly insufficient to simply eliminate the usernames from the data set, as the identity of a user may be easily derived from other facts mentioned in the tweets (including locations, other users, etc.).

A lot of data can be derived from Twitter data sets, and the larger the data set, the more detailed the user's profile. On the other hand, a case-by-case decision regarding each individual tweet is infeasible when dealing with Big Data. And there is a persistent threat that future, automatic, de-anonymisation technologies will make even those efforts futile (see Ohm, 2010; but contra Schwartz & Solove, 2011).

Again, a pragmatic approach is called for. Since data sharing is already limited due to the Rules of the Road (Twitter, 2012c) and Terms of Service (Twitter, 2012b), access to any data set used in research is only available to other researchers. The researchers must use best efforts to remove any personal references from the data sets. They have to remove user names or replace them with pseudonyms (including those used in the tweet's contents) and anonymise location data. Cleaning up personal data (like pet names, food eaten, and others) is infeasible (but would probably be required under the current legal regime). The issue is certainly in need of specific legislation, especially in the European Union. Yet, the current debate seems to ignore research issues to a large extent.

REFERENCES

Beurskens, M. (2012). Law: Friend or foe in scientific Internet use? In A. Tokar, M. Beurskens, S. Keuneke, M. Mahrt, I. Peters, C. Puschmann, T. van Treeck, & K. Weller (eds.), *Science and the Internet* (pp. 115–130). Düsseldorf, Germany: Düsseldorf University Press.

Bradley, C. (1997). Territorial intellectual property rights in an age of globalism. *Virginia Journal of International Law, 37*(2), 505–585.

Curtin, T. (2010). The name game: Cybersquatting and trademark infringement on social media websites. *Journal of Law & Policy, 19*(1), 353–394.

Database Directive. (1996). Directive 96/9/EC of the European Parliament and of the Council of 11 Mar. 1996 on the legal protection of databases. Official Journal L 077, 27/03/1996 P. 0020–0028.

Graham, J. (2011). Save the tweets: Library acquisition of online materials. *American Intellectual Property Law Association Quarterly Journal, 39*(2), 269–294.

Graham, N., & Anderson, H. (2010). Are individuals waking up to the privacy implications of social networking sites? *European Intellectual Property Review, 32*(3), 99–103.

Haas, R. (2010). Twitter: New challenges to copyright law in the Internet age. *John Marshall Review of Intellectual Property Law, 10*(1), 231–254.

Helman, L., & Parchomovsky, G. (2011). The best available technology standard. *Columbia Law Review, 111*(6), 1194–1247.

Jung, A. M. (2011). Twittering away the right of publicity: Personality rights and celebrity impersonation on social networking websites. *Chicago-Kent Law Review, 86*(1), 381–417.

McDonnell, J. (2012). The continuing viability of the hot news misappropriation doctrine in the age of Internet news aggregation. *Northwestern Journal of Technology and Intellectual Property, 10*(3), 255–276.

Monaghan, J. (2011). Social networking websites' liability for user illegality. *Seton Hall Journal of Sports and Entertainment Law, 21*(2), 499–532.

Nelson, A. S. (2012). Tweet me fairly: Finding attribution rights through fair use in the twittersphere. *Fordham Intellectual Property, Media and Entertainment Law Journal, 22*, 697–751.

Ohm, P. (2010). Broken promises of privacy: Responding to the surprising failure of anonymization. *UCLA Law Review, 57*, 1701–1760.

Schwartz, P. M., & Solove, D. J. (2011). The PII problem: Privacy and a new concept of personally identifiable information. *New York University Law Review, 86*, 1814–1894.

Sheppard, J. R. (2010). The thrill of victory, and the agony of the tweet: Online social media, the non-copyrightability of events, and how to avoid a looming crisis by changing norms. *Journal of Intellectual Property Law, 17*, 445–473.

Taylor, R. B. (2011). Consumer-driven changes to online form contracts. *New York University Annual Survey of American Law, 67*(2), 371–429.

Teebagy North, S. (2011). Twitteright: Finding protection in 140 characters or less. *Journal of High Technology Law, 11*(2), 333–364.

Terenzi, R. (2010). Friending privacy: Toward self-regulation of second generation social networks. *Fordham Intellectual Property, Media and Entertainment Law Journal, 20*(3), 1049–1106.

Twitter, Inc. & Google, Inc. (2010), Brief as Amici Curiae Supporting Reversal in Barclays Capital Inc. v. Theflyonthewall.com, No. 10-1372-cv (2d Cir. 22 June 2010). Retrieved from http://www.naa.org/Public-Policy/Legal-Affairs/Digital-Media/~/media/NAACorp/Public%20Files/PublicPolicy/LegalAffairs/google-brief.ashx

Twitter, Inc. (2012a, May). Twitter Privacy Policy. Retrieved from https://twitter.com/privacy on 1 May 2013

Twitter, Inc. (2012b, June). Terms of Service (Version 7). Retrieved from https://twitter.com/tos on 1 May 2013

Twitter, Inc. (2012c, September). Developer Rules of the Road. Retrieved from https://dev.twitter.com/terms/api-terms on 1 May 2013

Twitter, Inc. (2012d). FAQs about Verified Accounts. Retrieved from https://support.twitter.com/groups/31-twitter-basics/topics/111-features/articles/119135-about-verified-accounts on 1 May 2013

Weber Waller, S. (2012). Antitrust and social networking. *North Carolina Law Review, 90*, 1771–1805.

PART 2

Perspectives and Practices

From #FollowFriday to YOLO
Exploring the Cultural Salience of Twitter Memes

11
CHAPTER   Alex Leavitt

 flourishing memes like #ff or #occupy are affected both by social behaviour and technical features such as trending topics

I, FOR ONE, WELCOME OUR INTRODUCTION

Since the introduction of hashtags to Twitter in mid-2007 (Messina, 2007), these organic, categorical markers have become the primary means to mark, contextualise, and participate in the informational, social practices of the popular microblogging platform. Frequently, Twitter users use hashtags, though keywords, images, and URLs are also employed, in order to spread so-called "memes"—units of cultural information, akin to their biological equivalent, genes, that develop iteratively as they move from individual to individual, like jokes, rumours, and iconic artifacts of popular culture (Dawkins, 1976). While traditionally the meme concept has referred to any iterative piece of culture, Internet users and subcultures have adopted and adapted the term to apply to rapidly spreading, momentarily salient in-jokes; recognisable images (and image forms); and other artifacts like viral videos: all of which have collectively become known as "internet memes" (Burgess, 2008; Knobel & Lankshear, 2007; Shifman, 2012). On Twitter, memes generally take the form of a hashtag that gets passed around quickly, grows through participatory iteration as users encounter them in their feeds, gains high visibility (usually through Trending Topics), and achieves a state of recognition within the endogenous subculture of the platform, similar to a form of subcultural-capital-based knowledge (Thorton, 1996).

This chapter examines the concept of the meme and how these cultural forms evolve and exist within Twitter's vibrant ecosystem and network. I explore a handful of memes to illuminate current and future research areas at the intersection of technological infrastructure (Gillespie, forthcoming), networked publics (boyd, 2010b), and hyperactive, sociocultural phenomena critical for understanding memes' cultural salience in massive social media systems. I end by inquiring about the meme-as-meme, a metadiscussion of when the concept of *meme* becomes a meme itself, impacting how memes flourish in and across a platform, especially when users actively and knowledgably produce and spread them, rather than simply participate in their organic emergence.

I CAN HAS? INFRASTRUCTURE
AND THE POLITICS OF THE MEME

Twitter's rapid growth after 2009—following various televised promotions such as on CNN (with Ashton Kutcher) and on Oprah—fostered a boom in activity on the platform (Golder, 2009). This activity affected not only social behaviors and norms of use but also, recursively, the development of Twitter's technical infrastructure. With an immediate escalation in total users, the value of Twitter as a platform similarly increased. Network effects (Easley & Kleinberg, 2010) of course contributed to this social value, though Twitter, Inc. designed new infrastructural value, such as the automatic retweet button, in response to evolving social behaviors.

Scale, then, directly affects both social and technical attributes of this microblogging ecosystem, which in turn impacts cultural transformation as well. When we think about basic memes on Twitter, like popular jokes—such as #BindersFullOfWomen, a reaction to Mitt Romney's odd phrasing of gender policy during the 2012 U.S. presidential debates—the massive size of Twitter's networked publics, in combination with the ever-evolving platform, fosters unique social situations. For instance, visibility on Twitter (namely, if a user can see a particular message) is bounded by who follows whom, but if many different people spread the same message, whether through retweeting or copying their own version of the message into a new tweet, the visibility of the content increases, a process that Jenkins, Ford, & Green (2013) called "spreadability." The result of this process becomes especially manifest if users from strongly connected communities in the total network pass information to other communities: for example, the initial success of the Kony 2012 meme was largely due to the circulation of the #kony2012 hashtag on Twitter from initial seed networks like high-school teenagers (Lotan, 2012b).

The possibility of spreading memes on Twitter and the effects of that spread seem obvious, though the implications of scale become significant when we consider Twitter as infrastructure, as a technical platform where various behaviors are mediated, constrained, or facilitated. As memes move between individuals, the cultural salience of the meme increases: it becomes more meaningful for more people (Jenkins et al., 2013). But the evolution of Twitter's platform may have effects on the process of diffusion across the various global, interconnected networks of Twitter users.

To look at the impact of infrastructure on the salience of memes, we can look at the evolution of social behaviors around memes in relation to Twitter's technical platform. One example of the situated cultural salience of a meme is the Follow Friday hashtag. Follow Friday, demarcated in tweets by the hashtag #FollowFriday (later, sometimes marked at #FF), was started by technologist Micah Baldwin in January 2009, when he recommended that people use the hashtag at the end of the workweek to recommend a friend for others to follow. Now, each year, tens of thousands of #FF hashtags are used on Fridays. #FollowFriday developed organically from a core group of early Twitter users, where it was spread by word of mouth and visibility, as users reused and reshared the hashtag weekly.

More recently, many memes spread from a smaller network of users to a broader set of viewers with the aid of Twitter's technical architecture, the most notable element of which is Twitter's Trending Topics algorithm. Since September 2008, Twitter has promoted hashtags and keywords that "trend" according to a specified combination of measures, such as most tweets, time period, and exclusiveness. However, the algorithm's details remain undisclosed, and the specifics behind the algorithm have produced a number of tensions between Twitter, Inc. and its users (Lotan, 2011). For instance, due to the high amount of tweets that fans of young contemporary musician, Justin Bieber, produce, many of the global trending topics have contained Bieber-related keywords. As a large portion of non-fans view these trends as crowd-produced spam, Twitter responded in mid-2010 that they had updated the Trending Topics algorithm to expel the singer's constant presence (Parr, 2010).

In reaction to changes in and nondisclosure about the algorithm, certain trending topics (or lack thereof) have produced critical tensions for political and activist users. In late 2011, during the Occupy protests in the United States, various hashtags—such as #OccupyWallstreet or #OccupyBoston—did not materialise when users expected them to appear in the Trending Topics record. Many accused Twitter, Inc. of censoring the topics from reaching the global list. Similar accusations were proposed when various WikiLeaks-related terms did not trend in months prior (Indvik, 2010). These tensions fall in line with what Gillespie (2010) described as the "politics of platforms", where companies must negotiate specific uses of particular contexts in exceptional circumstances.

Although trending topics bring hypervisibility to messages in Twitter's ecosystem, they present a subjective take on what memes possess value and salience to Twitter's networked public. In the context of a global political moment, the Occupy hashtags represented a charged situation with significant consequences and inflected with meaning, but the process of them not trending subtracted from that value. Twitter's Head of Communications, Sean Garrett, stated in a reaction, "[Trending Topics] are the most 'breaking' and reward[ing] discussions that are new to Twitter. We are not blocking terms related to #occupywallstreet in any way, shape or form" (Jeffries, 2011). Of course, many protesters would probably disagree that an Occupy hashtag was not "rewarding" enough to make it to the top; although the movement's hashtags were certainly newsworthy and tactical for some, Twitter's algorithmic values around the promotion of topics diverge from those opinions.

More generally, the Occupy example demonstrates how Twitter as an infrastructure for communication impacts a meme. The technical architecture of the platform and the various decisions from Twitter's management directly affect how the meme develops and replicates, and therefore transform how users relate to the meme as a social and cultural artifact.

BUT WHO WAS AUDIENCE??

The dynamics of content circulation and *re*circulation on social media platforms like Twitter can generate novel circumstances around audience formation: as Jenkins (2007) argued, through this spreadable process, media can find new audiences and take on new meanings. Memes are particularly volatile in this circulatory activity: because memes maintain a unique aspect of recognition for a particular audience or community. As the meme spreads, the new meanings that users append to the bits of culture might shift away from that initial meaning, warping the meme and its significance. Thus, it is important to expound upon how audiences—and what kinds of audiences—work with and react to memes.

First, it is necessary to outline in categorical terms how users produce information within informational boundaries. Most scholars mark these boundaries as endogenous and exogenous information flows: that is, information coming from within the system or outside of it, respectively (Agrawal, Potamias, & Terzi, 2011; boyd, 2010a). In other words, a hashtag (such as #FF) might be created and fostered within Twitter's platform, while global news events (for example, such as #debate2012, for the 2012 U.S. presidential debates or the keywords, "Tohoku" and "earthquake", for the March 2011 Japanese natural disaster) prompt users to share information from other channels on Twitter. When we conceive of memes as constructed within particular contexts, exogenous memes behave differently than endogenous memes, and each have different meanings and purposes for the audiences that participate in them.

For instance, dozens of joke-based memes we encounter daily on Twitter request and require the participation of thousands of users to bring them to popularity. The most notable example of an endogenous, joke-based hashtag was "Cala Boca Galvão". In mid-2010, this phrase circulated within the (then rapidly growing) Brazilian Twitter user network. As non-Portuguese speakers saw the phrase begin trending on the platform, confused messages began to circulate. Two days later, though, an English-language video was uploaded to YouTube entitled "CALA BOCA GALVAO – Save Galvao Birds Campaign", which explained the phrase: the video petitioned individuals on Twitter to include the phrase in tweets to spread word about and raise money for a campaign to save a rare Brazilian bird.

However, a linguistic barrier between Brazilian and English-speaking users generated the opportunity for this meme to succeed: the reality hidden behind the phrase and its video explanation (a falsely translated trick) was that Brazilian users were reacting negatively to a national broadcast of the World Cup. The phrase actually translates to "Shut up, Galvão", calling out a local broadcast sports announcer's name, who held a disliked position in the hearts of Brazilian sports fans. Because the phrase spread widely outside of Brazil's local user networks and became misunderstood in the process of moving between audiences, Brazilian Twitter users took advantage of the situation to play a joke on the world. The meme's success, in that it trended for days on Twitter, led the

New York Times to call the joke "one of history's most successful cyberpranks" (Dwyer, 2010).

Of course, while trending topics do play a key role in the adoption of memes, especially endogenous ones, other social factors contribute to memetic spread and participation. For instance, human limitations such as sleep—perhaps more generalisable as behaviors around "attention"—play a large factor in the spread of any messages on a global scale.

The location of various potential audiences in diverse time zones, and the relative activity of users in those time zones thus contribute to a meme's dissemination, especially as memes begin to trend in particular, situated locations. Or, frequently, a user who initiates a meme might not have enough followers, or at least active, attentive ones, to help spread the tweet or hashtag around. Charting the distribution of various memes over time tends to then reveal particular users aiding in the broad propagation of the meme because of factors like audience size and time of day. For instance, you can see the relative amount of tweets in the spread of two memes that reached Trending Topic status. The hashtag meme

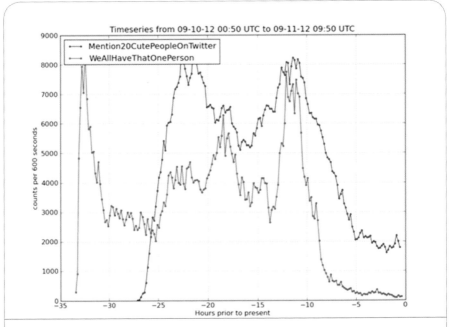

Figure 11.1: The Amount of Tweets over Time for Two Twitter Memes that Reached Trending Topic Status; Their Use over Time Is Likely Affected by Geographical Location of Adopting Participants

#Mention20CutePeopleonTwitter (where, to participate, a user appends the hashtag and names twenty darlings' usernames) rises and falls as it moves from its origin point in the Philippines and Indonesia, eventually reaching users on the East Coast of the United States, which kept its volume high (Figure 11.1). In comparison, it can also be noted that #Mention20CutePeopleonTwitter's gradual climb to trending status is markedly slower than #WeAllHaveThatOnePerson (participants included the hashtag while describing an individual they cannot forget, usually because of a detrimental reason), whose initial spike in tweets was much faster, but then was affected by geographic time differences and resultant behavioral patterns, dropping off quickly, but picking up again at a much later time.

In general, we also assume that Twitter users tend to demonstrate fairly established behavioral patterns: activity throughout the day with peaks during free times around lunch and dinner. However, various locales diverge from these assumptions, such as Japanese Twitter users, who—according to Rios & Lin (2012)—only tweet heavily in the late evening in Tokyo contrasted with other users in major cities around the world.

Further research is necessary to determine if such behaviors are a reflection of cultural attitudes toward social media use. But in compari-

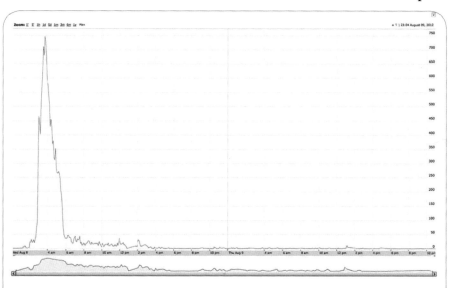

Figure 11.2: A Meme Created by Japanese Media Fans Spikes during the Night but Does Not Reemerge during the Day (Likely Due to Lower Japanese Twitter Use during Diurnal Hours)

son to the memes in Figure 11.1, the spread of another meme from Japan, #あなたをオタクへと導きだしたアニメは何ですか ("What is the Japanese anime show that led to be a hardcore fan?"), illustrates how the message becomes bounded by these sociocultural, local (and here, temporal) practices. The meme was started late one night in August 2012, peaked, and collapsed in the morning, but it did not regain a foothold amongst many other Japanese users (nor in non-Japanese linguistic networks) during the day, with only a few Japanese users barely sustaining the hashtag throughout the next 24 hours (Figure 11.2).

The situational context of audiences also plays a huge role in how a meme becomes accepted within a user network, and how that network decides to spread it. However, the infrastructural component discussed in the previous section heavily impacts how users ultimately interpret the contexts of various Twitter memes. In particular, when a hashtag hits the Trending Topics list (especially the handful of global trends), it immediately jumps into the public news spotlight. As people use the hashtag to talk *about* the hashtag, the meme's purpose becomes convoluted, as users mentioned the hashtag in a tweet to mark participation in the discourse rather than personal involvement in the meme.

As mentioned before, users have imagined audiences for the digital traces they leave on social media, and in a massive ecosystem like Twitter, frequently those audiences are paired with unintended audiences: what Marwick and boyd (2010) called "context collapse". One example of this occurred in 2011, when a clash occurred when one hashtag, #reasonstobeatyourgirlfriend, appeared in a global Trending Topics slot.

Started by two male users on 31 July, a small initial network participated in the hashtag, appending blunt opinions, crude jokes, and sexist commentary. Concurrently, a larger force of users grew in opposition to the hashtag's maintained presence on the Twitter main page, some even calling for Twitter, Inc. to remove the hashtag, as it was promoting domestic violence from such a visible position on the platform's website. But, as noted above, as users reacted to the hashtag by using the hashtag in their critical tweets, these reactions kept the hashtag trending in parallel with the criticism's increasing message-to-time velocity. As one user noted, "Most of the tweets in the #reasonstobeatyourgirlfriend stream are from people appalled at the trending topic, thus keeping it trending" (@devincf). Twitter CEO Dick Costolo even responded to one user's complaint about Twitter Inc.'s apparent preferential treatment for politically inflected trends, stating his faith in the algorithm, though also noting his displeasure with obscene trends and writing "the trends are algorithmic, not chosen by us but we edit out any w/ obscenities & I'd like to see clearly offensive out too" (@dickc).

As the #reasonstobeatyourgirlfriend hashtag illustrates, it is crucial to consider the coevolving roles that both social behaviors and technical mediation bring to the spread, and thus our understanding, of memes. We can see similar reactions emerge from conflicts between imagined audiences, unintended participants, and visibility brought about by technical features, such as when UK fashion shop Celeb Boutique tweeted the #aurora hashtag during the Aurora, CO movie theater shooting aftermath (Lotan, 2012a), after which the shop's Twitter account received a wave of hateful messages directed at the untimely mistake (as the shop had used the #aurora hashtag to advertise a new dress dubbed "Aurora").

SHOWS MEMES, ACTUALLY EXPLAINS THEM: MEME PARTICIPATION

This chapter, thus far, has established a foundation for understanding technology as infrastructure and audiences as the backbone for variations in communication on social media platforms. We can now explore the question of participation in memes (or lack thereof) and how users' consciousness of how a meme operates might affect that participation.

As stated throughout the chapter, memes rely on networks of users to spread them to others in a social media ecosystem. Participation, of course, is bounded by technical limitations (features and affordances), entrenched social behaviors, and inclusion in (or exclusion from) a shared understanding of the meme. Participation, too, is continually reworked as emergent practices restructure social behavior within Twitter. For instance, the technical implementation of the automatic retweet button subtly changed the way users both think about and, on a practical level, participate in resharing content on Twitter, such as issues around attribution, redundancy, and speed (Williams, 2009). Interestingly, this has implications for how certain memes spread.

For instance, Longcat is an Internet meme that originated in Japan of a picture of a cat being held up by its owner, stretched out by gravity (mona_jp, 2009). Eventually making its way onto the subcultural image board 4chan, and solidifying itself as an emblem of American Internet subculture, the phrase "Longcat is long" became associated with the meme and its imagery. In June 2010, three Twitter accounts were created—longcat111, longcat222, and longcat333—that each tweeted a part of the phrase—"Longcat", "is", and "long", respectively—and whose avatars also depicted Longcat broken across the three images (head, midsection, and feet). The joke was simple: retweet the three tweets in the correct order, and your followers would see Longcat appear in their personal Twitter feed. The meme became so popular via the act of retweeting that for some time, the accounts held the record of having the most retweeted tweets ever, with 311,380 (longcat111), 300,904 (longcat222), and 308,520 (longcat333) retweets.[1] The joke certainly continued despite users' failed participation to recreate the meme, as evidenced by the discrepancy in numbers of retweets across the three accounts, in accidentally retweeting the trio in the wrong order, or due to other endogenous circumstances like advertising (see Figure 11.3).

Sometimes various elements of the social media ecosystem

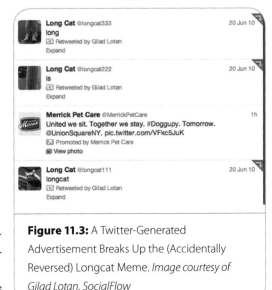

Figure 11.3: A Twitter-Generated Advertisement Breaks Up the (Accidentally Reversed) Longcat Meme. *Image courtesy of Gilad Lotan, SocialFlow*

also contribute to unexpected or unintentional noninvolvement in meme participation. For instance, with regard to Twitter's algorithm that detects hashtags and encodes a link to similarly hashtagged tweets, sometimes the hashtag identification fails, particularly with languages that rely on Unicode, an industry standard for representing text in non-Latin-character languages. Regarding the Japanese media fan meme previously mentioned in this chapter, a small number of users attempted to participate in the meme by copying the hashtag and appending their own answer to the question, but by not writing the tweet correctly with proper white space at the end of the hashtag's phrase so that the algorithm could detect it, these users cannot be seen as participating in the meme because the algorithm misses the identification. Figure 11.4 illustrates a network graph of the co-occurrence of entities—hashtags, mentions, and keywords—that emerge from the meme's collected tweets, but the nodes circling the core network component are attempts at creating the hashtag that failed.

Some memes, in contrast, are riddled with overactive participation. Sometimes these instances do not even involve human actors: occasionally, bots—automated entities that search for popular hashtags to join, usually to promote spam—will adopt memes for their own uses. Other social phenomena like the spread of rumours lead to overactive memes with little value. One example, from mid-2012, illustrates how a few celebrity gossip websites contributed to the spread of a rumour on Twitter regarding the supposed death of Reese Witherspoon, in which the name reached the Trending Topics list for a few hours. The rumour proved to be false . . . but even the existence of the rumour as reported by the gossip websites proved to be false! Data I collected on the keyword, *witherspoon*, around the time of the rumour—totalling 121,434 tweets across 86,263 users—shows hyperactive participation around the keyword, but the tweets did not mention the actress; instead, other rumours regarding the death of actor John Witherspoon (also untrue) cir-

#あなたをオタクへと導きましたアニメは何ですか

Figure 11.4: A Network Graph of a Japanese Meme, Where Nodes Represent Entities (Hashtags, Mentions, and Keywords) and Edges Represent Entity Co-Occurrence in a Tweet. Node Size Shows Occurrences of Each Entity (Thus the Meme's Hashtag Is the Largest). The Isolates around the Edge Represent Undetected Hashtags Due to Improper White Space Use.

culated to an explosive extent (Figure 11.5), while a few of those tweets were extended to include the actress. The apparent volume of tweets with the keyword, though, allowed it to trend, and based on a false assumption, the blogs incorrectly pinned such volume on the Reese Witherspoon rumour.

When discussing participation in Twitter memes, we must also contextualise the *who* and *how* of each instance, to build off points made earlier in this chapter. Issues of geography, class, race, linguistics: each produces various inconsistencies in how participation in memes occurs. For instance, so-called "black Twitter", along with the term *blacktags* (a racial othering of *hashtag*), have been used to categorise the community of African American Twitter users that frequently bump various meme hashtags to many local trending topics rankings, such as *#wordsthatleadtotrouble* or *#ghettobabynames* (Manjoo, 2010). Hargittai and Litt (2011) posited that adoption of Twitter amongst Black youth correlates with their interest in celebrity and entertainment news, and these racialised networks lead to above-average participation on the platform with regard to interest-based participation. Similarly, Brazilian and Japanese users

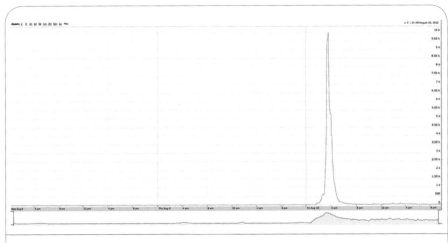

Figure 11.5: Tweet Volume for the Keyword *witherspoon* Shows a Gigantic Spike in Activity for a Short Period of Time, Demonstrating the Power of Rumours on Twitter

(as mentioned previously) participate in their respective, language-bounded memes, since other user networks tend to remain distinct and distant from Portuguese and Japanese information flows (or any language for that matter, as users simply do not understand the messages).

Ultimately, the particular, foundational audience networks on which memes move direct the extent of their flow. Future research might explore the exact reasons *why* Twitter users participate in memes, especially across their various typologies and contexts, and how the various reasons behind participation might illuminate various behaviors or even network structures that help propagate particular types of content like humour.

MINDBLOWN.GIF: MEMES AS A MEME

While the study of Twitter provides passing insight into the platforms' technical, social, and cultural processes, examining memes—single, iterative ideas that spread with a participatory audience—allows us to look at a bounded, digital artifact and how it impacts and is impacted by technical infrastructure, developing audiences, and various forms of and barriers to participation. However, researchers employing the concept of the meme as a form of participation must recognise that the popularity of the term *meme* over time has made participants self-aware of their practices. Essentially, how to participate in a meme has become commonplace knowledge—in fact, Knobel & Lankshear (2007) described it as a form of media literacy—particularly among individuals and groups that use the Internet as a primary means of communication and socialisation. In other words, the concept of the meme has become a meme, and this realisation leads to emergent forms of meta-participation in meme iteration and diffusion.

In order to understand the meme-as-meme, we might look at an exceptional meme that grew from Twitter and evolved in an exogenous fashion with particular, endogenous results: YOLO. An acronym for the phrase "You only live once", YOLO became a commonplace motto in American youth culture beginning in late 2011 and early 2012. While the phrase had emerged earlier in other contexts, a meme developed in autumn of 2011 when Canadian recording artist and rapper Drake released his hip hop single "The Motto", which includes the lyrics "Now she want a photo, you already know, though / You only live once: thats [sic] the motto, nigga, YOLO" in its refrain. On 23 October, Drake posted a picture of himself overlooking Las Vegas to his Twitter account with the YOLO acronym.

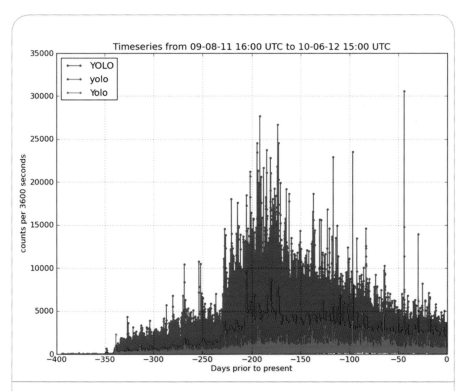

Figure 11.6: The Volume of Tweets over One Year for the Phrase YOLO

With his millions of followers on Twitter, Drake was able to spread the phrase rapidly and far across Twitter's global network. YOLO became particularly popular with American youth, who would post messages to Twitter describing irresponsible behavior and append YOLO as a flippant excuse, a *carpe diem* for the younger generation: for instance, "Pregamed my shower #yolo." Zimmer (2012) described its popularity for American teens, writing, "If you are over 25, YOLO likely means nothing to you. If you are under 25, you may be so familiar with YOLO that you're already completely sick of it." Figure 11.6 illustrates the spread of the acronym over one year on Twitter, showing the massive propagation of the keyword over time, particularly used in all caps.

As Twitter users spread the YOLO phrase in hundreds and then thousands of tweets, the phrase's adoption in American youth slang continued offline. More and more teens used it in school and with friends, and they replicated these uses online in tweets and Facebook posts; eventually, though, many individuals grew tired of the phrase's sudden rise to fame. Urban Dictionary, cataloger of popular slang terms, stores two popular entries for the phrase, the definitions of which read, "Also one of the most annoying abbreviations ever. . . ." and "In many cases, though, the term has been blown out of proportion and teenagers over use it by hash-tagging it in pictures and wall posts on Facebook because its [sic] become trendy" (Urban Dictionary, n.d.).

Irony around using the phrase emerged—for example, "Eating salad straight out the bowl #YOLO"—in which individuals participated in the meme for participation's sake, reflecting the recursivity of the meme-becoming-meme process. Eventually, though, the distaste for YOLO spread, and traces on Twitter containing the phrase or tagged with #yolo brought negative sentiments to the aggregate of meaning. Some users participated in what Godwin (1990) called a counter-meme, which Cheese (2012) used to describe similar practices around creating new memes or adapting current ones for oppositional purposes, where they would adopt the YOLO keyword and hashtag to complain about its spread.

Such mixture of varying practices around meme participation, though, poses particular challenges for qualitative researchers hoping to study meme participation in massive social media ecosystems. Future research on the dynamics of memes as represented in Twitter data needs robust methodologies for analysing the intricacies of social data, particularly in light of the various contexts discussed throughout this chapter. Figure 11.7, for instance, shows the entity co-occurrence network for a sample of Twitter data collected about "yolo" encompassing 629,786 tweets from 435,645 distinct users from the large spike that occurred in late August 2012. The dense mass of nodes in the center represents uses of the phrase in youth culture (the largest nodes, in yellow, represent instances of *YOLO* and *#yolo*), with other uses of *yolo* in particular conversations displayed around the graph (some instances appear contextually in other languages too, such as Dutch and Swedish). Interestingly, the dense, red clusters on the left represent Indonesian users' tweets, in which the keyword appears on its own or in local words. Such discrepancies from looking merely at keywords without exploring the underlying meanings and uses of those keywords thus introduce various issues for researchers entering this area.

Figure 11.7: An Entity (Hashtags, @ mentions, and Keywords) Network Graph Showing Co-Occurrences with the Keyword *yolo* for Almost 630,000 Tweets during a Massive Spike of the Term in August 2012

SOMEBODY SET UP US THE CONCLUSION

Memes are a useful lens for analysing not only how information spreads on social media platforms like Twitter, but also the implications of contexts of use, such as technological infrastructure, networked publics, and typologies of participation. As I have shown in this chapter through the exploration of various Twitter memes, these issues illuminate particular aspects of social media ecosystems that researchers should take into account when studying such spaces. By looking at the spread of single, iterative, participatory ideas like memes, we can demonstrate the importance of the structural and social particulars of the technical platforms we study, and we can identify trends of and exceptions to everyday information-producing and -circulating practices on social media that bring these memes to popularity through shared meaning-making. In other words, the contexts of use discussed in this chapter illuminate the processes around and barriers to what makes memes spread.

Current research already is examining the finer processes underlying the spread of memetic phenomena. Future research, especially in looking at the sociocultural effects of when information spreads in online, networked ecosystems, should look at how participatory moments like memes spread *across* social media platforms, the theoretical and empirical limitations to how (or if) those flows occur, and how memes—and information in general—are maintained in cross- or multi-platform ecosystems. By combining qualitative research with empirical, computational methods, we can begin to answer some of these questions to gain further knowledge about one of contemporary Internet culture's most recognisable media forms: Internet memes.

NOTE

1 This record was broken when in November 2012, retweets of an image of Barack Obama with his wife surpassed the retweets of the Longcat accounts after he won the 2012 U.S. Presidential election.

ACKNOWLEDGMENTS

I would like to thank Gilad Lotan and SocialFlow for providing resources and access to data for this book chapter (disclaimer: I was a Research & Development intern at SocialFlow while writing this essay). Additionally, many thanks go out to the editors for inviting me to participate in this book project.

REFERENCES

Agrawal, R., Potamias, M., & Terzi, E. (2011). Learning the nature of information in social networks (Microsoft Research Technical Report). Retrieved from http://research.microsoft.com/pubs/148652/natureofinformation.pdf

boyd, d. (2010a). 'Pep rally'—A truly exogenous trending topic on Twitter. *Apophenia*. Retrieved from http://www.zephoria.org/thoughts/archives/2010/10/08/pep-rally-a-truly-exogenous-trending-topic-on-twitter.html

boyd, d. (2010b). Social network sites as networked publics: Affordances, dynamics, and implications. In Z. Papacharissi (Ed.), *Networked self: Identity, community, and culture on social network sites* (pp. 39–58). Retrieved from http://www.danah.org/papers/2010/SNSasNetworkedPublics.pdf

Burgess, J. (2008). 'All your chocolate rain are belong to us?' Viral video, YouTube, and the dynamics of participatory culture. In G. Lovink & S. Niederer (Eds.), *Video vortex reader: Responses to YouTube* (pp. 101–109). Amsterdam, The Netherlands: Institute of Network Cultures.

Cheese, K. (2012). I spoke with An Xiao Mina for her story in *The Atlantic* today.... *Final Boss Form*. Retrieved from http://finalbossform.com/post/27135781606/i-spoke-with-an-xiao-mina-for-her-story-in-the

Dawkins, R. (1976). The selfish gene. Oxford, UK: Oxford University Press.

Dwyer, J. (2010, 15 June). A Brazilian Twitter campaign that really is for the birds. *The New York Times*. Retrieved from http://www.nytimes.com/2010/06/16/nyregion/16about.html

Easley, D., & Kleinberg, J. (2010). Networks, crowds, and markets: Reasoning about a highly connected world. Cambridge, UK: Cambridge University Press. Retrieved from http://www.cs.cornell.edu/home/kleinber/networks-book/networks-book-ch17.pdf

Gillespie, T. (2010). The politics of 'platforms.' *New Media & Society, 12*(3), 347–364.

Gillespie, T. (forthcoming). The relevance of algorithms. In T. Gillespie, P. Boczkowski, & K. Foot (Eds.), *Media technologies*. Cambridge, MA: MIT Press. Retrieved from http://www.tarletongillespie.org/essays/Gillespie%20-%20The%20Relevance%20of%20Algorithms.pdf

Godwin, M. (1990). Meme, counter-meme. *Wired*. Retrieved from http://www.wired.com/wired/archive/2.10/godwin.if_pr.html

Golder, S. (2009). Oprah, Iran and Twitter growth. Retrieved from http://scottgolder.wordpress.com/2009/07/02/oprah-iran-twitter/

Hargittai, E., & Litt, E. (2011). The tweet smell of celebrity success: Explaining variation in Twitter adoption among a diverse group of young adults. *New Media & Society, 13*(5), 824–842.

Indvik, L. (2010). Twitter: We are not keeping WikiLeaks out of Trending Topics. *Mashable*. Retrieved from http://mashable.com/2010/12/06/wikileaks-twitter-censorship/

Jeffries, A. (2011). Twitter: We are not blocking terms related to #OccupyWallStreet in any way, shape or form. *BetaBeat*. Retrieved from http://betabeat.com/2011/10/twitter-we-are-not-blocking-terms-related-to-occupywallstreet-in-any-way-shape-or-form/

Jenkins, H. (2007). Slash me, mash me, spread me. . . . Confessions of an Aca-fan: The official weblog of Henry Jenkins. Retrieved from http://henryjenkins.org/2007/04/slash_me_mash_me_but_please_sp.html

Jenkins, H., Ford, S., & Green, J. (2013). *Spreadable media: Creating value and meaning in a networked culture*. New York, NY: New York University Press.

Knobel, M., & Lankshear, C. (2007). Online memes, affinities, and cultural production. In M. Knobel & C. Lankshear (Eds.), *A new literacies sampler* (pp. 199–227). New York, NY: Peter Lang.

Lotan, G. (2011). Data reveals that "occupying" Twitter Trending Topics Is harder than it looks! *SocialFlow*. Retrieved from http://blog.socialflow.com/post/7120244374/data-reveals-that-occupying-twitter-trending-topics-is-harder-than-it-looks

Lotan, G. (2012a). Big data for breaking news: Lessons from #Aurora, Colorado. *SocialFlow*. Retrieved from http://blog.socialflow.com/post/7120245507/big-data-breaking-news-aurora-colorado

Lotan, G. (2012b). [Data Viz] KONY2012: See how invisible networks helped a campaign capture the world's attention. *SocialFlow*. Retrieved from http://blog.socialflow.com/post/7120244932/data-viz-kony2012-see-how-invisible-networks-helped-a-campaign-capture-the-worlds-attention

Manjoo, F. (2010). How Black people use Twitter. *Slate*. Retrieved from http://www.slate.com/articles/technology/technology/2010/08/how_black_people_use_twitter.html

Marwick, A., & boyd, d. (2010). I tweet honestly, I tweet passionately: Twitter users, context collapse, and the imagined audience. *New Media & Society, 13*(1), 114–133. Retrieved from http://educ333b.pbworks.com/w/file/fetch/53249911/marwick_boyd_twitter_nms.pdf

Messina, C. (2007). Groups for Twitter; or a proposal for Twitter tag channels. *FactoryCity*. Retrieved from http://factoryjoe.com/blog/2007/08/25/groups-for-twitter-or-a-proposal-for-twitter-tag-channels/

mona_jp. (2009). Longcat. KnowYourMeme. Retrieved from http://knowyourmeme.com/memes/longcat

Parr, B. (2010). Twitter improves Trending Topic algorithm: Bye bye, Bieber! *Mashable*. Retrieved from http://mashable.com/2010/05/14/twitter-improves-trending-topic-algorithm-bye-bye-bieber/

Rios, M., & Lin, J. (2012). Distilling massive amounts of data into simple visualizations: Twitter case studies. *Proceedings from ICWSM '12: The 6th International AAAI Conference on Weblogs and Social Media* (pp. 22–25). June 2012. Dublin, Ireland. Retrieved from http://www.aaai.org/ocs/index.php/ICWSM/ICWSM12/paper/view/4785/5095

Shifman, L. (2012). An anatomy of a YouTube meme. *New Media & Society, 14*(2), 187–203.

Thorton, S. (1996). *Club cultures: Music, media, and subcultural capital.* Middletown, CT: Wesleyan University Press.

Urban Dictionary. (n.d.). Yolo / YOLO. Retrieved from http://www.urbandictionary.com/define.php?term=yolo and http://www.urbandictionary.com/define.php?term=YOLO

Williams, E. (2009). Why retweet works the way it does. *EvHead*. Retrieved from http://evhead.com/2009/11/why-retweet-works-way-it-does.html

Zimmer, B. (2012, 26 Aug.). What is YOLO? Only teenagers know for sure. *Boston Globe*. Retrieved from http://www.bostonglobe.com/ideas/2012/08/25/what-yolo-only-teenagers-know-for-sure/Ids004FecrYzLa4KOOYpXO/story.html

Twitter and Geographical Location

12
CHAPTER   Rowan Wilken

 add a location to my tweets? only few
tweets are #geocoded, but locations
may be disclosed in different ways

*Twitter was meant to exploit a familiar blogging framework, one that
was quite locationless.*

—Erickson, 2010, p. 1198

The basic contention of this chapter is that the significance of the social media platform Twitter is further amplified when it is conceived of as a *locative platform*. Location, Fred Lukermann (1961) argues, is both a foundational concept for geography and one that carries important wider significance. From the Ancient Greek poets onwards, he writes, "how to describe 'where something is' becomes idiomatic in Western culture" (Lukermann, 1961, p. 197). Commensurate with the rise of mobile social—and especially location-based—media platforms is a renewal of interest in location. This reinvigorated concern for how we conceive of and experience location is encapsulated by de Souza e Silva and Frith (2012) as follows:

> The popularization of location-aware mobile technologies not only highlights the importance of location, but also forces us to re-think how location has been traditionally conceptualized. Locations are still defined by fixed geographical coordinates, but they now acquire dynamic meaning as a consequence of the constantly changing location-based information that is attached to them. (p. 9)

Thus, de Souza e Silva and Frith (2012) argue, where locations were once seen as "places deprived of meaning"—or more accurately, perhaps, whose meaning was dependent on other concepts and phenomena—they can now be seen as taking on "complex, multifaceted identities that expand and shift according to the information ascribed to them" (p. 10). Location-based services purportedly "comprise the fastest growing sector in web technology business" (p. 9). In addition, and of crucial importance in the present context, is that questions of location and location-awareness are increasingly central to our contemporary engagements with the Internet and mobile media. According to Gordon and de Souza e Silva (2011), "unlocated information will cease to be the norm" and location will become a "near universal search string for the world's data" (pp. 19–20). In the words of McCullough (2006), "information is now coming to you . . . wherever you are", and is "increasingly *about* where you are" (p. 26).

Twitter is interesting in this context insofar as it has not generally been thought of as a locative platform (Erickson, 2010, p. 1198). However, this is changing with growing acceptance of Twitter as a large-scale, data-rich platform of considerable social network significance (Cohen, 2009)—one, what is more, that is associated increasingly with a "re-awakened" understanding of "the importance of location" (Field & O'Brien, 2010, p. 5). Further recognition of the importance of Twitter's locational capabilities followed with the release in 2009 of its geotagging functionality, which, among other things, allows for the automatic (or selective) tagging of tweets with location information. Indeed, when considered as part of the larger communications ecosystem referred to as the geospatial web, or "geoweb" (Scharl & Tochtermann, 2007; see also Crawford & Goggin, 2009), Twitter emerges as a distinctly locative platform. As Elwood and Leszczynski (2011) explain, while the term "geoweb" is most closely associated with user-generated content and "the 'geotagging' of online content, or the assignation of place names, latitude/longitude coordinates, or any other locational information" (p. 6), it also implies much more than this:

> The geoweb consists of hardware (mobile devices), software objects (applications and services) and programming techniques (such as "mashing up" content) that include

virtual globes, interactive mapping platforms, spatial application programming interfaces (APIs), and technical standards (such as GPX) that guide its curation, aggregation, and dissemination. (Elwood & Leszczynski, 2011, p. 6)

It is within this context—that is, the "recent phenomenon of the merg-
ing of web content with locational referents" (Elwood & Leszczynski, 2011, p.
7)—that I examine Twitter in this chapter. Twitter developed geotagging capabilities in large part to encourage a richer user experience and more contextually relevant, finely granulated data (Cohen, 2009): as Twitter tells its users,
"twittering 'Earthquake!' alone is not as informative as 'Earthquake!' coupled
with your current location" (Twitter Blog, 2009). Noteworthy, though, is the
speed with which access to this data has been subsequently restricted: within
the space of only three years (2009–2012) Twitter has gone from opening up
geolocational operability to users and to developers via their API, to moving
to restrict third-party access to this data and, in the process, shoring up their
own corporate (and corporate partner) control of this increasingly rich, geotagged information source. I explore these developments in three steps. The first
gives a brief account of the opportunities provided to end users for disclosing
location information within Twitter. The second examines researcher interest
in geocoded Twitter information, and the challenges that are encountered in
extracting this data. The third discusses these data-extraction issues within the
context of the strategic and discursive framing of services and technologies that
Tarleton Gillespie (2010) labels "platform politics".

TWITTER AND LOCATION DISCLOSURE

There are a variety of means by which a Twitter user's location can be disclosed.
First, it can be conveyed by the information entered in the predetermined data
fields in Twitter's settings. These range from the most precise, such as checking
the "add a location to my Tweets" box, to the less precise, such as completing
the location box in a user's profile (a notoriously inaccurate measure of location,
as we shall see below), and selecting a country and a time zone. Geolocational
information can also be disclosed in tweets themselves as part of what Erickson
(2010) called "citizen microbroadcasting" (p. 1201): that is, "microblogging"
practices that focus on "sending timely, location-tagged bits of information to
the members of a community" (p. 1202). Erickson suggested that this form of
location disclosure is most apparent during "critical events" (p. 1201), such as a
natural disaster or other significant incident (see also Bruns & Burgess, Chapter
28 in this volume). More subtle forms of geolocational "broadcasting" via Twitter

are also possible, and are part and parcel of tweets associated with micro-scale daily mobilities that involve subtle negotiations of what Hjorth (2012) terms "mobile intimacy", which she defines as "the overlaying of the geographic and the electronic with the emotional and socio-cultural" (p. 199).

On a larger, arguably less intimate, scale, location is also disclosed through Twitter subscribers' use of third-party applications that connect with the Twitter interface. For example, in May 2010, Twitter announced plans to launch a feature, initially dubbed Points of Interest and subsequently relabeled Twitter Places (Ingram, 2010), which involved alliances with the location-based, mobile, social networking services Gowalla (now defunct) and Foursquare (Ingram, 2010). This development permitted users to click on a geotagged tweet and see other recent tweets from, or in close proximity to, that particular location (Twitter Blog, 2010). The data used to generate this location information was made possible by strategic partnerships with TomTom (Cowan, 2010) and Localeze (Siegler, 2010) and as a result of the acquisition in December 2009 by Twitter of Mixer Labs, the creator of the GeoAPI service (Ingram, 2010). These deals were in addition to content indexing licensing arrangements Twitter struck in 2009 with Google, Microsoft, and Yahoo! (Van Couvering, 2011), which gave these companies access to Twitter's "firehose" (high volume) data stream and opportunities for real-time search of Twitter content, including of geocoded tweets. Two years later, on 3 July 2011, Google cancelled its real-time search offering, and chose not to renew its deal with Twitter. Significantly, the expiration of this deal coincided with the launch, the previous week, of Google's own social networking service Google+ (Rosoff, 2011a). Meanwhile, a further two months later, in September 2011, Twitter renewed its licensing deal with Microsoft (Rosoff, 2011b). Along with its prior relationship with Yahoo!, Twitter is said to have similar "fire hose deals" with NTT Docomo and Yahoo Japan (Gannes, 2011). However, the Microsoft Bing deal is especially significant, given that, in the words of one commentator, it means Microsoft "will have preferred access to social signals from both Twitter and Facebook, while Google will only have access to a more limited and indirect supply of publicly available data" (Gannes, 2011).

These forms of location disclosure are significant on at least three levels. First, at the level of everyday practice, specific socio-technical competencies have emerged around user engagement with geolocational services like Twitter. For example, the ability displayed by Twitter subscribers to rapidly parse their Twitter feeds and perceive within them information that carries "geographical saliency" has been described by Erickson (2010, p. 1202) as "a form of *socioloc-*

ative topography". Following Crawford (2009a, 2009b), I would argue that this information-processing capacity of users is both honed through, and intrinsic to, the different forms of "listening" we practice and "ambient intimacy" we develop through our engagements with social media (see also van Dijck, 2011).

Second, location disclosure is also significant in the context of wider public debate over and anxiety around privacy (see de Souza e Silva & Frith, 2010). As de Souza e Silva and Frith (2012) write elsewhere, once a social network platform user's location "becomes a crucial determinant of the type of data accessed", then, "consequently, privacy issues become more directly interconnected with location" (p. 118). Drawing on the influential work of Solove (2008), de Souza e Silva and Frith (2012) argue that transparency, and exclusion and aggregation, are key issues attending the disclosing of location data in social media. With respect to the first of these, their argument is that the privacy policies of popular, location-based services "rarely delineate if they share location information with third parties, how they share the information, or if location information is stored" (p. 128). With respect to the second, interrelated concerns of exclusion and aggregation, the issue here, they suggest, is that

> as companies collect more and more data to build increasingly robust profiles, people have little recourse to access what information has been collected or whether that information is correct . . . [and consequently] they have little control over what is done with their own locational information. (de Souza e Silva & Frith, 2012, pp. 128–129)

De Souza e Silva and Frith (2012, p. 119) go on to make two further important points. The first is that, ultimately, "locational privacy needs to be understood contextually". Location information is not inherently private—indeed, as Elmer (2010) has argued, all location-based social media platforms operate around a tension, continuously negotiated by their users, between "finding" (someone or something) and "being found". Given this, de Souza e Silva and Frith (2012) suggest that "the loss of privacy occurs when the context shifts away from how the information was originally intended" (pp. 119–120). Their second point pertains to medium specificity, and the way that locational privacy must be understood as shifting from platform to platform: "Sharing location with a small group of friends in Foursquare is different from allowing anyone that uses Whrrl to see one's location, which is also different from sharing one's location via Twitter openly on the Web" (p. 131). In response to both of the above points, the key issue, de Souza e Silva and Frith argue, is that users' negotiations of locational privacy is, and ought to be, "intimately related to the ability to control the context in which one shares locational information" (p. 129).

In light of the above considerations of locational privacy, it is worth briefly considering Twitter's own policies. In terms of users' ability to control the context in which location information is shared, the Twitter Help Center provides very detailed instructions, along with clear explanations of associated risks ("About the Twitter Location Feature", 2012). Meanwhile, in terms of what locational information is stored and why, the Information Collection and Use section of Twitter's Privacy Policy states the following:

> We may use and store information about your location to provide features of our Services, such as Tweeting with your location, and to improve and customize the Services, for example, with more relevant content like local trends, stories, ads, and suggestions for people to follow. (Twitter Privacy Policy, 2012)

Beyond this, however, the privacy policy is rather less forthcoming about what the company does with this information, especially in terms of providing third-party access to and analysis of it.

The third reason location disclosure in Twitter is significant is because the accumulation of geotagged information via services like Twitter generates an informationally rich data pool. As Lapenta (2011) explains it, "geomedia transform the geolocation of their users, their geosphere, into data, and connect these data to existing information that describe users' online activities and identities (and their infosphere)" (p. 20). This data carries immense potential commercial value (Miller, 2012), most obviously in relation to possibilities for location-aware advertising (Cheng, Caverlee, & Lee, 2010; Davis, Pappa, Rennó Rocha de Oliveira, & de L. Arcanjo, 2011), and other initiatives that tap into the practices that bind location to identity construction and performance—what de Souza e Silva and Frith (2012, pp. 162–184), after Erving Goffman, term the "presentation of location". The richness of Twitter data also makes it of great research interest, although researchers face numerous challenges in extracting geolocational information from the platform.

TWITTER AS A LOCATIVE PLATFORM AND THE CHALLENGES OF DATA EXTRACTION

Research interest in geotagged Twitter data spans a variety of fields, including (to name a few): crisis communication (Bruns & Liang, 2012); disaster prediction (Earle, Bowden, & Guy, 2011; Gelernter & Mushegian, 2011); population health (Cheng et al., 2010); geography and geographical information systems (GIS) (Cheng et al., 2010; Field & O'Brien, 2010); media and communication research, especially the study of social movements (Burns & Eltham, 2009; Lysenko & Desouza, 2012; Reips & Garaizar, 2011); urban studies, where, to cite one example, Twitter data has been used as a diagnostic tool in order to understand the "processes that underpin the function and changing nature of urban space and place" (Pettit et al., 2012, p. 153) and the complexities of these processes (see also, Wakamiya, Lee, & Sumiya, 2011); and in understanding and mapping social network dynamics (Takhteyev, Gruzd, & Wellman, 2012; Yardi & boyd, 2010; Davis et al., 2011; Gonzalez, Cuevas, & Guerrero, 2011; Toole, Cha, & González, 2012). Within this chapter, geotagged Twitter data is generally retrieved in the following ways: from the tweet itself; from the Twitter profile location field; using qualitative data analysis software such as NVivo10's NCapture facility; or, from the geocode functionalities associated with the Twitter search API or the Twitter Geolocation API, a process performed either by the researcher or, increasingly, on their behalf, by an API aggregation company, such as Gnip.

Despite the introduction of Twitter's geolocation user settings and associated API search facilities, "location sparsity" in Twitter (Cheng et al., 2010, p. 760) remains a commonly remarked-upon research issue. The problem, in short, is that "most tweets are not associated with a geographic location" (Davis et al., 2011, p. 739), and researchers report considerable difficulties in obtaining data that holds any degree of granular location information. There are a number of possible explanations for this. For instance, user embrace of the geotagging function since its introduction has been slow, with the percentage of tweets containing locational information reported to be "increasing but still very small", rising from 0.23% in January 2010 to 0.6% in June 2010 (Reips & Garaizar, 2011, p. 636). A further explanation (as flagged in the previous section) concerns the vagaries of self-disclosed location by users in their profile settings. To illustrate, in one study, Cheng and colleagues accessed the Twitter API to collect and analyse 1,074,375 user profiles. Of these, only 21% listed "a location as granular as a city name", and only 5% provided "a location as granular as a latitude/longitude coordinate" (Cheng et al., 2010, pp. 760–761; cf. Takhteyev et al., 2012, who report better results than this while still acknowledging the limitations of this approach to locational data gathering).

These "location sparsity" issues have led the research community to develop more experimental methods for extracting geocoded Twitter data. One of the more notable approaches, applied across many Web-based platforms (not just Twitter), is that of location estimation based on analyses of Web content using a process known as "geoparsing": that is, automated place-name recognition within text strings which converts these names into longitude/latitude coordinates (Janowicz & Keßler, 2008, p. 1129). There are, however, particular challenges in applying this technique to Twitter. As Cheng et al. (2010, p. 761) point out, it is often difficult to determine whether "clear location cues [are] embedded in a user's tweets at all" due to a low signal-to-noise ratio and due to the use of textual abbreviations and non-standard vocabulary. Thus, once geocoded data is retrieved, considerable additional labour is involved in verifying and "disambiguating" locational information using "toponym resolution" techniques. Generally, this involves reference to a digital gazetteer. Gazetteers, as Janowicz and Keßler (2008, p. 1129) define them, are "place name directories containing names, spatial references, feature types and additional information for named geographic places". The principal objective in using digital gazetteers is not just to provide information on named features, but, crucially, to translate between informal and formal systems of place referencing and their possible locations in order to accurately determine and match a given feature's name with its location and its type (road, hill, etc.) (Goodchild & Hill, 2008; Hastings, 2008). While these and other related methods of location inference are considered to show promise (for discussion, see Davis et al., 2011), they are, evidently, time and resource intensive.

Access to geocoded data has also been further hindered by two controversial, recent changes by Twitter in how it manages access to its data. As Burgess and Bruns (2012) explain:

> First, the company locked out developers and researchers from direct "firehose" (very high volume) access to the Twitter feed; this was accompanied by a crackdown on free and public Twitter archiving services . . . and coincided with the establishment of what was at the time a monopoly content licensing arrangement between Twitter and Gnip, a company which charges commercial rates for high-volume API access to tweets . . . . A second wave of controversy . . . occurred in August 2012 in response to Twitter's release of its latest API rules . . . , which introduce further, significant limits to API use and usability in certain circumstances.

The second of these controversies is merely the most recent in a series of moves by Twitter to adjust access to and control of its API, especially its geolocation resources. In 2011, for instance, Twitter shut down the GeoAPI it acquired with Mixer Labs in 2009, rolling much of its functionality into its own Twitter API, while continuing to use the GeoAPI internally for its own applications (Schonfeld, 2011). Thus, in the space of only three years (2009–2012), Twitter has gone from opening up geolocational operability via their API to users and to developers, to moving to restrict third-party access to this functionality and its data. Political economic analyses of social and search media provide some productive insights into the motives for this changed stance (for further discussion of these issues, see Puschmann & Burgess, Chapter 4 in this volume).

CONCLUSION: GEOMEDIA AND PLATFORMS POLITICS

The many challenges faced in extracting and capturing geocoded data are becoming further pronounced in light of recent moves by Twitter to limit third-party access to and development for its API, and, in the case of researchers, by encouraging them to use commercial (and expensive) data access and analytics services (Burgess & Bruns, 2012).

Twitter was born into what Lewis (2001) would term a post-Netscape era of fluid business models, where tech start-ups are backed by venture capital and other forms of investment, while the developers and investors go about figuring out how best to monetise their assets. This is the broader context in which Twitter's continuous API adjustments need to be understood. As van Dijck (2011) explains, over the course of its development, "Twitter's pursuit of reaching a large, worldwide user base prompted the modification of its hardware to become interchangeable with other global platforms; changing its interface to promote follower lists in turn accommodated the insertion of sponsored content" (p. 343).

Such changes are strategically significant for a number of reasons. They are, for instance, increasingly crucial means by which social media platforms like Twitter work "discursively to frame their services and technologies" (Gillespie, 2010, p. 348; for discussion, see van Dijck, 2011, and Burgess & Bruns, 2012). In addition, ongoing alterations to Twitter's interface are significant insofar as they are part and parcel of emergent forms of economic logic that Van Couvering (2011) suggested are indicative of various new media platforms. Collectively labeling them "navigational media", Van Couvering suggests that search and social media services tend to be characterised by media platforms that facilitate exchange between producers and audiences, and which operate across "complex content pools that are large in size, extremely varied in terms of producers, and frequently refreshed" (Van Couvering, 2011, p. 198). Due to the complexity of the "content pool", the platform thus "becomes the central way to mediate connections between audiences and producers" (p. 198). Significantly, "if the content pool is the network", then "audience traffic, enabled through the platform, are the connections within the network" (p. 198). It is these "connections" which form the "core, saleable asset" for the owners of the platform (p. 198). In this way, Twitter's APIs, as key gateways to the platform's "audience traffic", form

vital instruments "enabling the capitalization" (Lapenta, 2011, p. 22) of its network data. The deals struck with Google and Microsoft in 2009, for example,
were said to be worth US$25 million for Twitter (Ante, 2009). Furthermore, in
its 2011 renegotiations of its licensing arrangements with Microsoft, Twitter was
reportedly seeking "about [US]$30 million per year for its exhaustive real-time
stream, a doubling of the previous fee", as well as "more user interface control,
a larger cut of ads sold next to its tweets [on Bing] and more linking back to
Twitter" (Gannes, 2011).

Herein lies the specific commercial significance of Twitter's geocoded data:

> Information within these [geomedia] systems is not only linked back to [users'] local
> referents (the physical space and the body of the user), but users themselves (and their
> surrounding space) are transformed into information—a commodified image—which
> is once again embedded in a controlled as well as socially and economically struc-
> tured system. (Lapenta, 2011, p. 22; for detailed discussion of the wider implications
> of the capitalization of geocoded data, see Barreneche, 2012)

By enabling geolocational functionality, van Dijck (2011, p. 343) points out, that
Twitter users "could be monitored more precisely; hence, certain revenue options
became more viable." As the commercial value of this data increases, so, too, it
would seem, will the economic costs associated with gaining research access *to*
this data (boyd & Crawford, 2012, pp. 673–675), and the wider social and other
costs associated with control *of* this data (Barreneche, 2012).

ACKNOWLEDGMENT

This chapter is an output of the Australian Research Council (ARC) funded project "The
Cultural Economy of Locative Media" (DE120102114). I wish to thank Emily van der Nagel for
her invaluable research assistance.

REFERENCES

About the Twitter location feature. (2012). *Twitter Help Center*. Retrieved from https://support.
twitter.com

Ante, S. E. (2009, 21 Dec.). Twitter is said to be profitable after making search agreements.
Bloomberg.com. Retrieved from http://www.bloomberg.com

Barreneche, C. (2012). Governing the geocoded world: Environmentality and the politics of
location platforms. *Convergence: The International Journal of Research Into New Media
Technologies, 18*(3), 331–351.

boyd, d., & Crawford, K. (2012). Critical questions for Big Data. *Information, Communication & Society, 15*(5), 662–679.

Bruns, A., & Liang, Y. E. (2012). Tools and methods for capturing Twitter data during natural disasters. *First Monday, 17*(4). Retrieved from http://firstmonday.org/htbin/cgiwrap/bin/ojs/index.php/fm/article/view/3937/3193

Burgess, J., & Bruns, A. (2012). Twitter archives and the challenges of "Big Social Data" for media and communication research. *M/C Journal, 15*(5). Retrieved from http://journal.media-culture.org.au/index.php/mcjournal/article/viewArticle/561

Burns, A., & Eltham, B. (2009, November). Twitter free Iran: An evaluation of Twitter's role in public diplomacy and information operations in Iran's 2009 election crisis. Paper presented at CPRF, Sydney, Australia. Retrieved from http://vuir.vu.edu.au/15230/1/CPRF09BurnsEltham.pdf

Cheng, Z., Caverlee, J., & Lee, K. (2010, Oct.). *You are where you tweet: A content-based approach to geo-locating Twitter users.* Paper presented at CIKM'10, Toronto, Ontario, Canada. Retrieved from http://infolab.cse.tamu.edu/static/papers/cikm1184c-cheng.pdf, 759–768.

Cohen, N. (2009, 9 Nov.). Refining the Twitter explosion. *The New York Times.* Retrieved from http://www.nytimes.com

Cowan, R. A. (2010, 17 June). Twitter Places: A tweet deal for TomTom? *The Wall Street Journal.* Retrieved from http://blogs.wsj.com

Crawford, A., & Goggin, G. (2009). Geomobile Web: Locative technologies and mobile media. *Australian Journal of Communication, 36*(1), 97–109.

Crawford, K. (2009a). Following you: Disciplines of listening in social media. *Continuum: Journal of Media & Cultural Studies, 23*(4), 525–535.

Crawford, K. (2009b). These foolish things: On intimacy and insignificance in mobile media. In G. Goggin & L. Hjorth (Eds.), *Mobile technologies: From telecommunications to media* (pp. 252–265). New York, NY: Routledge.

Davis C. Jr., Pappa, G., Rennó Rocha de Oliveira, D., & de L. Arcanjo, F. (2011). Inferring the location of Twitter messages based on user relationships. *Transactions in GIS, 15*(6), 735–751. doi: 10.1111/j.1467-9671.2011.01297.x

de Souza e Silva, A., & Frith, J. (2010). Locational privacy in public spaces: Media discourses on location-aware mobile technologies. *Communication, Culture & Critique, 3*(4), 503–525.

de Souza e Silva, A., & Frith, J. (2012). *Mobile interfaces in public spaces: Locational privacy, control, and urban sociality.* New York, NY: Routledge.

Earle, P. S., Bowden, D. C., & Guy, M. (2011). Twitter earthquake detection: Earthquake monitoring in a social world. *Annals of Geophysics, 54*(6), 708–715.

Elmer, G. (2010). Locative networking: Finding and being found. *Aether: The Journal of Media Geography, Spring,* 18–26. Retrieved from http://geogdata.csun.edu/~aether/pdf/volume_05a/elmer.pdf

Elwood, S., & Leszczynski, A. (2011). Privacy, reconsidered: New representations, data practices, and the geoweb. *Geoforum, 42*(1), 6–15.

Erickson, I. (2010). Geography and community: New forms of interaction among people and places. *American Behavioral Scientist, 53*(8), 1194–1207. Retrieved from http://abs.sagepub.com/

Field, K., & O'Brien, J. (2010). Cartoblography: Experiments in using and organising the spatial context of micro-blogging. *Transactions in GIS, 14*(s1), 5–23.

Gelernter, M., & Mushegian, N. (2011). Geo-parsing messages from microtext. *Transactions in GIS, 15*(6), 753–773.

Gannes, L. (2011, July 15). With Google gone (for now), Twitter tries to come to terms with Microsoft's Bing. *All Things Digital.* Retrieved from http://allthingsd.com

Gillespie, T. (2010) The politics of "platforms". *New Media & Society, 12*(3), 347–364.

Gonzalez, R., Cuevas, R., & Guerrero, C. (2011, 18 May). Where are my followers? Understanding the locality effect in Twitter. arXiv.org, arXiv:1105.3682v1 [cs.SI]. Retrieved from http://arxiv.org/pdf/1105.3682v1.pdf

Goodchild, M. F., & Hill, L. L. (2008). Introduction to digital gazetteer research. *International Journal of Geographical Information Science, 22*(10), 1039–1044.

Gordon, E., & de Souza e Silva, A. (2011). *Net locality: Why location matters in a networked world.* Chichester, UK: Wiley-Blackwell.

Hastings, J. T. (2008). Automated conflation of digital gazetteer data. *International Journal of Geographical Information Science, 22*(10), 1109–1127.

Hjorth, L. (2012). iPersonal: A case study of the politics of the personal. In L. Hjorth, J. Burgess, & I. Richardson (Eds.), *Studying mobile media: Cultural technologies, mobile communication, and the iPhone* (pp. 190–212). New York, NY: Routledge.

Ingram, M. (2010, 14 June). A place for every tweet and every tweet in its place. *GigaOM.* Retrieved from http://www.gigaom.com

Janowicz, K., & Keßler, C. (2008). The role of ontology in improving gazetteer interaction. *International Journal of Geographical Information Science, 22*(10), 1129–1157.

Lapenta, F. (2011). Geomedia: On location-based media, the changing status of collective image production and the emergence of social navigation systems. *Visual Studies, 26*(1), 14–24.

Lewis, M. (2001). *The new new thing.* London, UK: Penguin.

Lukermann, F. (1961). The concept of location in classical geography. *Annals of the Association of American Geographers, 51*(2), 194–210.

Lysenko, V. V., & Desouza, K. C. (2012). Moldova's Internet revolution: Analyzing the role of technologies in various phases of the confrontation. *Technological Forecasting & Social Change, 79*(2), 341–361.

McCullough, M. (2006). On the urbanism of locative media. *Places, 18*(2), 26–29.

Miller, R. (2012, September 17). Data is the real business model for social. *O'Reilly Media.* Retrieved from http://strata.oreilly.com

Pettit, C., Widjaja, Russo, P., Sinnott, R., Stimson, R., & Tomko, M. (2012). *Visualisation support for exploring urban space and place.* Paper presented at XXII ISPRS Congress, Melbourne, Australia. Retrieved from http://www.isprs-ann-photogramm-remote-sens-spatial-inf-sci.net/I-2/153/2012/isprsannals-I-2-153-2012.pdf

Reips, U., & Garaizar, P. (2011). Mining Twitter: A source for psychological wisdom of the crowds. *Behaviour Research Methods, 43*(3), 635–642.

Rosoff, M. (2011a, 4 July) Google doesn't need Twitter anymore, so it just let their deal expire. *Business Insider.* Retrieved from http://www.businessinsider.com

Rosoff, M. (2011b, 6 Sep.) Microsoft and Twitter renew their search deal. *Business Insider.* Retrieved from http://www.businessinsider.com

Scharl, A., & Tochtermann, K. (Eds.). (2007). *The geospatial Web: How geobrowsers, social software and the Web 2.0 are shaping the network society.* London, UK: Springer.

Schonfeld, E. (2011, 4 Mar.). Twitter will shut off GeoAPI to developers. *TechCrunch.* Retrieved from http://techcrunch.com/2011/03/04/twitter-shuts-geoapi/

Siegler, M. G. (2010, 17 June). Facebook also said to have a deal with Localeze for Facebook Places. *TechCrunch.* Retrieved from http://www.techcrunch.com

Solove, D. (2008). *Understanding privacy.* New York, NY: New York University Press.

Takhteyev, Y., Gruzd, A., & Wellman, B. (2012). Geography of Twitter networks. *Social Networks, 34*(1), 73–81.

Toole, J. L., Cha, M., & González, M. C. (2012). Modeling the adoption of innovations in the presence of geographic and media influences. *Public Library of Science ONE, 7*(1), 1–9.

Twitter Blog. (2009, December 23). Mixing it up at 795 Folsom St. *Twitter Blog.* Retrieved from http://blog.twitter.com

Twitter Blog. (2010, 14 June). Twitter Places: More context for your tweets. *Twitter Blog.* Retrieved from http://blog.twitter.com

Twitter Privacy Policy. (2012, 17 May). Information collection and use. Retrieved from https://twitter.com/privacy

Van Couvering, E. (2011). Navigational media: The political economy of online traffic. In D. Winseck & Y. Yin (Eds.), *Media political economies: Hierarchies, markets and finance in the global media industries* (pp. 183–200). London, UK: Bloomsbury.

van Dijck, J. (2011). Tracing Twitter: The rise of a microblogging platform. *International Journal of Media and Cultural Politics, 7*(3), 333–348.

Wakamiya, S., Lee, R., & Sumiya, K. (2011). *Urban area characterization based on semantics of crowd activities in Twitter.* Paper presented at the 4th International Conference on Geospatial Semantics, Brest, France. Retrieved from http://www.springerlink.com/content/8r175343882678k1/fulltext.pdf

Yardi, S., & boyd, d. (2010, May). *Tweeting from the town square: Measuring geographic local networks.* Paper presented at the 4th International AAAI Conference on Weblogs and Social Media, Washington, DC. Retrieved from http://www.aaai.org/ocs/index.php/ICWSM/ICWSM10/paper/view/1490/1853

Privacy on Twitter, Twitter on Privacy

13
CHAPTER   Michael Zimmer and Nicholas Proferes

while Twitter's privacy settings appear binary (#public / #restricted), there's much in between and little is known about users' perceptions

"What are you doing right now?" is the compelling question that greeted users for years at the website Twitter.com. Twitter's prompt served as the most prominently displayed instructional message on the homepage that suggested to individuals how they should use the service, and while some simply answer the simple prompt with an equally simple (and often mundane) description of their current activities, most largely ignore the question, and instead find a "myriad ways to share pretty much anything they wanted, be it information, relationships, entertainment, citizen journalism, and beyond" (Dybwad, 2009, para. 2). This sharing of "information, relationships, entertainment, citizen journalism, and beyond" has made Twitter a cultural phenomenon. Yet, as Twitter's popularity increases, so do privacy concerns with regard to personal or sensitive information shared by users and stored on the platform. The unauthorised sharing or misuse of personal information can result in harm to one's reputation (Solove, 2007), impact employment (Weiss, 2006), lead to identity theft, or fuel various forms of discrimination (Lyon, 2003). In contrast to the granular privacy controls provided by Facebook and Google+, Twitter offers a simple binary in terms of privacy control: either a user's Twitter activity is public to everyone, or restricted, requiring authorisation before access (to all tweets) is granted to particular users. Since Twitter's default privacy setting is that all messages are public—and the simple binary of public versus restricted accounts offers little room for ambiguity—arguments are commonly made that the 90% of users of the service who maintain public account settings (see Moore, 2009) have minimal expectations of privacy (Crovitz, 2011), and as a result, deserve little consideration in terms of possible privacy harms (Fitzpatrick, 2012). Seen this way, Twitter offers little in terms of large-scale privacy concerns or controversies.

This chapter, however, will argue that there are justifiable concerns over privacy on Twitter. While the technical controls which Twitter provides appear to provide simple and clear means for users to manage their information flows, personal and sensitive information routinely is shared—and leaked—beyond users' intended audience, while Twitter's own data security and data-sharing practices add new threats to user privacy. Thus, *privacy on Twitter* is a clear and present issue. Additionally, Twitter's own organisational rhetoric regarding the "ephemerality" of the platform shapes users' expectations of privacy, increasing the likelihood of the sharing of personal and sensitive information. Interrogating the very language Twitter uses to describe itself suggests that the majority of its own rhetoric focusses on the real-time nature of the communication exchange that Twitter provides, while often remaining silent or ambiguous about the permanence of tweets, and the privacy threats such permanence brings. Thus, examining *Twitter on privacy* reveals how its own rhetoric about a false ephemerality of tweets intensifies the overall privacy concerns of the platform.

PRIVACY ON TWITTER

Privacy on Twitter is, at first sight, simple to understand and to manage. Most popular, online, social networking platforms, such as Facebook and Google+ on the one hand, provide users with highly detailed and customisable privacy settings that allow users to specify levels of access to certain content (e.g., posts, photos, videos, comments) based on user-defined groups (e.g., friends, friends of friends, coworkers, family, etc.). Twitter, on the other hand, offers a simple binary of public versus restricted, as described above. Accounts are public by default. When a new user signs up for a Twitter account, no mention of the public nature of the account is provided (although links to the site's Privacy Policy and Terms of Service are provided), and no settings are immediately available to manage the visibility of the account. Unless a user takes steps to manage their account settings and opts to create a protected account, all Twitter activity will be publicly accessible. The power of this default setting and interface design is undeniable, as studies have shown that fewer than 10% of Twitter users have taken steps to gain privacy through restricting their accounts (Meeder, Tam, Kelley, & Cranor, 2010; Moore, 2009).

Twitter encourages users to "follow" other accounts, thereby creating a live feed of tweets to monitor and engage with. Consequently, some Twitter users might assume that only their "followers" will have access to and read their messages, when in fact any message posted by a user who has not changed their default privacy settings may be accessed by any other users, as well as by search engines and third-party applications using authorised APIs. Similarly, the use of the "@" symbol before a username allows Twitter users to direct public messages toward a specific user. When a Twitter message includes an @username, the tweet message will be posted in the receiver's timeline as well as potentially on the timeline of other users (the sender and receiver of the @mention, as well as any followers they have in common, if the @mention is at the beginning of the tweet; or the sender of the @mention as well as all of the sender's followers, as well as the receiver of the @mention, if the @mention is anywhere else in the tweet than at the very beginning), despite the possible expectation that such a tweet is viewable by the target only. Any mistaken sense of private communication between the sender and receiver of @mentions might lead some Twitter users to share information intended for a more restricted audience, not realising that their tweets are viewable by all. Research has shown that between 40% and 50% of tweets included information about the author herself (Honeycutt & Herring, 2009; Naaman, Boase, & Lai, 2010), which might include contact data, other personally identifiable information, locational data, health information, and the like (see, for example, Mao, Shuai, & Kapadia, 2011), posing potential privacy threats to users unaware of the fully public nature of their activity.

Users seeking privacy can restrict availability of their tweets by setting their accounts to "private" and therefore accessible only to authorised followers, as well as by revoking such authorisations, using Twitter's block feature. Taking this step affords control over the visibility of one's tweet stream to ensure privacy, but leakages can easily occur. Twitter users frequently share other users' tweets with their own followers through the mechanism of retweeting. This was originally accomplished through informal means of copying tweets appearing in one's feed into a new tweet, with various conventions to indicate that the content originated elsewhere. Eventually the syntax standardised into a simple "RT @[username]" to denote the origin (boyd, Golder, & Lotan, 2010; see also Chapter 3 by Halavais in this volume). The act of retweeting can extend the immediate visibility of a public tweet beyond the original author's expected audience. More importantly, users who have been granted access to restricted accounts can easily retweet protected tweets by copying and pasting into their own, unprotected feed, violating the privacy protections enacted by the original author. In a study of over 80 million Twitter accounts, nearly 250,000 protected accounts had at least one restricted tweet retweeted by a public user (Meeder et al., 2010).

Twitter eventually formally introduced functionality to make retweeting a feature of the service. These "official" retweets are treated differently from those created by the copy-and-paste method: the author of the original tweet is displayed prominently in the timeline, and a small footnote link indicates which person retweeted the message; the retweeting user is not able to add any commentary or annotation. Most importantly, the official retweet functionality implemented by Twitter respects user privacy settings, in that a tweet originating from a restricted account cannot be retweeted. However, many users continue to use the copy-and-paste retweet convention, as it allows them to add comments of their own, and to overcome the restrictions placed on the original tweet. Additionally, the use of third-party clients, such as Seesmic or Echofon, enables users to retweet protected tweets, albeit with a warning that such an action might violate the original account's privacy (Meeder et al., 2010). Thus, the practice of retweeting represents a sizeable risk for the leakage of tweets that had been intended for a restricted audience, thereby generating a considerable privacy threat.

Regardless of whether a tweet is public or private, and of whether the act of retweeting respects the privacy of the original account, Twitter retains copies of all tweets, logging all related data and metadata such as hashtags, page views, links clicked, geolocational data, searches, and relationships between Twitter users and their followers. The service also requires users to provide their full name and a valid email address to create an account. Although this data collection and retention is necessary for Twitter to operate effectively, it also intensifies privacy concerns stemming from the archiving and possible release or exposure of personal or sensitive information. Twitter, of course, provides a Privacy Policy describing the information it collects and how it might be used or shared with third parties (Twitter, 2012a). Despite such assurances and related security measures, breaches of Twitter's vast databases of user information and activity remain possible, with numerous vulnerabilities, exploits, and attacks threatening user privacy. Twitter also provides detailed information for law-enforcement agencies seeking to request information about Twitter users or particular activity (Twitter, 2012c). While Twitter has shown the willingness to fight subpoenas for access to its data (Cohen, 2011; Zetter, 2012), the general possibility of law enforcement gaining access to its logs represents a continued privacy threat for users.

To summarise, privacy concerns on Twitter range from users lacking a sufficient understanding of how publicly viewable their tweets actually are, through the retweeting of restricted tweets by public users, to the general threat of the release—whether intended or not—of the vast amounts of user data stored by Twitter itself. In a recent case where Twitter, Inc. was ordered to release data about a user arrested at an Occupy Wall Street protest, the judge ruled partially on the grounds that Twitter users who tweet publicly have "no reasonable expectation of privacy" (Fitzpatrick, 2012). Indeed, the simplicity of Twitter's privacy settings, the ease of retweeting private tweets, and the inherently public nature of the platform fuel arguments that it is unreasonable for anyone to expect any privacy while participating on the platform, while millions of users continue to use the platform and tweet private and potentially sensitive information. Yet, when confronted with the reality that tweets—even those knowingly made public—might become archived digitally at the Library of Congress (Raymond, 2010), many individuals were shocked by the announcement. Some comments on the Library of Congress's Web version of the announcement of the archive expressed surprise and frustration about the seemingly newfound permanence of tweets, and about the privacy threats fostered by such an archive. Three examples follow:

> So with no warning, every public tweet we've ever published is saved for all time? What the hell. That's awful.

> I can see a lot of political aspirations dashed by people pulling out old Tweets. I've always thought of the service as quite banal and narcissistic, but I've had a Twitter account to provide feedback to a college and a couple of vendors. I think I'll close my account now. I don't need to risk Tweeting something hurtful or stupid that will be around for all recorded time.

> Now future generations can bear witness to how utterly stupid and vain we were—1. for creating this steaming mountain of pointless gibberings, and 2. for preserving it for posterity. LOC, you nimrods. (as quoted in Raymond, 2010)

Even in broadcasting the news, the language *Wired* magazine chose underscored the apparent transition from a fleeting existence for tweets to a newly instilled sense of permanence when it stated that, "while the short form musings of a generation chronicled by Twitter might seem ephemeral, the Library of Congress wants to save them for posterity" (Singel, 2010).

This perceived ephemerality of Twitter prompted surprise by many users, who, in the face of the Library of Congress announcement, suddenly were confronted with the fact that Twitter had, of course, been saving every message sent through the service in the first place. Indeed, Twitter's own language used to describe the platform contributes to this false belief that privacy might exist due to the fleeting nature of one's tweets.

TWITTER ON PRIVACY

Despite the fact that Twitter maintains copies of all tweets and related user activity—leading to many of the privacy threats outlined above—its organisational rhetoric frames Twitter as a "real-time" and ephemeral service. Through an analysis of the descriptive language present in interview comments made by Twitter's founders, and on the Twitter homepage itself during Twitter's early years of operation, this section—exploring Twitter's own rhetoric *on* privacy—addresses how the language used to describe Twitter, and the language of Twitter, helped construct user expectations and experiences regarding the service. Our focus in the following pages is on foregrounding the conflicting messages about the temporality and permanence of tweets in order to reveal how this particular conflict—between the rhetoric used to describe a technology and the operation of the technology itself—may have disempowered users by helping to instil certain expectations of privacy.

Rhetoric can have a profound impact on understanding (Scott, 1967). As such, the concern of this analysis is specifically the presence of language that describes the temporality of Twitter and the permanence of tweets within Twitter's organisational rhetoric. Within this descriptive language, there is an inherent attempt to influence the knowledge of an audience who, during the early days of Twitter's operation, may have been encountering Twitter for the first time. The organisation of Twitter, Inc. functions rhetorically through the communicative practices of its organisational leaders (such as CEOs, founders, and public relations representatives); therefore these messages are the first object of concern for this analysis. Twitter's website itself is also home to a number of rhetorical messages that are representative of the organisation or, as Gallant and Boone (2008) put it, "Internet sites are inherently rhetorical" (p. 185). The instructional language on Twitter.com that orients users and visitors to the technical operation of the site similarly serves as an argumentative description regarding the temporal properties of Twitter and the permanence of tweets. Therefore, messaging present on Twitter.com is the second instance of organisational rhetoric included in this analysis.

Twitter founders Evan Williams, Biz Stone, and Jack Dorsey have been active in discussing their service in the media. They have each, and sometimes collectively, given interviews in a variety of news outlets, talk shows, and at a variety of technology conferences. The language that this group uses to describe Twitter functions as an argument for how to conceptualise and view the service. The descriptive language used within the rhetoric can be split into three categories: language that suggests that Twitter maintains an archive of tweets, language that suggests that Twitter does not maintain an archive of tweets, and language that focusses an audience on the real-time nature of the service while neglecting any description of the permanence of tweets. Eight news interviews, recorded during the period of 2006–2011, were used as part of this analysis. In six of the interviews, the founders only used descriptive language that focussed on the real-time nature of the medium while neglecting any description of the permanence of tweets. For instance, in a 2010 interview with Wolf Blitzer on CNN's "The Situation Room", Biz Stone was asked if he could sum up the real point of Twitter. He responded: "the real point of Twitter is to help people discover and share what it is that is happening around them in the world . . . it really has become an information network that is focussed on real-time" (Blitzer, 2010). Language such as this positions Twitter as being a tap into what is happening right now. Similarly, in an interview with *AgoraNews*, Jack Dorsey stated that Twitter "brings a lot of immediacy to the conversation, it allows people to interact in real-time, and it allows a great mass of people to interact and report from wherever they are and whatever they are doing. . . . I think a tool like this allows people to get immediately into something and then share it" (*AgoraNews*, 2009). Seldom found within these messages about what Twitter is are descriptions that detail how this constant stream of real-time information was being stored for the long term. Instead, listeners are invited to focus on the immediacy of Twitter as a medium. Recipients of this sort of rhetorical message are invited to consider Twitter as something in the moment, "a constant babble of thoughts by users" (Kinzie, 2009, para. 7), "about instantaneous notification" (Kinzie, 2009, para. 8), and an "up-to-the-minute venue" (Ody, 2009, para. 7).

In none of the interviews did the founders ever describe Twitter as explicitly maintaining a permanent record of all messages sent through the system. However, in two of the interviews, the founders described Twitter as "being like" other technologies which may internally maintain permanent archives, though this descriptive language is often muddied by incomplete or conflicting analogies. Analogy to older technologies is particularly important for meaning-making, as Lipartito (2003) wrote, because "when confronted with a truly new technology that had not been an option before, consumers must find some way to match the unexpected with previous experience" (p. 56). The first example of this is from the 2006 interview with "@LunchMeet". In this interview, when asked to describe what Twitter is, the founders referred to Twitter as being like "a chatroom", and then seconds later, described the service as "like Livejournal" (Slutsky & Codel, 2006). Here, the analogy to a chatroom is inherently problematic, as chatrooms are a technology which, depending on their technical structure, may or may not have a centralised storage of messages. This analogy is further muddied with the additional comparison to LiveJournal.com, a blogging/diary platform substantively different from a chatroom, with an extensively different temporality and permanence. LiveJournal maintains an accessible archive of posts made to its servers; chatrooms may or may not do so. So while the recipients of organisational rhetoric are occasionally invited to conceptualise Twitter as maintaining a permanent archive because Twitter is "like" other technologies that do, this message is diluted with parallels to technologies that have ambiguous message-retention policies.

The Twitter website itself is home to numerous rhetorical messages that guide users in the sense-making process. This chapter approaches these messages as they appear to a user who is using the site for the first time in 2011, on a laptop or desktop computer through a Web browser. This distinction is necessary, as Twitter offers a mobile version of their site, and as there are numerous applications for various mobile devices that also interface with Twitter. Again, this analysis relies on the categorisation of rhetoric as being either descriptive of the archival practices for tweets, focussing users on the real-time nature of the technology, or describing tweets as ephemeral.

Visitors are oriented towards the real-time nature of Twitter the second the landing page loads. The first page at which a visitor to the Twitter homepage arrives contains large text on the right-hand side of the screen stating, "follow your interests"; underneath, "instant updates from your friends, industry experts, favorite celebrities, and what's happening around the world" (Twitter, 2011a). This is the only language that appears on the landing page that is not sign-up or sign-in specific. Sign-up information appears immediately to the right of this statement. Immediately, users are oriented towards the real-time, "instant", and global nature of the medium.

The Twitter interface itself has changed somewhat since its original design in 2006. The question that appeared at the top of the screen in 2006, "What are you doing right now?" has been replaced by "What's happening?" (Twitter, 2011b). Underneath, there is a text box in which users may choose to enter a response, with a button next to it marked "tweet". Clicking this button sends the message off into the world of Twitter. A message just sent shows up in a user's "Timeline," the area directly underneath the input box. The Timeline displays, in reverse chronological order, both the messages of the user and the messages that have been sent by the individuals whom a user follows. On the right-hand side of the screen is information about whom a user is following, who is following that user, suggestions for more people to follow, and an area marked "trends," along with a search bar. Within the realm of this interface there are several rhetorical messages about the way that users should experience the site, and about the historicity of messages. The question "What's happening?"—while not as obviously as "What are you doing right now?"—invites a user to form a response tweet that is of the moment. It is a question that Twitter seems to be asking of users (or perhaps, one's followers are asking of the user). When the user enters a response, it is immediately populated within the chronological timeline on the user's page. A small bit of text under each tweet appears in the timeline that describes how long ago that message was posted. The twenty most recent tweets appear in the timeline as a default. Only when a user scrolls down further and further on the page do older messages appear. Despite the fact that these older messages appear, as of 2012, there is a technical limit on the number of tweets that can be accessed (Owens, 2011). A user can only "go back" so far into their history before the site will load no more older tweets, allowing users to plausibly draw the conclusion that once a certain number of tweets are populated, the old ones disappear. Of course, this is not the technical reality. Twitter retains the tweets beyond the cutoff point; users are simply unable to access them without knowing the exact URL of the original message.

Twitter's Terms of Service and Privacy Policy are the documents that govern user access and use of the Twitter service. While anyone who has ever set up a Twitter account has agreed to these conditions, a 2011 survey found that "only 18 per cent of social media users surveyed said that they read the terms and conditions for posting to the sites they use" (Dugan, 2011, para. 7). By agreeing to these conditions, "you consent to the collection and use (as set forth in the Privacy Policy) of this information" (Twitter, 2012b, para. 7)—that is, of any information provided to Twitter. Despite their length, the Terms of Service and

Privacy Policy never explicitly state that Twitter maintains a permanent record of tweets, nor do they state that older tweets are removed from the site. Instead, the Terms of Service includes statements such as "what you say on Twitter may be viewed all around the world instantly" (Twitter, 2012b, para. 3), and

> By submitting, posting or displaying Content on or through the Services, you grant us a worldwide, non-exclusive, royalty-free license (with the right to sublicense) to use, copy, reproduce, process, adapt, modify, publish, transmit, display and distribute such Content in any and all media or distribution methods. (Twitter, 2012c, para. 12; see also the discussion of official Twitter documents by Puschmann & Burgess, Chapter 4 in this volume)

In these quotes, there again appears language that invites users to consider the real-time nature of the medium, but absent from this language are terms like "forever", "in perpetuity", or "archive". While this licence may grant Twitter the legal right to archive tweets in perpetuity, there does not appear to be any rhetorical language that would invite a reader to understand that this was happening.

In summary, the majority of the organisational rhetoric of Twitter focusses on the real-time nature of the communication exchange that Twitter provides, while often remaining silent or ambiguous about the permanence of tweets. This ambiguity contributes to the potential for the rhetorical "ephemerality" of the platform to shape users' expectations of privacy.

CONCLUSION

This chapter has argued that Twitter presents a unique privacy challenge for its users. More than the simple decision of whether to create a public or restricted account, concerns over privacy play out on Twitter through a complex assemblage of potentially mistaken user expectations, leakages of restricted tweets, and the persistent logging of vast amounts of user information—what we refer to as *privacy on Twitter*. This combines with a powerful organisational rhetoric by Twitter itself which can lead users into embracing the "ephemerality" of the platform, thereby shaping users' expectations of privacy—what we refer to as *Twitter on privacy*. These two spheres of influence—how privacy actually exists on Twitter and how Twitter frames overall privacy concerns—have significant impact on how users engage with the platform, the information they choose to share, and the expectations they bring to the context of being a part of the Twitter phenomenon.

The privacy threats to Twitter users are real, and can occur in a number of places along the path of tweet production, distribution, and consumption.

However, there are a number of practical steps that could be taken by Twitter to buffer against some of the potential privacy harms. Twitter's own privacy model of the simple "protected/unprotected" binary could be altered to support a more robust and granular set of user privacy controls. Greater emphasis could be placed on new user education. In particular, prominently displayed, descriptive language that clearly articulates how tweets are stored by Twitter in the long term and how those tweets are shared with third parties could help alleviate some of the privacy problems identified by this chapter. Having new users be confronted with the choice to keep their tweets protected or unprotected as part of the sign-up process could further boost the visibility of these options.

Alongside these recommendations to be addressed by Twitter, there are numerous opportunities for additional scholarly research to better understand and address the privacy threats that stem from Twitter's rhetoric and practices. First, measuring users' actual understandings and expectations regarding the privacy and relative visibility of their tweets can help assess the impact of Twitter's rhetoric stressing the ephemerality of tweets. Additionally, further exploration into the Library of Congress's plans to archive public tweets is needed, addressing key variables, such as whether users can opt out of the archive, delete tweets, or request the removal of any protected tweets that happened to be retweeted publicly. Similar research should take place to explore the increased archiving and use of Twitter data by data brokers, such as Gnip or DataSift. The knowledge produced by such research, in tandem with increased control choices and data-flow transparency, could yield greater protection for users answering the question "What are you doing right now?"

REFERENCES

AgoraNews. (2009, 8 Apr.). Interview with Jack Dorsey. *AgoraNews.* Retrieved from http://www.youtube.com/watch?v=qt9eH74PmVw

Blitzer, W. (2010, 9 Nov). Twitter co-founder, Biz Stone, talks with CNN. *CNN.* Retrieved from http://edition.cnn.com/video/?/video/tech/2010/11/09/tsr.twitter.biz.stone.cnn

boyd, d., Golder, S., & Lotan, G. (2010). Tweet, tweet, retweet: Conversational aspects of retweeting on Twitter. *Proceedings of the 43rd Hawaii International Conference on System Sciences (HICSS-43)* (pp. 1–10), Kauai, HI: IEEE.

Cohen, N. (2011, 9 Jan.). Twitter shines a spotlight on secret F.B.I. subpoenas. *The New York Times.* Retrieved from http://www.nytimes.com/2011/01/10/business/media/10link.html

Crovitz, G. (2011). Are we too hung up on privacy? *The Wall Street Journal.* Retrieved from http://online.wsj.com/article/SB10001424052970204138204576604503117575260.html

Dugan, L. (2011). 52 percent of Twitter users do not consider legal implications of their tweets. *All Twitter*. Retrieved from http://www.mediabistro.com/alltwitter/twitter-legal_b15407

Dybwad, B. (2009, November 19). Twitter drops 'what are you doing?' Now asks 'what's happening?' *Mashable*. Retrieved from http://mashable.com/2009/11/19/twitter-whats-happening/

Fitzpatrick, A. (2012). Judge: Public tweets have no 'reasonable expectation of privacy.' *Mashable*. Retrieved from http://mashable.com/2012/07/03/twitter-privacy/

Gallant, L. M., & Boone, G. M. (2008). Communicative informatics: A social media perspective for online communities. In *Proceedings of the 2nd International Conference on Weblogs and Social Media (ICWSM)* (pp. 184–185). 30 Mar. –2 Apr. 2008. Seattle, WA. Menlo Park, CA: AAAI Press.

Gallant, L. M., & Boone, G. M. (2011). Communicative informatics: An active and creative audience framework of social media. *tripleC—Cognition, Communication, Co-operation*, 9(2), 231–246.

Honeycutt, C., & Herring, S. (2009). Beyond microblogging: Conversation and collaboration via Twitter. *Proceedings from the Forty-Second Hawai'i International Conference on System Sciences, (HICSS-42)* (pp. 1–10). doi: 10.1109/HICSS.2009.89

Kinzie, S. (2009, 26 June). Some professors losing their Twitter jitters. *The Washington Post*. Retrieved from http://www.washingtonpost.com/wp-dyn/content/article/2009/06/25/AR2009062504027.html

Lipartito, K. (2003). Picturephone and the information age: The social meaning of failure. *Technology & Culture*, 44(1), 50–81. doi: 10.1353/tech.2003.0033

Lyon, D. (Ed.). (2003). *Surveillance as social sorting: Privacy, risk, and digital discrimination*. London, UK: Routledge.

Mao, H., Shuai, X., & Kapadia, A. (2011). Loose tweets: An analysis of privacy leaks on Twitter. *Proceedings of the 10th Annual ACM Workshop on Privacy in the Electronic Society* (pp. 1–12). Chicago, IL. doi: 10.1145/2046556.2046558

Meeder, B., Tam, J., Kelley, P. G., & Cranor, L. F. (2010, May). *RT@ IWantPrivacy: Widespread violation of privacy settings in the Twitter social network*. Paper presented at the Web 2.0 Privacy and Security Workshop, IEEE Symposium on Security and Privacy, Oakland, CA. Retrieved from http://www.cs.cmu.edu/~./bmeeder/papers/Meeder-SNSP2010.pdf

Moore, R. (2009). Twitter data analysis: An investor's perspective. Retrieved from http://techcrunch.com/2009/10/05/twitter-data-analysis-an-investors-perspective-2/

Naaman, M., Boase, J., & Lai, C. H. (2010, February). *Is it really about me?: Message content in social awareness streams*. Paper presented at CSCW 2010 conference, Savannah, GA. Retrieved from http://infolab.stanford.edu/~mor/research/naamanCSCW10.pdf

Ody, E. (2009). Market news, delivered by Twitter. *The Washington Post*. Retrieved from http://www.washingtonpost.com/wp-dyn/content/article/2009/04/25/AR2009042500101.html

Owens, M. (2011). Why the 3200 tweet user_timeline limit, and will it ever change? | Twitter Developers. *Twitter.com*. Retrieved from https://dev.twitter.com/discussions/276

Raymond, M. (2010, 14 Apr.). How tweet it is!: Library acquires entire Twitter archive. *Library of Congress Blog*. Retrieved from http://blogs.loc.gov/loc/2010/04/how-tweet-it-is-library-acquires-entire-twitter-archive/

Scott, R. L. (1967). On viewing rhetoric as epistemic. *Communication Studies, 18*(1), 9–17.

Singel, R. (2010). Library of Congress archives Twitter history, while Google searches it. *Wired Magazine.* Retrieved from http://www.wired.com/epicenter/2010/04/loc-google-twitter/

Slutsky, I., & Codel, E. (2006, 5 Dec). A look at the early days of Twitter. *LunchMeet.* PodTech. Retrieved from http://www.youtube.com/watch?v=93NGaicjHnE

Solove, D. (2007). *The future of reputation: Gossip, rumor, and privacy on the Internet.* New Haven, CT: Yale University Press.

Twitter. (2011a). Twitter. *Twitter.com.* Retrieved from https://twitter.com/

Twitter. (2011b). Twitter / Home. *Twitter.com.* Retrieved from https://twitter.com/

Twitter. (2012a, May). Twitter / Twitter Privacy Policy. *Twitter.com.* Retrieved from https://twitter.com/privacy

Twitter. (2012b, June). Twitter / Twitter Terms of Service. *Twitter.com.* Retrieved from https://twitter.com/tos

Twitter. (2012c). Twitter / Guidelines For Law Enforcement. *Twitter.com.* Retrieved from https://support.twitter.com/entries/41949-guidelines-for-law-enforcement#

Weiss, P. (2006, 19 Mar.). What a tangled Web we weave: Being Googled can jeopardize your job search. *New York Daily News.* Retrieved from www.nydailynews.com

Zetter, K. (2012). Twitter fights back to protect 'Occupy Wall Street' protester. Retrieved from http://www.wired.com/threatlevel/2012/08/twitter-appeals-occupy-order/

Automated Twitter Accounts

14
CHAPTER   Miranda Mowbray

not all #bots are evil: how to distinguish between spammers, entertainers and tweeting #toasters

Twitter was designed for communication between human beings. Evan Williams, the co-founder of Twitter, has said in a television interview that the purpose of Twitter is "about humans connecting with each other, and often in ways that they couldn't otherwise" (Williams, 2009). However, Twitter is also used by automated accounts. In 2009, Sysomos, Inc. estimated that 24% of all public tweets were sent by automated accounts tweeting at least 150 times a day (Sysomos, Inc., 2009). Chu, Gianvecchio, Wang, and Jajodia (2010) have estimated human, semi-automated, and automated Twitter accounts to be in the proportion 5:4:1, and Zhang and Paxson (2011) found that 16% of active accounts show a high degree of automation, judging by the regularity of tweet timestamps. At a conservative estimate, tens of millions of automated tweets are sent every day.

The term *bot* will be used in this chapter to mean an account that is at least semi-automated. Typically, a Twitter bot will send automated tweets or direct messages (DMs), although its human owner may also send non-automated tweets through the account. Many bots also automatically make and accept friend requests. The content of a bot's tweets may be pre-written by its owner or may be algorithmically generated, in some cases responding to other Twitter users' behaviour.

Twitter encourages the use of automation, and does not require that bots identify themselves to be such. However it forbids some types of automated behaviour abused by spammers.

A BESTIARY OF TWITTER BOTS

This section describes the most common types of Twitter bots (these types overlap).

MARKETING BOTS

The most common use of Twitter bots is for marketing. Marketing bots may simply tweet marketing messages, or may allow interaction. For example, @KLMfares replies with airfares to structured queries such as "@KLMfares Johannesburg to Manchester April".

USEFUL BOTS

Many bots are designed to be informative or otherwise useful. There are Twitter bots that tweet public holidays (@whatholidayisit); weather forecasts (@AccuWx); earthquake information (@earthquakebot); and arrests in Knoxville, Tennessee (@knoxarrests). Services offered via @t411 include looking up stock prices, rail times, maps, and Bible verses. @twisst alerts you when the International Space Station is overhead at your location. @gcal lets you add events to your Google calendar through Twitter. The OKITE alarm clock app sends slightly embarrassing tweets (in Japanese) through the accounts of users who hit snooze (eureka, Inc., 2011).

ENTERTAINING BOTS

There is a well-developed subculture of Twitter bots designed to entertain. These may tweet works of literature line by line, play interactive games, or converse with human Twitter users. @pentametron tweets rhyming iambic pentameters made entirely out of retweets. @Betelgeuse_3 responds when anyone tweets "beetlejuice" three times. @Yoda_Bot's Twitter page says "Only once per day, tweet to you can I. Follow me if you like, yes. Tweet to me 'optout', bother you no more I will!"

THE INTERNET OF TOASTERS

Twitter is a natural communication channel for the Internet of Things (Slavin, 2009). Things that tweet include a toaster (@mytoaster—"The Internet of Toasters" comes from the toaster's Twitter bio); shoes (@ramblershoes); and a catflap (@GusAndPenny)—the catflap's tweets say which cat is going out or in, and link to a security camera photo. A house may tweet its water consumption, room temperatures, or energy use (@tweetawatt; Cellan-Jones, 2009). Remote human users have used commands over Twitter to operate a coffee machine (InstructablesTV, 2011) and a robot (@tweetnoid), and to water plants that tweet when they need watering (Fahner, 2009).

ANTISOCIAL BOTS

Not all Twitter bots are benign. Spambots (aggressive marketing bots that break Twitter's Terms of Service) use various tricks that decrease Twitter's signal-to-noise ratio for human users. Some bots market malware, for example, persuading Twitter users to download an infected application that allows spammers to use their Twitter account, or criminals to use their computer as part of a botnet. Some botnet computers receive control information via Twitter.

Twitter bots have also been used in underhand ways for political motives, for example to spread smears about candidates (Ratkiewicz et al., 2011). It seems likely that the approximately 90,000 bot followers acquired by 2012 U.S. presidential candidate Mitt Romney's account in one day were bought by someone else to embarrass him (Green, 2012): there is a black market in Twitter followers, priced at a few U.S. cents each. More seriously, in 2011 to 2012 tens of thousands of coordinated bots attempted to drown out political conversations on Twitter by protesters in Russia (Thomas, Grier, & Paxson, 2012) and Mexico (Santiesteban, 2012).

Twitter's security team detects and suspends accounts that break their rules. Most are bots: the economics of spamming, online fraud, and targeted follower-selling require automation.

DO NOT LOOK LIKE A BOT

A blogger used to communicate frequently on Twitter with two other Twitter users. They wished him good morning every day. He regarded them as close friends. After some time, he discovered that in fact they were not human beings, but software programs (Coconutsfine, 2009). (Twitter is not the only environment where this happens. An expert on conversational bots was fooled for about four months by one that he encountered on a dating website; see Epstein, 2007.)

To gain some idea of how common it is for Twitter bots' profiles to appear human, in Matwyshyn & Mowbray (2012) we examined the profiles of 727 accounts (sampled by Nazareno Andrade) that sent tweets in April–May 2010 using unregistered clients. Manual inspection confirmed that these accounts were almost certainly bots. However, 273 (37%) of the profiles contained some human-like indicator and gave no indication that the account was a bot. The human-like indicators were that names were recognisably human, with a given name and surname; profile images were human photos; or text in bios implied that the account was human, for example, "*Love Travelling and Learning*".

This analysis only considered profiles, and did not examine tweet content or the accounts' behaviours over time. Some automated accounts' behaviours may also make them appear human, such as automatically replying to tweets, making targeted friend requests, or sending human-like tweets such as "off 2 bed". The website for the TweetBuddy bot software advertised, "Don't look like a bot to twitter they don't like that. Our custom settings delay message and responses to give the impression it is a human doing all the work" (as quoted in *Twitter, Inc., v. Skootle Corp. et al.*, 2012, p. 11).

BOT OR NOT?

Because some bots can appear to be human, methods are needed to identify whether or not a Twitter account is a bot. This information is of interest to Twitter users in deciding whether or not to follow an account, or how much to trust a product endorsement or political message on Twitter.

Tweet-based predictions about widely tweeted topics can be impressively accurate (Asur & Huberman, 2010). However, bot behaviour may seriously distort Twitter analytics connected with a less common word or hashtag, and therefore reliable bot identification is also important for marketers, politicians, and social scientists using Twitter for research into human opinions and behaviour. Judging by a 24-hour sample, about 10% of all public tweets containing the word *communist* are sent by the entertaining @RedScareBot, for example. A campaign using multiple bots might cause a particularly large distortion.

A body of research has been published on identifying Twitter spambots or other antisocial bots (e.g., Benevenuto, Magno, Rodrigues, & Almeida, 2010; Chu, Widjaja, & Wang, 2012; Gayo-Avello & Brenes, 2010; Lee, Caverlee, Kamath, & Cheng, 2012; Lee, Caverlee, & Webb, 2010; Lee, Eoff, & Caverlee, 2011; Song, Lee, & Kim, 2011; Stringhini, Kruegel, & Vigna, 2010; Wang, 2010; Yang, Harkreader, & Gu, 2011), and a few articles have addressed identifying Twitter bots in general (Chu et al., 2010; Laboreiro, Sarmento, & Oliveira, 2011; Zhang & Paxson,

2011). There is consensus that the way to build a bot-identification algorithm is to use standard machine learning techniques to determine a weighted combination of different measures that provides good bot/non-bot discrimination. Identification algorithms with a reported accuracy of 93% or more have been constructed using this method.

There are five broad classes of measures mentioned in these articles and/or currently used for bot identification:

1. *Spamming method indicators*
   Indicators of spamming methods include the presence of words associated with spam in tweets or profile content; frequent use of hashtags, @mentions, or trending topics, which increase the visibility of tweets; frequent URLs in tweets, which may indicate web page promotion; account age, because 77% of suspended accounts are suspended within a day of their first tweet (Thomas, Grier, Song, & Paxson, 2011); and the *ff ratio*, which is the number of followers divided by the number of friends. More sophisticated versions of these include the number of different URLs tweeted divided by the number of tweets (Lee et al., 2010), and the number of @mentions of non-followers divided by the total number of @mentions (Song et al., 2011). Spamming method indicators are used by various Twitter anti-spam apps (e.g., Emerge2 Digital Inc., 2009–2012; Joi Company, 2009–2010; 97th Floor, 2010) and by Twitter's own anti-spam engine: for instance, Twitter temporarily misidentified @twisst's alert-sending accounts as spambots, because they made many @mentions of non-followers.

   These measures are only designed to identify spambots, and may be useless for identifying other kinds of bot. Moreover, their effectiveness may decrease as spamming techniques evolve. The history of the ff ratio illustrates this. A technique commonly used by spammers and account sellers to obtain followers is to automatically follow many users, some of whom will follow back. This technique used to produce a low ff ratio, until bot designers programmed their software to automatically unfollow those who did not follow back. Twitter banned automated unfollowing in response. The ff ratio now has low discrimination power (Chu et al., 2010). Measures of follower and friend dynamics (Lee et al., 2011) can detect automated unfollowing, but these measures may also lose effectiveness over time.

2. *Social graph measures*
   Some bot-identification measures use properties of accounts' social graphs. The local clustering coefficient of the graph measures how

densely it is connected. The betweeness centrality of an account in its social graph measures whether it appears on many of the shortest paths between its neighbours. These are robust indicators of spambots (Yang et al., 2011), because spambots tend to target socially unconnected users. Similarly, an account contacting users socially distant from it may be suspect (Peri, 2009; Song et al., 2011). Quercia, Capra, & Crowcroft (2012) used a combination of the social graphs' reciprocity and geographic span. Some measures rate accounts as a function of scores of their friends (passivity in Romero, Galuba, Asur, & Huberman, 2010; taste in Chu et al., 2012), others as a function of scores of their followers (e.g., TunkRank (Findable, 2010) and other prestige or influence measures).

Social graph measures can be subverted by bot designers. For instance, spammers consistently beat non-spamming marketers in prestige ratings (Gayo-Avello & Brenes, 2010). Measures relying on differences between bots' and humans' social graphs may be vulnerable to bots that mimic human social patterns by following their followers' social contacts. Although this tactic does not currently appear to be widely used by Twitter bots, this may just be a matter of time, as it has been shown to be highly effective for gaining Facebook friends (Boshmaf, Muslukhov, Beznosov, & Ripeanu, 2011). Influence measures do not distinguish bots that are not influential from the many human accounts that are not influential; a study of 11.5 million accounts found that 21% had never tweeted (Sysomos, Inc., 2009). (Some algorithms used to identify fake followers, e.g., Calzolari, 2012, have the same weakness.)

3. *Automated tweeting indicators*
Measures used to detect algorithmically generated tweet content vary from the simple (percentage of identical tweets) to the sophisticated (format similarity, identifying text, URLs, and numbers, see Laboreiro et al. 2011; similarity modulo synonym substitution, see Chu et al. 2012). Tests for algorithmically generated text can also be applied to profile information to detect accounts auto-registered in bulk for sale and/or antisocial use. Tweet stylistics may be used: Laboreiro et al. (2011) found emotional tweets and incorrect grammar to be good indicators of a human account. The Twitter API reports the client used to send a tweet. Human accounts usually use either the Web or a mobile client, and some clients are used mostly for automated tweets. Regularly sent tweets indicate automated scheduling. Zhang and Paxson (2011)

identified potential bots by a measure of the regularity of the distribution of seconds and hours in tweet timestamps, and manually inspected dozens of these accounts; nearly all showed strong indications of being automated.

Measures based on tweet content can be evaded by communicating through DMs. Some bots that do this do not tweet at all; others tweet text copied from humans, as camouflage. Some spambots copy other users' tweets that contain a URL, but change the URL to one that the bot is promoting. Client information has some discrimination power, however, there are many different clients, not all easily categorisable, and some bot software can send "Web" tweets. TweetAttacks, a bot software vendor, advertised that its bots' automated tweets were "posted via the WEB NOT THE API and it will look like being posted by a REAL HUMAN" (as quoted in *Twitter, Inc. v Skootle Corp. et al.*, 2012, p. 8). Timing regularity measures may not detect bots whose automated tweeting behaviour is triggered by human actions.

4. *Blacklists*
Some Twitter anti-spam apps use a blacklist of accounts, URLs, or DMs shared by users of the app (SocialOomph.com, 2008–2011; Stay N' Alive Productions, 2012; Whitlock, 2009). Shared blacklists are useful for spotting trends and testing algorithms, but for filtering use it can be challenging for them to keep up with the rate of creation of new accounts and message content.

5. *CAPTCHAs*
CAPTCHAs (von Ahn, Blum, Hopper, & Langford, 2003) are automated challenge tests that most humans should pass, but current machines should fail; for example, the challenge may be to identify distorted letters in an image. TrueTwit uses CAPTCHAs to validate Twitter accounts (TrueTwit LLC, 2010), and Twitter requires a CAPTCHA solution for account registration.

Automated solutions have been found for some types of CAPTCHA; ironically, some Web bots have been detected because they were solving CAPTCHAs too quickly to be human. More fundamentally, machines can solve CAPTCHAs by forwarding them to human solvers. The price for a thousand CAPTCHA solutions, solved using human labour in low-wage countries, is a few U.S. dollars (Motoyama et al., 2010). Variants of the Koobface malware entice or scare the owners of infected machines into solving CAPTCHAs for free (Vaas, 2007).

SUMMARY OF BOT IDENTIFICATION

In summary, it appears that all currently known bot-identification measures could be bamboozled by a determined bot designer. However, several bot-identification algorithms have been reported to have good accuracy. Different measures require different evasion techniques, so a good approach is to use a combination of a large number of measures in different classes (except perhaps for CAPTCHAs, because they annoy humans). To counter bot design evolution, the algorithm should be regularly updated by re-running machine learning on recent data, and by adding new measures if they are found. The TWASE engine (Chivers & Hampson, 2011; in beta) takes this general approach, but is focussed on identifying spambots, rather than any bots.

NON-DECEPTIVE BOT DESIGN

In Mowbray (2002), I advocated that ethical bot designers should avoid making their bots deliberately deceptive. More recently, in Matwyshyn & Mowbray (2012), Andrea Matwyshyn and I have suggested mandatory labelling for bots, to reduce deception and for owner accountability, by analogy with dog ownership laws. A similar principle has been advocated for physical robots (ESPRC, 2010).

The importance of this requirement depends on the purpose of the deception. For example, it may be ethically justifiable to use a deceptive bot in security research to improve defences against antisocial bots, but not to increase customers' satisfaction with a customer service bot by deceiving them into thinking it is human.

I suggest that owners of Twitter bots should either indicate in their bots' bios and/or screen names that they are in fact bots, or alternatively (for semi-automated accounts and apps that send automated tweets through users' accounts), include a brief indication that messages are automated within the messages themselves. A reviewer of this chapter suggested an official Twitter bot badge, similar to the verified account badge.

Bot designers and owners should also follow Twitter's automation rules and best practices (Twitter, 2012). Among other things, these forbid certain types of content and abusive use, mass creation of automated accounts with overlapping use, buying or selling followers, sending automated @mentions to many users unless the recipients pre-approved them, providing services with no opt-out option, and any auto-following behaviour other than followback. They also discourage automated tweeting to trending topics, and automated retweeting based on keywords.

CONCLUSION

Although this chapter has concentrated on antisocial bots, these are in the minority. There are very many bots on Twitter which are useful, entertaining, or simply delightful. It is my hope that research in this area will help to maintain the ecosystem within which these good bots can thrive.

ACKNOWLEDGMENT

Parts of the content of this chapter previously appeared in Mowbray (2011) and Matwyshyn & Mowbray (2012). My thanks to the ETHICOMP organisers, Andrea Matwyshyn, and the creators of the bots named here. This chapter is dedicated to Harriet Hill, Olivia Hill, and Keely Taylor-Hall.

REFERENCES

97th Floor. (2010). Clean tweets. Retrieved from https://addons.mozilla.org/en-US/firefox/addon/clean-tweets/

Asur, S., & Huberman, B. (2010). Predicting the future with social media. *Proceedings of the 2010 IEEE/WIC/ACM International Conference on Web Intelligence and Intelligent Agent Technology* (pp. 492–499). doi:10.1109/WI-IAT.2010.63

Benevenuto, F., Magno, G., Rodrigues, T., & Almeida, V. (2010, July). *Detecting spammers on Twitter.* Paper presented at Collaboration, Electronic Messaging, Anti-abuse and Spam (CEAS) Conference, Redmond, WA. Retrieved from http://ceas.cc/2010/papers/Paper%2021.pdf

Boshmaf, Y., Muslukhov, I., Beznosov, K., & Ripeanu, M. (2011, Dec.). *The SocialBot network: When bots socialize for fame and money.* Paper presented at the Annual Computer Security Applications Conference, Orlando, FL. Retrieved from http://lersse-dl.ece.ubc.ca/record/272

Calzolari, M. C. (2012, 8 June). Analysis of Twitter followers of leading international companies. Retrieved from http://www.camisanicalzolari.com/MCC-Twitter-ENG.pdf

Cellan-Jones, R. (2009, 23 June). Things that tweet. Retrieved from http://www.bbc.co.uk/blogs/technology/2009/06/things_that_tweet.html

Chivers, J., & Hampson, M. (2011). TWASE: An anti-spam engine for Twitter. Retrieved from http://twase.com/

Chu, Z., Gianvecchio, S., Wang, H., & Jajodia, S. (2010, Dec.). *Who is tweeting on Twitter: Human, bot, or cyborg?* Paper presented at the Annual Computer Security Applications Conference, Austin, TX. doi: 10.1145/1920261.1920265

Chu, Z., Widjaja, I., & Wang, H. (2012, June). *Detecting social spam campaigns on Twitter.* Paper presented at the 10th International Conference on Applied Cryptography and Network Security, Singapore. doi: 10.1007/978-3-642-31284-7_27

Coconutsfine. (2009, 9 Mar.). The people I've been friends with for ages on Twitter were bots [in Japanese]. Retrieved from http://d.hatena.ne.jp/coconutsfine/20090309/1236611519

Emerge2 Digital Inc. (2009–2012). TwitSweeperTM. Retreived from http://www.twitsweeper.com

Epstein, R. (2007). From Russia, with love. *Scientific American MIND, 18*, 16–17.

ESPRC. (2010). Principles of robotics. Retrieved from http://www.epsrc.ac.uk/ourportfolio/themes/engineering/activities/Pages/principlesofrobotics.aspx

eureka, Inc. (2011). OKITE. Retrieved from http://itunes.apple.com/us/app/okite/id457818344

Fahner, C. (2009, 12 Oct.). #garden [Video file]. Retrieved from http://vimeo.com/7036697

Findable, LLC. (2010). TunkRank. Retrieved from http://www.tunkrank.com

Gayo-Avello, D., & Brenes, D. J. (2010, June). *Overcoming spammers in Twitter—A tale of five algorithms.* Paper presented at Congreso Español de Recuperación de Información (CERI 2010), Madrid, Spain. Retrieved from http://ir.ii.uam.es/ceri2010/papers/ceri2010-gayo-avello.pdf

Green, Z. (2012, 23 July). Is Mitt Romney buying Twitter followers? Retrieved from http://140elect.com/2012-twitter-politics/is-mitt-romney-buying-twitter-followers/

InstructablesTV. (2011, 5 Jan.). *Tweet-a-pot* [Video file]. Retrieved from http://www.youtube.com/watch?v=_Y-F9Zdk_qM

Joi Company. (2009–2010). StopTweet. Retrieved from http://stoptweet.com

Laboreiro, G., Sarmento, L., & Oliveira, E. (2011, Oct.). *Identifying automatic posting systems in microblogs.* Paper presented at Progress in Artificial Intelligence Conference, Lisbon, Portugal. doi: 10.1007/978-3-642-24769-9_46

Lee, K., Caverlee, J., Kamath, K. Y., & Cheng, Z. (2012, April). *Detecting collective attention spam.* Paper presented at WebQuality conference, Lyon, France. doi: 10.1145/2184305.2184316

Lee, K., Caverlee, J., & Webb, S. (2010, July). *Uncovering social spammers: Social honeypots + machine learning.* Paper presented at ACM SIGIR Conference on Research and Development in Information Retrieval, Geneva, Switzerland. doi:10.1145/1835449.183552

Lee, K., Eoff, B. D., & Caverlee, J., (2011, July). *Seven months with the devils: A long-term study of content polluters on Twitter.* Paper presented at AAAI International Conference on Weblogs and Social Media (ICWSM '11), Barcelona, Spain. Retrieved from http://citeseerx.ist.psu.edu/viewdoc/download?doi=10.1.1.230.412& rep=rep1&type=pdf

Matwyshyn, A., & Mowbray, M. (2012, 24 Mar.). Bot or not?: Digital augmentation and personhood. Presentation at Internet Law Work-in-Progress, New York Law School, New York, NY.

Motoyama, M., Levchenko, K., Kanich, C., McCoy, D., Voelker, G. M., & Savage, S. (2010, Aug.). *Re: CAPTCHAs—Understanding CAPTCHA-solving services in an economic context.* Paper presented at the Nineteenth USENIX Security Symposium, Washington, DC. Retrieved from http://static.usenix.org/event/sec10/tech/full_papers/Motoyama.pdf

Mowbray, M. (2011, September). *A rice cooker wants to be my friend on Twitter.* Paper presented at ETHICOMP, Sheffield, UK. Retrieved from http://www.hpl.hp.com/techreports/2011/HPL-2011-175R1.pdf

Mowbray, M. (2002). Ethics for bots. In I. Smit & G. E. Lasker (Eds.), *Cognitive, emotive and ethical aspects of decision making and human action* (pp. 24–28). Baden-Baden, Germany: International Institute for Advanced Studies in Systems Research and Cybernetics.

Peri, C. (2009, 15 Mar.). Friendfilter: Replace your Twitter email with a scored new follower email. Retrieved from http://www.perivision.net/wordpress/2009/03/friendfilter-replace-your-twitter-email-with-a-scored-new-follower-email/

Quercia, D., Capra, L., & Crowcroft, J. (2012, June). *The social world of Twitter: Topics, geography, and emotions.* Paper presented at the 6th International AAAI Conference on Weblogs and Social Media (ICWSM '12), Dublin, Ireland. Retrieved from http://www.aaai.org/ocs/index.php/ICWSM/ICWSM12/paper/download/4612/4996

Ratkiewicz, J., Conover, M., Meiss, M., Gonçalves, B., Patil, S., Flammini, A., & Menczer, F. (2011, Mar.–Apr.). *Truthy: Mapping the spread of astroturf in microblog streams.* Paper presented at World Wide Web Conference (WWW 2011), Hyderabad, India. doi: 10.1145/1963192.1963301

Romero, D. M., Galuba, W., Asur, S., & Huberman, B. (2010), Influence and passivity in social media. Retrieved from http://papers.ssrn.com/sol3/papers.cfm?abstract_id=1653135

Santiesteban, I. (2012). Goodbye bots. Retrieved from http://santiesteban.org/adiosbots/en.html

Slavin, K. (2009,16 June). *Things that twitter at the 140 character conference—Day 1* [Video file]. Retrieved from http://www.blip.tv/file/2267845

SocialOomph.com. (2008–2011). SocialOomph. Retrieved from http://www.socialoomph.com

Song, J., Lee, S., and Kim, J. (2011, September). Spam filtering in Twitter using sender-receiver relationship. Paper presented at the 14th International Symposium on Recent Advances in Intrusion Detection (RAID), Menlo Park, CA. doi: 10.1007/978-3-642-23644-0_16

Stay N' Alive Productions. (2012). SocialToo. Retrieved from http://www.socialtoo.com

Stringhini, G., Kruegel, C., & Vigna, G. (2010, December). *Detecting spammers on social networks.* Paper presented at the Annual Computer Security Applications Conference, Austin, TX. Retrieved from http://www.cs.ucsb.edu/~gianluca/papers/socialnet-spam.pdf

Sysomos, Inc. (2009, June). Inside Twitter: An in-depth look inside the Twitter world. Retrieved from http://www.sysomos.com/insidetwitter/

Thomas, K., Grier, C., & Paxson, V. (2012, April). *Adapting social spam infrastructure for political censorship.* Paper presented at the USENIX Workshop on Large-Scale Exploits and Emergent Threats (LEET), San Jose, CA. Retrieved from https://www.usenix.org/system/files/conference/leet12/leet12-final13_0.pdf

Thomas, K., Grier, C., Song, D., & Paxson, V. (2011, November). Suspended accounts in retrospect: An analysis of Twitter spam. Paper presented at the Internet Measurement Conference, Berlin, Germany. doi:10.1145/2068816.2068840

TrueTwit, LLC. (2010). TrueTwit. Retrieved from http://www.truetwit.com/truetwit/welcome/index

Twitter. (2012). Automation rules and best practices. Retrieved from http://support.twitter.com/articles/76915#

Twitter, Inc. v Skootle Corp. et al. (2012). US District Court, Northern District of California, San Francisco Division, Case number 3: 12-cv-0 1721, filed 5 Apr. 2012.

Vaas, L. (2007, 31 Oct.). Striptease used to recruit help in cracking sites. *eWEEK.* Retrieved from http://www.eweek.com

von Ahn, L., Blum, M., Hopper, N. J., & Langford, J. (2003, May). *CAPTCHA: Using hard AI problems for security.* Paper presented at EUROCRYPT, Warsaw, Poland. doi: 10.1007/3-540-39200-9_18

Wang, A. H. (2010, July). *Don't follow me: Spam detection in Twitter.* Paper presented at Security and Cryptography (SECRYPT 2010), Athens, Greece. Retrieved from http://www.personal. psu.edu/hxw164/files/SECRYPT2010_Wang.pdf

Whitlock, T. (2009, 3 Aug.). TwitBlock spam ratings explained. Retrieved from http://web.2point1. com/2009/08/03/twitblock-spam-ratings-explained/

Williams, E. (Interviewee). (2009, August 5). *Newsnight* [Television broadcast]. London, UK: BBC.

Yang, C., Harkreader, R. C., & Gu, G., (2011, September). *Die free or live hard? Empirical evaluation and new design for fighting evolving Twitter spammers.* Paper presented at the 14th International Symposium on Recent Advances in Intrusion Detection, (RAID), Menlo Park, CA. doi: 10.1007/978-3-642-23644-0_17

Zhang, C. M., & Paxson, V. (2011, March). *Detecting and analyzing automated activity on Twitter.* Paper presented at Passive and Active Measurement Conference (PAM 2011), Atlanta, GA. Retrieved from http://pam2011.gatech.edu/papers/pam2011--Zhang.pdf

Information Retrieval for Twitter Data

15
CHAPTER

Ke Tao, Claudia Hauff,
Fabian Abel, Geert-Jan Houben

what is #relevant? searching for specific pieces
of information within millions of daily tweets
is a challenging task for #ir

Broadly speaking, given a corpus of documents and a user's information needs (such as *Who won the 2012 U.S. presidential elections?*), together with a subsequent search request which is submitted to a search engine ("2012 US elections"), information retrieval (IR) is concerned with the efficient retrieval and ranking of the documents in response to the query. A ranking of documents is considered to be of high quality if the top-ranked documents are *relevant*, that is, if they aid the user in answering their information needs.

Traditionally, text-based information retrieval research has focussed on corpora, which—though diverse in the type of documents (such as news articles, Web pages, or patents)—have a number of commonalities:

1. Each document consists of at least a few hundred words,
2. The content is (mostly) correctly spelled and grammatically sound, and
3. The information contained in each document is expressed in multiple ways.

These factors are one reason for the success of IR techniques for text corpora which is evident in search engines such as Google and Bing.

In the case of microblogs such as Twitter, the need for an effective search engine to find and filter relevant messages is evident, as Twitter receives more than a billion queries a day (Twitter Engineering Blog, 2011). Considering this from an IR perspective, each Twitter message can be viewed as a document and for each user query, all messages that were created up to that point in time can then be considered for retrieval.

In this chapter, we first describe what makes Twitter different from established document corpora. Based on that analysis, we then present a number of IR techniques that have been shown to perform well in this context. The last part of this chapter discusses event detection and analysis: an important use case of Twitter, where users react in real time to events occurring in the world around them. This reaction can be analysed and exploited for a range of applications, such as public safety.

WHAT MAKES TWITTER SPECIAL?

Although IR techniques are already used in many scenarios, its characteristics make Twitter different when applying these techniques. The main reason for this is the limited length of tweets. Further, users exhibit behaviours during their information searches on microblogging sites which differ markedly from established search behaviours elsewhere on the Web (Teevan, Ramage, & Morris, 2011).

We distinguish three categories of user activity in which these differences are most pronounced: user behaviour, posting style, and search style.

USER BEHAVIOUR

Due to the limited length of tweets, and the ease of posting them, users show behavioural characteristics in Twitter usage which differ from other text-based platforms such as personal blogs. Kwak, Lee, Park, and Moon (2010) investigated the intentions of Twitter users worldwide and found that Twitter resembles much more a news media platform than a social network, since 85% of tweets are related to news. This makes Twitter a significant source of information about emerging events. However, Twitter is quite different from traditional media since its users act as "social sensors" (Sakaki, Okazaki, & Matsuo, 2010) who can provide first-hand information on various aspects of news events.

With hundreds of millions of tweets being generated every day, IR techniques are needed to enable users to fulfil their information needs when searching.

POSTING STYLE

Users rely on different methods to adhere to the 140-character limit while conveying their intended meaning and emotions:

- Users tend to use abbreviations, remove vowels, drop articles, or use acronyms (Gouws, Metzler, Cai, & Hovy, 2011).
- Interjections and long sequences of repeated letters are widely used to express emotions.
- Hashtags are used both to engage in a discussion on a particular topic and to promote oneself (Laniado & Mika, 2010).
- Mentions are used when explicitly notifying others and when replying to others.
- URL shortening services are used to save characters when users include links. Such shortening removes valuable information such as the URL's domain name (often used as quality indicator).

SEARCH STYLE

In addition to these platform-specific practices in posting information, users' behaviours in searching for information on Twitter also differ from established Web searching practices. For example, Teevan et al. (2011) revealed that on Twitter, people typically use an average of 1.64 words in their search queries, while on the Web they use 3.08 words. This can be explained by Twitter's 140 characters per message limitation: since long keyword queries easily become too restrictive, people tend to use broader and fewer keywords for searching. The queries issued to Twitter also contain more references to people. In addition, it has been observed that people repeatedly search on Twitter using the same queries in order to monitor new(s) content, such as developments in popular news stories and upcoming events.

WHAT ARE IMPORTANT INFORMATION RETRIEVAL TECHNIQUES FOR TWITTER DATA?

Since Twitter content is different from the documents traditionally used in IR, researchers have encountered several problems when applying existing techniques to Twitter-based search and retrieval tasks. In the following, we first introduce the standard Twitter benchmark used in academic research—the so-called TREC Microblog Track (NIST, 2011). We then explain the most important IR techniques that researchers have adopted for microblog corpora.

TREC MICROBLOG

In 2011, the Text REtrieval Conference (TREC)—an annual conference where IR researchers and practitioners come together and evaluate their algorithms on common data sets—introduced a new benchmark based on Twitter, the Microblog Track. The goal of the benchmark is to investigate the prevalent search tasks on Twitter, and the evaluation methodologies applicable to these tasks. In 2011, the search task was defined as follows: given a keyword query Q and the query's timestamp, retrieve the *interesting and relevant* tweets for Q that are at least as old as Q (to simulate Twitter's streaming nature). The corpus used for this benchmark contains sixteen million tweets that cover a duration of two weeks starting from 24 January 2011. The corpus was derived by sampling Twitter in order to create "a reusable, representative sample of the Twitter sphere".

Fifty search topics were released for the corpus in 2011 for the real-time, *ad hoc* task. For each search query (an example being "Jintao visits US"), a ranked list (in time-descending order) of tweets was expected as retrieval result. Ideally, all of the retrieved tweets are relevant and interesting with respect to the query.

A variety of information retrieval techniques may be used to address such challenges. The following are descriptions of some of these techniques.

NAMED ENTITY RECOGNITION (NER)

Given the limited length of tweets, and the posting styles that this limitation generates, information which would be valuable for retrieval purposes is rarely available in a tweet's surface form. The challenge is, then, how to extract useful semantics from the tweet.

To tackle this problem we make use of Named Entity Recognition (NER) services, which semantically enrich the tweet's content by linking terms and phrases to knowledge bases.

The same procedure can be applied to the query. NER services can be utilised to identify names and their synonyms, as well as to expand abbreviations from text snippets, where the snippets can be queries or tweets. One example from the TREC 2011 Microblog Track is the personal name "Jintao", which refers to the former President of the People's Republic of China. However, in tweets he

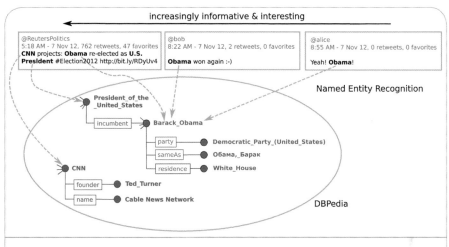

Figure 15.1: Tweets Are Semantically Enriched by Linking Recognised Entities to the Linked Open Data Cloud (here: DBpedia)

may also be referred to as "President Hu" or "Chinese President". If these variants of a person's name and titles are considered when searching the corpus, a wider variety of tweets can be found. Once NER has been applied, the text snippets will be annotated with named entities as well as the complete names.

EMBEDDING INTO THE LINKED OPEN DATA (LOD) CLOUD

The LOD cloud (Bizer, Heath, Berners-Lee, & Hausenblas, 2010) integrates various data sets that are available under open licenses in order to take advantage of knowledge across different fields. In particular, DBpedia (Mendes, Jakob, Garcia-Silva, & Bizer, 2011) is one of the key interlinking hubs in this cloud. Once a query or tweet term is coupled to an entity in the LOD cloud (Figure 1), additional information can be extracted from this cloud by following the relationships between different entities in the entity graph. Consequently, a greater understanding of both the tweets and the queries can be gained. It should be noted, though, that such embedding is only useful when the NER services perform with a high degree of accuracy—erroneous entity assignments will harm the search effectiveness considerably.

QUERY AND DOCUMENT EXPANSION

In addition to applying NER, there is another way to extract additional, meaningful information from tweets. Since a large percentage of tweets contains hyperlinks that may point to more complete information—such as news reports, blog entries, bulletin board posts, etc.—it is possible to leverage the information acquired by following the URLs and by performing information extraction on these documents. It has also been found that tweets containing URLs are more likely to be relevant for search queries, and to be interesting to the users searching for information (Tao, Abel, Hauff, & Houben, 2012a).

FEATURE COMBINATIONS

Due to the high variance in the content quality of tweets (i.e., the amount of useful information in them), it is not sufficient to only consider algorithms that compute scores which indicate how closely the tweet text matches the search query. Rather, a more accurate measure of the relevance of a given tweet may be derived from a combination of features, such as the tweet-query similarity, tweet content quality, tweet language, and user trustworthiness. In combining these individual scores into one overall score, various weighting parameters can be applied in order to represent the relative importance of each feature.

For example, in the learning to rank (LTR) framework (Liu, 2009), the ranking model is learnt from a number of training examples where each example consists of a user query and a ranked list of tweets with the most relevant tweets ranked at the top (Metzler & Cai, 2011). The LTR approach is not the only machine learning approach, however; logistic regression has also been successfully employed to evaluate the influence of different features on the learnt model and their impact on the ability to detect interesting and relevant tweets (Naveed, Gottron, Kunegis, & Alhadi, 2011; Tao et al., 2012a; Tao, Abel, Hauff, & Houben, 2012b). In contrast to LTR, logistic regression is a classification approach which does not produce a ranking, but a set of relevant tweets. Overall, the challenge is to develop a useful and diverse set of features, as well as to accurately estimate the weighting parameters.

NOVELTY AND DIVERSITY

Due to the high similarity between tweets that report or comment on popular and breaking news, and the heavy usage of retweets in this context, the diversification of search results on Twitter is especially meaningful when aiming to supply users with information about a certain topic from a variety of angles. One of the problems in diversification is how to measure the similarity between tweets—in other words, the extent to which a pair of tweets can be considered as duplicates of each other. Gadiraju (2012) proposed a six-level measurement to define the duplicity scale, and outlined several strategies for the extraction of various sources of evidence for determining the duplicity between tweet pairs. Although the diversification of search results has also been extensively studied in Web search (Agrawal, Gollapudi, Halverson, & Ieong, 2009; Drosou & Pitoura, 2010), corresponding research into the diversification of Twitter search results is still lacking.

EVENT DETECTION AND ANALYSIS

IR techniques are not only used in search and retrieval applications, they are also employed in specific scenarios, such as (real-time) event detection and analysis in the stream of Twitter messages for the purpose of public safety. Twitter exhibits substantial spikes in activity as major events unfold; examples are (natural) disasters (Earle, Bowden, & Guy, 2011; Iyengar, Finin, & Joshi, 2011; see also Chapter 28 by Bruns & Burgess, in this volume); the spread of diseases (Culotta, 2010; Sadilek, Kautz, & Silenzio, 2012); or political (Hu, John,Wang, & Kambhampati, 2012), entertainment (Benson, Haghighi, & Barzilay, 2011; Iyengar et al., 2011), or sports events (Nichols, Mahmud, & Drews, 2012). Here, users act as social sensors (Sakaki et al., 2010), and their tweets can be considered as imperfect sensory information.

One of the first studies which explored the possibility of real-time disaster detection on Twitter was presented by Sakaki et al. (2010), who detected earthquakes and notified users immediately. Due to the speed of detection, users who live 100 km away from the epicentre of an earthquake have a twenty-second time window after notification before the earthquake arrives. A later study (Earle et al., 2012) found that this result only holds for strong earthquakes originating or passing through densely populated areas. According to Sakaki et al. (2010), events whose development can be observed and analysed based on tweets need to fulfil at least the following attributes:

1. The events are large in scale (to generate enough tweets for analysis),
2. The events influence users in some way (users have an incentive to tweet about this influence on their lives), and
3. The events have spatial and/or temporal boundaries.

To provide an example for the amount of data that can be generated in response to an important event, consider the first presidential TV debate between Barack Obama and Republican candidate Mitt Romney on 3 October 2012: within a time span of ninety minutes, more than ten million tweets had been posted (Twitter Blog, 2012).

Research in this area usually focusses on one of three major elements in the event detection and analysis pipeline:

1. Detection of the type or types of events (e.g., earthquakes),
2. Detection of events of the specified type in the Twitter stream (as well as of their sub-events and boundaries), and
3. Analysis of the tweets associated with the events identified.

Each element of the pipeline can be processed either manually or automatically. The type of event to be detected is usually a choice made by the researcher. Only very recently have the first studies appeared which can address any type of event (Jackoway, Samet, & Sankaranarayanan, 2011), or are able to automatically determine event types based on tweets (Ritter, Mausam, Etzioni, & Clark, 2012). In cases where the detection of particular events (step 2) is not the focus of the work, simple filters based on tweet content, user location, and time are used. Finally, step 3 often involves qualitative as well as quantitative approaches.

In the following discussion, we outline the technical challenges and the design decisions that need to be considered with respect to the underlying, specific research questions.

TIME OF ANALYSIS

The analysis of the tweets can occur in *real time*, that is at the time of the event's occurrence, or *post-event*, that is after the event has occurred. Real-time analysis is particularly useful as a monitoring tool to recognise potential dangers and to warn people or alert the authorities (Abel, Hauff, Houben, Stronkman, & Tao, 2012). Post-event analysis is often employed to learn more about the dynamics of events (for example, how diseases spread), and as a learning tool for organisations and governments. Depending on the time of analysis, different types of algorithms have to be considered. A real-time analysis needs to process a stream of tweets (at each point in time, new tweets are posted and the algorithm merges the new information with existing information), whereas a post-event analysis can use more traditional approaches, as all tweets are considered to be known at once.

DATA COLLECTION

Since Twitter does not provide complete access to its message archive, a design decision has to be made with respect to the data collection process (see also Chapter 5 by Gaffney & Puschmann, in this volume). Through Twitter's

Streaming API, tweets with particular characteristics—specified manually—can be collected over time. Depending on the event type investigated, different methods have been used in the past: a stream of tweets that contain particular keywords or hashtags—for example, "grass fire" (Vieweg, Hughes, Starbird, & Palen, 2010), "#patriots" (Chakrabarti & Punera, 2011), "earthquake" (Earle et al., 2012)—or tweets that have been posted by users within a particular geographic area, or all tweets by a set of pre-defined Twitter users. Jackoway et al. (2011) proposed an approach to future event detection through which potential future events (described by their time frame and a list of keywords) are extracted from news articles; once such future events are extracted, the data collection process can be tuned so that the tweets which are subsequently collected are likely to be about the desired event.

EVENT-RELATED TWEETS

Even using such targeted data collection approaches, not all of the tweets posted at the time or location of an event are going to be related to it. Thus, it is necessary to classify each tweet as being or not being about the event under investigation. Classification can be done manually (Vieweg et al., 2010), though this is only feasible for a small amount of tweets. It can also be based on filters that remove tweets which are retweets, or are directed at another user and contain URLs (Earle et al., 2012)—such types of tweets are assumed not to refer to the event in question. More complex, and often more successful, are classification schemes that employ machine learning to determine for each tweet whether it belongs to an event (e.g., Sakaki et al., 2010). While this type of classification is common, it may also be necessary to distinguish between the roles of users who are tweeting: are these users eyewitnesses, or are they removed from the situation and simply commenting on it based on information gathered from other sources? Finally, we note that it has also been shown that the tweet behaviour in response to events differs according to the *user type*; in particular, De Choudhury, Diakopoulos, & Naaman (2012) investigated the different tweeting behaviours of individuals, journalists, and organisations.

CONCLUSION

We have outlined a number of IR techniques that are particularly useful when dealing with Twitter data. We first described the differences between Twitter and more traditional document corpora, and then presented strategies that allow us to compensate for the peculiarities of tweets. We then described the type of search algorithms that are successful on Twitter data, and finally outlined the specific IR challenges for the case of event detection and analysis.

Usage of Twitter is growing continuously: it is a portal which is used by people across the globe who have different information needs, different ways of expressing themselves, and different views on the same events. This diversity makes the utilisation of Twitter data in search and retrieval applications difficult, but at the same time the ubiquitous nature of Twitter, and its widespread use, allow us unprecedented insights into people's day-to-day activities.

REFERENCES

Abel, F., Hauff, C., Houben, G. J., Stronkman, R., & Tao, K. (2012). Semantics+Filtering+Search =Twitcident. Exploring information in social Web streams. In *Proceedings of the 23rd ACM Conference on Hypertext and Social Media* (pp. 285–294), Milwaukee, WI.

Agrawal, R., Gollapudi, S., Halverson, A., & Ieong, S. (2009). Diversifying search results. In *Proceedings of the 2nd ACM International Conference on Web Search and Data Mining* (pp. 5–14), Barcelona, Spain.

Benson, E., Haghighi, A., & Barzilay, R. (2011). Event discovery in social media feeds. In *Proceedings of the 49th Annual Meeting of the Association for Computational Linguistics: Human Language Technologies—Volume 1* (pp. 389–398).

Bizer, C., Heath, T., Berners-Lee, T., & Hausenblas, M. (Eds.). (2010). *Proceedings of the WWW2010 Workshop on Linked Data on the Web*, Raleigh, NC.

Chakrabarti, D., & Punera, K. (2011). Event summarization using tweets. In *Proceedings of the Fifth International AAAI Conference on Weblogs and Social Media* (pp. 66–73), Barcelona, Spain.

Culotta, A. (2010). Towards detecting influenza epidemics by analyzing Twitter messages. In *Proceedings of the 1st Workshop on Social Media Analytics* (pp. 115–122), Washington, DC.

De Choudhury, M., Diakopoulos, N., & Naaman, M. (2012). Unfolding the event landscape on Twitter: Classification and exploration of user categories. In *Proceedings of the ACM 2012 Conference on Computer Supported Cooperative Work* (pp. 241–244), Seattle, WA.

Drosou, M., & Pitoura, E. (2010). Search result diversification. *SIGMOD Record, 39*(1), 41–47.

Earle, P., Bowden, D., & Guy, M. (2011). Twitter earthquake detection: Earthquake monitoring in a social world. *Annals of Geophysics, 54*(6), 708–715.

Gadiraju, U. K. (2012). Detection of duplicate content on Twitter (Technical report). Delft, The Netherlands: Delft University of Technology.

Gouws, S., Metzler, D., Cai, C., & Hovy, E. (2011). Contextual bearing on linguistic variation in social media. In *Proceedings of the Workshop on Languages in Social Media* (pp. 20–29), Portland, OR.

Hu, Y., John, A., Wang, F., & Kambhampati, S. (2012). ET-LDA: Joint topic modeling for aligning events and their Twitter feedback. In *Proceedings of the 26th AAAI Conference on Artificial*

Intelligence, Toronto, Canada. Retrieved from http://www.public.asu.edu/~yuhenghu/paper/aaai12.pdf

Iyengar, A., Finin, T., & Joshi, A. (2011). Content-based prediction of temporal boundaries for events in Twitter. In *Proceedings of the Third IEEE International Conference on Social Computing* (pp. 186–191). Retrieved from http://ebiquity.umbc.edu/_file_directory_/papers/576.pdf

Jackoway, A., Samet, H., & Sankaranarayanan, J. (2011). Identification of live news events using Twitter. In *Proceedings of the 3rd ACM SIGSPATIAL International Workshop on Location-Based Social Networks* (pp. 25–32), Chicago, IL.

Kwak, H., Lee, C., Park, H., & Moon, S. (2010). What is Twitter, a social network or a news media? In *Proceedings of the 19th International Conference on World Wide Web* (pp. 591–600), Raleigh, NC.

Laniado, D., & Mika, P. (2010). Making sense of Twitter. In *ISWC '10: Proceedings of the 9th International Semantic Web Conference* (pp. 42–51), Shanghai, China.

Liu, T.-Y. (2009). Learning to rank for information retrieval. *Foundations and Trends in Information Retrieval, 3*(3), 225–331.

Mendes, P. N., Jakob, M., Garcia-Silva, A., & Bizer, C. (2011). DBpedia Spotlight: Shedding light on the web of documents. In *I-Semantics: Proceedings of the 7th International Conference on Semantic Systems* (pp. 1–8), Graz, Austria.

Metzler, D., & Cai, C. (2011). USC/ISI at TREC 2011: Microblog track. In *Proceedings of the 2011 Text REtrieval Conference*, Gaithersburg, MD. Retrieved from http://don-metzler.net/papers/usc-microblog11.pdf

National Institute of Standards and Technology (NIST). (2011, 30 Aug.). Tweets2011. Retrieved from http://trec.nist.gov/data/tweets/

Naveed, N., Gottron, T., Kunegis, J., & Alhadi, A. C. (2011). Bad news t fast: A content-based analysis of interestingness on Twitter. In *Proceedings of the 3rd International Conference on Web Science*. Koblenz, Germany. Retrieved from http://journal.webscience.org/435/1/50_paper.pdf

Nichols, J., Mahmud, J., & Drews, C. (2012). Summarizing sporting events using Twitter. In *Proceedings of the 2012 ACM International Conference on Intelligent User Interfaces* (pp. 189–198), Lisbon, Portugal.

Ritter, A., Mausam, Etzioni, O., & Clark, S. (2012). Open domain event extraction from Twitter. In *Proceedings of the 18th ACM SIGKDD International Conference on Knowledge Discovery and Data Mining* (pp. 1104–1112), Beijing, China.

Sadilek, A., Kautz, H. A., & Silenzio, V. (2012). Predicting disease transmission from geo-tagged micro-blog data. In *Proceedings of the 26th AAAI Conference on Artificial Intelligence*, Toronto, Canada. Retrieved from http://www.cs.rochester.edu/~sadilek/publications/Sadilek-Kautz-Silenzio_Predicting-Disease-Transmission-from-Geo-Tagged-Micro-Blog-Data_AAAI-12.pdf

Sakaki, T., Okazaki, M., & Matsuo, Y. (2010). Earthquake shakes Twitter users: Real-time event detection by social sensors. In *Proceedings of the 19th International Conference on World Wide Web* (pp. 851–860), Raleigh, NC.

Tao, K., Abel, F., Hauff, C., & Houben, G. J. (2012a). Twinder: A search engine for Twitter streams. In *ICWE '12: Proceedings of the 12th International Conference on Web Engineering* (pp. 153–168), Berlin, Germany.

Tao, K., Abel, F., Hauff, C., & Houben, G. J. (2012b). What makes a tweet relevant for a topic? In *Proceedings of the 2nd Workshop on Making Sense of Microposts* (pp. 49–56), Lyon, France.

Teevan, J., Ramage, D., & Morris, M. R. (2011). #TwitterSearch: A comparison of microblog search and Web search. In *Proceedings of the 4th International Conference on Web Search and Web Data Mining* (pp. 35–44), Hong Kong.

Twitter Blog. (2012, 4 Oct.). Dispatch from the Denver debate. Retrieved from http://blog.twitter.com/2012/10/dispatch-from-denver-debate.html

Twitter Engineering Blog. (2011, 31 May). The engineering behind Twitter's new search experience. Retrieved from http://engineering.twitter.com/2011/05/engineering-behind-twitters-new-search.html

Vieweg, S., Hughes, A. L., Starbird, K., & Palen, L. (2010). Microblogging during two natural hazards events: What Twitter may contribute to situational awareness. In *Proceedings of the 28th International Conference on Human Factors in Computing Systems* (pp. 1079–1088), Atlanta, GA.

Documenting Contemporary Society by Preserving Relevant Information from Twitter

16
CHAPTER

Thomas Risse, Wim Peters,
Pierre Senellart, and Diana Maynard

can we preserve tweets and their contexts
and still understand what they mean in 10+
years? #archive

WHY ARCHIVE TWITTER?

In recent years, Twitter has changed from a medium for posting personal updates or status information to a channel for sharing and distributing information of all kinds. Its increasingly pervasive nature is encouraging more and more people to give insights into their daily life and to stay in contact with friends. This also attracts many companies and media agencies, attempting to establish a more or less constant flow of information to their customers. The limitation to 140 characters reduces efforts, and focusses the tweet on the core information. The ease of use of Twitter and its availability on every smartphone also encourages people to act as citizen journalists and immediately report the events they witness. Twitter can thus be seen as the foremost channel for "breaking news", where information about events appears before being distributed via traditional channels. Follow-up messages on Twitter complement news articles with sentiments, opinions, and related information. Prominent examples for citizen journalism are the Arab Spring (Mourtada & Salem, 2011) or the "Miracle on the Hudson" ("Twitter First Off the Mark with Hudson Plane Crash Coverage", 2009).

As a side effect of its active and pervasive usage, Twitter documents contemporary society in rich detail. Tweets give valuable insights into individuals, groups, and organisations, and enable an understanding of the public perception of events, people, products, or companies, including the flow of information. While in the past reports about society were written by individuals and were therefore biased, today Twitter and other Social Web applications create the possibility of a live documentation of our society. It gives unprecedentedly rich and detailed insights into the day-to-day process of public communication. This will allow later generations to understand how topics were spreading, how sentiments and opinions were developing, or to better understand the impact of technological developments like Twitter on the evolution of culture and society as it is possible today.

The long-term preservation of public Twitter content and its accessibility for research is thus becoming a cultural necessity. For short-term usage, the probability that Twitter content remains accessible at Twitter itself can be assumed to be high. In the long-term perspective—meaning more than 10 years—no prediction can be made, as the experience with past popular Internet sites such as GeoCities shows. GeoCities—founded in 1994, bought by Yahoo! in 1999, and closed down in 2009—was a popular Web service for hosting free user homepages. Nowadays, some of these homepages are preserved in a Web archive, thanks to some last-minute crawling activities, while others are lost forever. To avoid such a loss of valuable information for Twitter, capturing its content and preserving it for future generations is necessary.

The aim of the capturing effort for Twitter should be to preserve the content, the presentation, and the social context scope of a tweet. According to Middleton (2012), "social context scoping is a critically important scope because it collects the subject alongside the social commentary for a more complete historical record". The U.S. Library of Congress (LoC) is currently archiving all tweets since Twitter's inception in 2006, but their accessibility is unclear (see also Chapter 13 by Zimmer & Proferes, in this volume). While on the one hand this archive is already a big achievement, on the other hand the access limitations constrain its usability. In addition, the LoC archive only holds the tweets, but not necessarily their social context.

Contextualised Twitter capturing goes beyond the pure collection of tweets. The limitation of 140 characters per tweet forces the poster to be very focussed and brief. Hence, there is little or no room for any introduction or explanation to help understand the tweet. To assist the reader of a tweet in the future, it is necessary to give them more information about its context. The context within Twitter is defined by the person who tweets, the topic defined by the hashtag (if one is present), but also by the answering and re-tweeting chain in which a tweet may participate. In addition, some tweets have links to external pages, which can provide more details about the topic of a tweet. An interesting application that highlights this requirement is Speak2Tweet (Speak2Tweet, 2012), set up in January 2011 during the Egyptian revolution, which allows the tweeting of a URL to voice recordings for those without an Internet connection by making a phone call to a designated number. On the one hand, to simply capture the tweets which contain such links might lead to a loss of highly valuable information about the Arab Spring. On the other hand, following links present in a tweet, and gathering other tweets with the same hashtag or from users @mentioned within a tweet allows preserving a more comprehensive context of the tweet.

For implementing the described contextualised crawl strategy, the European-funded project ARCOMEM (ARchive Communities' MEMories) (ARCOMEM, 2012) follows a two-step philosophy. In the first step, Twitter content is captured and analysed to extract semantic and contextual information. This information triggers in a second step the Web crawler to collect relevant content from the Web.

During the capturing of tweets and the crawling of their context, future usage should be taken into account. Bearing in mind the large number of tweets and related pages generated per day, efficient access mechanisms are mandatory. One means of going beyond standard, full-text search is to enrich each tweet with descriptive meta-information. This meta-information consists of (a) information directly gathered from Twitter (e.g., user, creation date, geolocation) and (b) information extracted from the content, such as topics, events, and sentiments. This meta-information can be used in conjunction with a full-text search to select the appropriate content from the archives.

EXTRACTING INFORMATION FROM TWITTER

To create incrementally enriched Web archives which allow access to all sorts of social media content in a structured and semantically meaningful way, we need to extract relevant information from the tweets (which can point to related information on the Web). Semantic technologies have the potential to help people cope better with social media-induced information overload, by making use of content from the social Web that is relevant to a specified topic, event, or entity that researchers and archivists may be interested in. From this information, we can also identify opinions, and track opinion changes over time, both of which help gauge public interest.

The content-collection process which we describe in the remainder of this chapter, in the form of information-extraction methodologies and crawling techniques and strategies, is under continuous development within the ARCOMEM project. Extraction covers the initial identification and structured representation of knowledge about events and entities from previously unstructured material. This process faces issues arising from diversity in the nature and quality of Web content, in particular when considering social media and user-generated content, where further issues are posed by informal use of language. Since archiving has to consider the evolution of content and metadata over time, temporal and dynamic aspects are of special importance. We aim to extract relevant information from tweets in order to answer questions such as:

- How did people talk about the issue or event?
- How are opinions distributed in relation to demographic user data?
- Who are the most active Twitter users?
- Who are the opinion leaders?
- Where did they come from?
- What did they talk about?
- How has the public opinion on a key person evolved?

ENTITY AND EVENT EXTRACTION

Information extraction from tweets involves the use of natural language processing (NLP) techniques to extract events, entities, and other kinds of information from the (unstructured) text of the tweets. The extracted information can then be used for targeted Web crawling, allowing the crawling strategy to be gradually refined according to some specification of the entities and events. A further challenge is then to make appropriate use of these outcomes to create focussed archives. Recognising occurrences of named entities (such as persons, locations, etc.) within a text can be broken down into two main phases: ontology-based entity annotation (or candidate selection) and entity linking (also called reference disambiguation or entity resolution). This is useful so that the entities extracted can be linked together (co-referenced), even when they appear in different documents, and disambiguated when multiple meanings are possible. For example, the word *Paris* could refer to the entities *Paris, Texas,* or *Paris, France,* or even *Paris Hilton*: we want to ensure that each time it occurs in a tweet, we know which of these it is referring to. We can also then group together all tweets that talk about Paris, Texas separately from those which talk about Paris, France. Ontology-based entity annotation identifies all mentions in the text of classes and instances from an ontology (such as DBpedia. org). The entity-linking step then uses contextual information from the text, as well as knowledge from the ontology, to choose the correct entity, associated with a unique identifier (uniform resource identifier, or *URI* in Semantic Web speak) in the case of ambiguity.

There are many tools and methods for extracting information from text, using both rule-based and statistical techniques. The extraction techniques used in the ARCOMEM project are all developed in GATE (Cunningham, Maynard, Bontcheva, & Tablan, 2002), an architecture for language engineering which contains a number of components for language processing and text mining. The extraction task can be broken down into the following tasks:

- document preprocessing (document format analysis, content detection);
- linguistic preprocessing (language detection, separating the text into words and sentences, annotation with simple grammatical features such as part-of-speech categories (nouns, verbs, etc.), and dependencies between them such as subject and object);
- entity and event recognition (ontology-based lookup, annotation using specific sets of rules, extraction of important terms and phrases, entity linking).

Traditionally, named entities are of the types Person, Location, Organisation, Date, Time, and Money. However, in some cases, we also want to extract entities specific to the domain in question. For example, for tweets about music events (rock concerts and so on), we might want to extract entities such as band names; for political tweets we might want to specifically extract political parties as a subtype of Organisation. Similarly, event types may be dependent on the domain: for example, music festivals have events such as performances, and sub-events such as incidents that happen during a band's performance. Usually, these specific types of event and entity will be predetermined, but there are also possibilities for creating and extracting new types on the fly, using techniques such as clustering of similar examples. And finally, event extraction involves the recognition of domain-important happenings or situations within which entities are related to each other.

OPINION EXTRACTION

Extracted entities and events can be used to drive the extraction of opinions from tweets. It is not enough in this case to simply know whether a tweet is positive or negative in general, but to know what exactly it is positive or negative about. It is thus important to relate the opinion to a target (topic); for example, a tweet may be negative overall (e.g., sadness about the death of a famous person) but positive about the actual person. We therefore use the entities and events as possible targets to which the opinions are anchored. Opinions and sentiments are first gathered at the sentence and word level from text-based documents, based on the recognition of sentiment referring to the entities and events previously identified; more information on how to do this is given by Maynard, Bontcheva, and Rout (2012). Opinions can then be aggregated over wider elements such as whole documents or individual blog posts, and stored along with the individual sentiments.

ISSUES WITH ANALYSING SOCIAL MEDIA

The analysis of tweets is challenging for text-mining systems because of their informal use of language and style. Typically, tweets are rich in abbreviations, slang, domain-specific terms, and spelling and grammatical errors. NLP techniques are usually developed to deal with standard language, and therefore tend to produce lower-quality results on this kind of informal text. For example, shortened or misspelled words increase the variability in the forms for expressing a single concept. One solution to this is the normalisation of text before processing, but this is not possible here because we wish to preserve the content in its original form. For example, misspelled entities need to be recognised as such, but also to be connected with the correctly spelled versions of the same entity. The quality of the text affects not only the actual recognition of entities but also all the linguistic processing components, such as part-of-speech (POS) taggers and so on, mentioned in the previous section. Degraded performance on any of these components may have a negative effect on any other components which rely on these, because they are run in series, with each depending on the results of the next. So the higher up the chain the error, the worse the knock-on effect; in particular, errors in tokenisation and POS tagging can severely hamper the entity and opinion extraction. For preprocessing, we can adopt a number of techniques, such as retraining the components specifically on tweets; using techniques from SMS processing; adding lists of emoticons; recognising artefacts such as smileys, @mentions, and hashtags separately; replacing common abbreviations with their full words (e.g., tnx = thanks); and so on. We can also adopt backoff strategies for dealing with informal text, such as using more flexible grammar rules and additional use of co-reference techniques: see Maynard et al. (2012) for a description and discussion of these.

CONTEXTUALISED TWITTER ARCHIVING

The previous section has mentioned a number of interesting features that can be extracted from the content of Twitter posts and the social networks they exist in. We now explain how to leverage this extracted information in the construction of focussed, contextualised archives of Twitter and Web data. This is done by using these features to guide a Web crawler.

The first step is for a Web archivist to specify the *scope* of the archive, with the scope specification relying on information extracted from Twitter (entities, social context, etc.) in addition to more traditional URL-based features. Once this is done, the archiving process can be launched. In contrast to more traditional Web crawling approaches, archiving Twitter requires using the Web APIs provided by Twitter, rather than conventional Web page crawls. Feedback from information extracted is then used to guide the crawler. We do not stop at capturing Twitter data—it is also important to crawl the content of the Web context of Twitter posts, in particular the URLs that Twitter posts point to. Finally, the archive can be enriched with a more in-depth analysis of its content.

ARCHIVE SCOPE

Traditionally, Web archivists and crawl engineers, when they use an archival Web crawler such as Heritrix (Mohr, Kimpton, Stack, & Ranitovic, 2004) to archive a part of the Web, express the scope of the indented crawl as a *crawl specification*. This is a document specifying a set of seed URLs, from which the crawl should be started, and a description of in-scope URLs, based on a whitelist and blacklist of URL patterns (typically described by regular expressions) and file formats (described by patterns on file extensions or MIME types). Such a specification may, for example, express that, for a given crawl, only resources under the *.gov.uk* domain name hierarchy should be archived, and that only HTML content together with some associated files (scripts, stylesheets, images) should be retrieved, excluding other kinds of content such as videos and PDF documents.

When one moves from regular Web archiving to the archiving of Twitter and associated social web content, this kind of crawl specification becomes too limited to express the scope of the archival process. Instead, in addition to regular URL seeds and URL patterns, an archival specification should consist, on the one hand, of keywords and key phrases relevant to the archive scope (e.g., "barack obama", "U.S. Politics") serving as *seeds* to search for on Twitter, and on the other hand, of a description of which structured entities (e.g., *Barack Obama*) or social network features (e.g., "users from the United States that are opinion leaders") are relevant. Essentially, anything that can be detected by the information-extraction components mentioned above can be added as a filter. All such components come with a *score* (a number between 0 and 1) quantifying relevance to the scope of the archive; this scope is then used to prioritise the crawler.

CAPTURING API CONTENT

Like any other Web site, Twitter can be crawled using a regular archival Web crawler. However, this is not the most efficient way to capture Twitter data, and it is usually preferable to access Twitter using its rich HTTP Application Programming Interface (see https://dev.twitter.com/ for the documentation). Indeed, the regular Web interface, which makes heavy use of AJAX to present information (presenting only a list of 20 or so recent tweets by default, with more being loaded asynchronously as the user scrolls down the Web page) is more cumbersome to use for retrieving content. The API, which provides 200 tweets at a time (for a user's timeline) as structured records of information, offers a wealth of different querying methods, such as *search* to discover tweets containing keywords and key phrases, or *streaming* to get a continuously updated list of tweets on a given topic (see Chapter 5 by Gaffney and Puschmann, in this volume).

The Twitter API restricts the number of requests that can be performed per hour (the precise amount depends on the method used). A Twitter API archiving system needs to be aware of this policy limitation in order to automatically adapt its rate of crawl. For this purpose, we have developed a general Social Web API crawler tool, *API Blender* (Gouriten & Senellart, 2012), that eases the burden of developing API-specific capturing tools that manage authentication, adapt to policy limits, and even transform the specific schema of the information presented by different social networking platforms (Twitter, Google+, Facebook, etc.) into a common, unified schema.

GUIDING THE ARCHIVING

Information extraction and social network analysis components drive the Twitter capture, in conjunction with the archive specification: once tweets are captured (starting with a search from the key phrases), they are analysed as described above (for example, to extract named entities) and their relevance to the crawl is assessed by a *prioritisation* module that decides whether to explore more of their context (social data, retweets, Web links), and determines the ordering of further requests to the API. Archival guidance can also come in a more indirect manner: once a capture has been made and the archive enriched and annotated (see Archive Enrichment below), the archive specification can be refined, and another capture can be launched, to focus more on those parts of the Twitter social network that were judged important.

CRAWLING THE CONTEXT

Building an archive of Social Web content is more than just building an archive of tweets: it is also critical to crawl the Web context of these tweets, in the form of the Web resources referenced in tweets, and possibly neighbouring pages thereof. The Twitter API capturing system thus needs to extract all hyperlinks found in tweets, if deemed relevant to the archive specification, and hand them over to a regular Web crawler. This regular Web crawler, in turn, uses these URLs as seeds, and crawls the corresponding Web content, also applying the scoring and filtering criteria defined by the crawl specification.

Conversely, once the Web crawler encounters the URL of a Twitter user, it makes sense for it to delegate the capture of the corresponding content to API Blender, by transforming the URL into the corresponding API method call.

One technical problem is raised by the common use of URL shorteners (HTTP redirection services that replace a long URL with a shorter one such as http://bit.ly/dG6yFL). Indeed, it is often the case that URLs make use of a chain of shorteners: they use a generic URL shortener in addition to Twitter's own, mandatory shortener http://t.co/. The use of these URL shorteners makes it harder for the information-extraction components to estimate the relevance of a given link to an archive specification, since nothing in the URL indicates its content. The URL must therefore be resolved before assessing it.

ARCHIVE ENRICHMENT

The result of the focussed archiving guided by information extraction described above is an archive that can be further enriched with metadata on attributes such as entities and opinions. (The extraction of these attributes has been described above.) Archive enrichment is an important aspect of social media preservation, because it enhances the quality and usefulness of the content. It enables different perspectives on the data to be encoded and searched, since archive users can search not just by the content of texts but also by the metadata attributes assigned to them. For example, they can investigate particular opinions about certain entities, look for changes to these over time, and perform other complex, information filtering processes, thus inferring new knowledge from the captured, enriched, and contextualised Web content.

ARCHIVE USAGE

Given the fast growth of social media exploitation, it is to be expected that the use of social media in general, and of Twitter content in particular, will rapidly extend to all areas of professional activity where organisations profit from gaining insight into the social repercussions of issues that are closely related to these organisations' interests. Social networks are a rich information source, whose structures can be exploited to acquire knowledge about facts and opinions (Dietze et al., 2012) as well as social connections and interactions (Agichtein, Castillo, Donato, Gionis, & Mishne, 2008). This is recognised by an increasing number of stakeholders in a wide variety of application domains for Web archiving. In addition to their traditional sources (news agencies, PR material, or library content), professionals such as archivists and journalists want support for selecting and archiving relevant, user-generated content from tweets, in order to preserve this ephemeral content, and to enable the retrieval of relevant, tweet-derived source material. Stakeholders with an interest in the appraisal of their products in social media environments, such as media organisations and political actors, will be able to mine and follow societal feedback for short-term purposes. Beyond this, although short-term storage is required when immediate use is important, additional storage strategies are necessary for longer-term preservation.

LONG-TERM PRESERVATION

The long-term usage of Twitter archives, and of Web archives more generally, raises a range of issues. Archived content should be kept accessible and usable well into the future. Also, access to the archive, as well as to a contextual understanding of the content at the time of publication, should be supported.

First challenges arise when the technological development concerns the accessibility and interpretability of the content (CCSDS, 2012). In the worst case, there are no tools available to present the content of the archive in an intelligible form. To avoid this situation, the usage of standards that are supported by a wide range of tools and maintained over time is an obvious necessity. Capturing Twitter results in a substantial number of JSON files (IETF, 2006). JSON (JavaScript Object Notation) is a lightweight, text-based, language-independent, data interchange format standardised by the Internet Engineering Task Force (IETF). The documentation is publicly available and widespread. Therefore, the semantics of the information items within a JSON file are well-documented for future usage.

However, preserving the JSON files alone is not sufficient. There are also many different forms of technical and descriptive meta-information that need to be preserved. The Web ARChive file format (WARC) has been standardised by the International Organisation for Standardisation (ISO, 2009). This choice of format for the long-term preservation of Web resources of all kinds is an accepted standard in the archive community. WARC archives aggregate multiple resources into a single file. Besides the content, it also stores related meta-information.

To ensure the technical accessibility and usability of the archive content is one step for the long-term usage of Twitter archives. As discussed above, individual tweets are limited in their length and contain very little information, which complicates the intelligibility of the content at a later time. A Twitter message such as "The new #ipod is cool http://bit.ly/NWou" will hardly be understandable in 50 years without additional knowledge. It is impossible to predict whether future users of the archive will have information on what an iPod was in 2012. While traditional materials, such as papers or books, often provide enough contextual information to be intelligible, this is rarely the case for user-generated content on the Social Web. Therefore, as much context information as possible—like descriptions of major entities and concepts (such as the concept of a portable media player, in the iPod example)—should be kept together with the tweet. This will not guarantee full intelligibility, but it is an important step in that direction.

We have outlined above how identified entities and concepts could be connected to the linked data cloud, for example by referencing DBpedia. When searching across long-term archives, different instances of a concept such as "portable media player" might occur: for example, Walkman, Watchman, Discman, MP3 Player, iPod. URIs linking to DBpedia or Wikipedia as references to an entity can help to identify information objects with similar semantics. The benefit for the reader is that they can access contextual information in terms of related documents as well as the description of an entity. This ensures the long-term semantic interpretability of the content.

CONCLUSION

Capturing tweets together with their context (if a context link or other context information is provided) allows for a better understanding of individual messages and groups of tweets. Therefore, a comprehensive mechanism for the archiving of Twitter content must consist of capturing API content as well as regular Web crawling, in order to collect both types of information. The enrichment of the captured data enhances the subsequent access and usage of the archives which this mechanism creates.

REFERENCES

Agichtein, E., Castillo, C., Donato, D., Gionis, A., & Mishne, G. (2008). Finding high quality content in social media, with an application to community-based question answering. In *Proceedings of Web Search and Data Mining (WSDM)* (pp.183–194). Stanford, CA: ACM Press.

ARCOMEM. (2012). ARchive Communities' MEMories (ARCOMEM). Retrieved from http://www.arcomem.eu/

CCSDS. (2012). Reference model for an Open Archival Information System (OAIS). Magenta Book. Issue 2 June 2012. Retrieved from http://public.ccsds.org/publications/archive/650x0m2.pdf

Cunningham, H., Maynard, D., Bontcheva, K., & Tablan, V. (2002). GATE: A framework and graphical development environment for robust NLP tools and applications. In *Proceedings of the 40th Anniversary Meeting of the Association for Computational Linguistics (ACL'02)* (pp. 168–175) Philadelphia, PA.

Dietze, S., Maynard, D., Demidova, E., Risse, T., Peters, W., Doka, K., & Stavrakas, Y. (2012). Entity extraction and consolidation for social web content preservation. In *Proceedings of 2nd International Workshop on Semantic Digital Archives (SDA)* (pp. 18–29), Pafos, Cyprus.

Gouriten, G., & Senellart, P. (2012). API Blender: A uniform interface to social platform APIs. In *Proceedings of the 21st World Wide Web Conference (WWW 2012), Developer Track*. Retrieved from http://www2012.wwwconference.org/proceedings/nocompanion/DevTrack_039.pdf

IETF. (2006). The application/json media type for JavaScript Object Notation (JSON). Retrieved from http://www.ietf.org/rfc/rfc4627.txt

ISO. (2009). Information and documentation— The WARC file format (ISO/DIS 28500). Retrieved from http://www.iso.org/iso/iso_catalogue/catalogue_tc/catalogue_detail.htm?csnumber=44717

Maynard, D., Bontcheva, K., & Rout, D. (2012). Challenges in developing opinion mining tools for social media. In *Proceedings of @NLP can u tag #usergeneratedcontent?! Workshop at*

LREC 2012, May 2012, Istanbul, Turkey. Retrieved from http://gate.ac.uk/sale/lrec2012/ugc-workshop/opinion-mining-extended.pdf

Middleton, M. (2012). *Defining Web archive scope*. Retrieved from http://web.hanzoarchives.com/bid/90416/Defining-web-archive-scope

Mohr, G., Kimpton, M., Stack, M., & Ranitovic, I. (2004). Introduction to Heritrix, an archival quality Web crawler. *Proceedings of the 4th International Web Archiving Workshop (IWAW 2004)*. Retrieved from http://www.iwaw.net/04/Mohr.pdf

Mourtada, R., & Salem, F. (2011). Civil movements: The impact of Facebook and Twitter. *Arab Social Media Report*, 1(2). Retrieved from http://www.dsg.ae/En/Publication/Pdf_En/DSG_Arab_Social_Media_Report_No_2.pdf

Speak2Tweet. (2012). https://twitter.com/speak2tweet

Twitter first off the mark with Hudson plane crash coverage. (2009). Retrieved from http://www.editorsWeblog.org/2009/01/19/twitter-first-off-the-mark-with-hudson-plane-crash-coverage

The Perils and Pleasures of Tweeting with Fans

17
CHAPTER   Nancy K. Baym

 #artists and #fans connect in new ways on Twitter, but this is both an opportunity and a challenge

You could see the progress from MySpace to Facebook to Twitter. Everyone just loses their minds at the latest thing, and says "No, this is how you do it." And there's never any sort of consensus. I mean as corrupt and horrible as the old record industry was, at least it was a barely stable way to get the word out about music.

— Gary Waleik, guitarist/singer, Big Dipper

The collapse of the music industry has sort of caused this knee jerk reaction in a lot of musicians. It's like, "Oh my God, I got to Twitter!" "Oh my God, I got to Facebook!" And they think all these things. "I got to be on 27 sites, I got to have my music on every single site." But nobody stops to think, "do I really need to do that?"

— Kate Schutt, jazz singer/guitarist

Musicians have always needed to gain attention in order to build and keep audiences for their recordings and live performances. However, as Waleik says in the opening quote, the consensus about how that attention should be gained and who is responsible for gaining it has disappeared. If artists were able to gain the attention of and be signed to labels, it was the job of the labels, their publicists, and the music press to promote artists and see that they got attention. This division of duties—and relatively centralised press with stringent gatekeepers—left musicians free to disappear for long intervals, emerging periodically to promote their latest albums and tour.

But in 1999, the recording industry reached its peak and began a rapid decline (IFPI, 2009). As file-sharing grew and major labels fumbled the transition to online sales throughout the early 2000s (Goldman, 2010), social media gained currency as a potential fix. The launch of MySpace in 2002 made it possible for musicians and audiences to reach one another directly online. Musicians' ability to take charge of self-presentations through social media is in many ways liberating, but it also has brought pressure to create compelling identities that attract attention. Artists are increasingly expected to spend time seeking attention and building relationships with audiences directly in order to earn income. "It's a lot of work to build a career in the era of digital creativity", wrote Kirsner in his book of advice for creatives:

> But there are huge benefits. . . . The on-going conversation with your audience can be a source of inspiration, motivation, and ideas. It is this powerful new link with the audience that the old power players don't understand. They still live in a world of press releases, flashy billboards in Times Square, and expensive-but-never-changing Web sites. (Kirsner, 2009, p. 4)

Twitter seems ideal for those in careers that are increasingly dependent on audience relationships, including musicians, other kinds of artists, entertainers, brand managers, and public figures. Twitter combines broad reach, a tendency toward interest-driven, "ambient affiliation" (Zappavigna, 2011), easy mobile use, and does it all in epistles of only 140 characters.

Yet Twitter's simplicity belies its challenges, as this chapter will show. As Marwick and boyd (2012a) have argued, Twitter has a site-wide norm of "authenticity" created through both official rhetoric and user practices. Twitter's official developer blog offers explicit advice for musicians using the platform that begins with this:

> For music fans, Twitter is the next best thing to being backstage. And for performers, connecting with your fans in an authentic way is one key to your success. A Twitter connection tells fans how much you appreciate them, and it also enables you to tailor

your messages. **The fact is, Twitter provides more authenticity and creative control than any other online medium**. Tweets come straight from you, and go right to your followers all over the world, in real-time. (Twitter, n.d.)

For all Twitter users, "authenticity" is in tension with other pressures toward self-commodification and away from privacy (Marwick, 2010; Marwick & boyd, 2011a). These tensions are amplified for musicians, who are already engaged in a field which positions authenticity as central to creativity and audience relationships, and views commodification as authenticity's opposite. This chapter examines musicians' perspectives on Twitter, looking at the tensions at play as they decide whether and how to use it.

Musical authenticity is often equated with freedom from industry, especially among independent musicians (Fonarow, 2006). As Hesmondhalgh (2007) explained:

> Romantic conceptions of art in 'Western' societies established the idea that art is at its most special when it represents the original self-expression of a particular author. At one level, this is a mystification, so to set creativity too strongly against commerce— as a great deal of romantic and modernist thought about art did—is silly. [But] it has had the long-term effect of generating very important tensions between creativity and commerce. (p. 20)

Fans identify with musicians because of the felt authenticity of the connection forged through music. As a result, authenticity can paradoxically be an important branding strategy. This is particularly true in indie music, which, as described by Fonarow (2006), is positioned as opposition to the mainstream, connoting "small, personal, and immediate", in contrast to "all that is enormous, distant and unspecialized" (p. 63). However, the use of "authentic" as a branding strategy is by no means limited to independent music, or even to music. Banet-Weiser (2012) described a "transformation of culture of everyday living into brand culture" that "signals a broader shift, from 'authentic' culture to the branding of authenticity" (p. 5). In this culture, "building a brand is about building an affective, authentic relationship with a consumer, one based—just like a *relationship* between two people—on the accumulation of memories, emotions, personal narratives, and expectations" (p. 8).

Despite the importance of authenticity in the artist-audience relationship, "there is no straightforward or intrinsic link between the lives of fans, the meaning of musical texts and the identity of a particular artist" (Negus, 1996, p. 133). Instead, identification, authenticity, and relationship are situated and socially constructed negotiations between performers and audiences (e.g., Marwick & boyd, 2011a; Negus, 1996). Musicians' identity performances on the new stages of social media can enhance the relationship forged through music, but performing authenticity through tweets poses very different challenges from performing it through music. Musicians must make sense of their audiences on Twitter, they must make choices about how to communicate with them there, and they must figure out how to fit Twitter into a communication system comprised of many other media and face-to-face encounters. These understandings and choices can be complicated, contradictory, and confusing.

This chapter draws on interviews about audience interaction that I conducted between 2009 and 2012 with 37 musicians from seven countries and more than a dozen genres. Most were either 'legacy artists' who had been in the business since at least the 1980s, or what Norwegian rock star Sivert Høyem called 'the last generation of analogue musicians', who found audiences in the late 1990s. I also spoke to musicians who got their start after MySpace. (A list of interviewed musicians is in Appendix I.) Everyone quoted and named here consented to being identified. Most, but not all of the musicians I spoke with used Twitter. Their follower counts ranged from 86 to 1,271,783. Most had fewer than 10,000. Some had joined in its first year. Others had joined more recently. Their experiences with and attitudes towards Twitter varied widely. In what follows, I address musicians' understandings of their Twitter audience; of communication on Twitter as broadcasting, listening, being real, and interacting; and of the place of Twitter in their multimodal communication systems.

THE TWITTER AUDIENCE

Senft (2008), Marwick (2010), and Marwick & boyd (2011a, 2011b) have studied micro-celebrities who built audiences for themselves primarily through social media. The musicians I interviewed built audiences primarily through recordings and live performance. While social media audiences are often taken to be more imagined than the audiences addressed in embodied encounters (e.g., Litt, 2012; Marwick & boyd, 2011a), for people with mass-mediated audiences, social media can personalise an otherwise anonymous group. Visible follower lists can help artists understand who pays attention to them. "It was interesting to see who follows you", genre-hopping Afropunk and electronica musician Honeychild Coleman told me. "I think it really tells a lot, because they weren't all people that I knew".

However, an audience on Twitter is not the same as an audience of listeners. Social media present a new kind of audience—one that is neither live in person in front of the performance stage nor the silent unseen people listening to recordings. Musicians with huge Twitter followings may have far fewer listeners and vice versa. Zoë Keating is an ambient solo cellist. For months, Twitter recommended her to all new users as someone to follow. As a result, she has more than a million followers, far surpassing the audience for her music. She does not "put a lot of weight" on her follower count:

> I know that all of those people are not really following me. It's like what it is is it's really a chance for me to win people over. That's what I see it as. It's like here's this medium where I have some number of my fans are here listening and I can talk to them. And then there's this much, much larger chunk of people who are wondering "Who the hell is @zoecello?" And if some of them are drawn in and start listening great, that's awesome. But I'm certainly under no illusions that they are all my fans.

In contrast, Høyem is one of Norway's most popular musicians. Unlike Keating, he can count on earning a living from his music. He has more than 60,000 Facebook followers, but has trouble understanding how to reach his large audience through other sites. When we spoke, he had fewer than 1,000 Twitter followers. We discussed the music site Blip.fm, which he had been using:

> I don't seem to get a lot of listeners on Blip.fm. I don't know how people do that. Probably they communicate with these blips like they do on Twitter. But I don't really get the Twitter thing either. I don't know how that happens.

To some extent, Høyem's difficulty in reaching an audience on Twitter stems from his own discomfort with the medium, to which I will return below. It may also be a matter of whether his listeners are on Twitter. There are many times more music listeners than there are Twitter users. As solo bass player Steve Lawson pointed out:

> There are a whole lot of people who treat the entire internet like Facebook. And they very rarely go outside of it. It's Facebook or Google. So they'll Google something to look it up, use it, back to Facebook. The idea of getting on Twitter to them is like 'Why would I need that? I'm not curious about that. I don't have that, I don't have the need for it. I get that side of what I need for that from Facebook.'

Despite their lack of correspondence, Twitter followers are often conflated with listening audiences by third parties who value "the ability to strategically appeal to broad audiences and retain the attention of others", and thus interpret follower counts as status signs (Marwick & boyd, 2011a, p. 127). For better or worse, the collapse of traditional metrics such as Soundscan sales in the music industry inflate the importance of follower counts. Singer-songwriter Erin McKeown explained:

> These social networks come along and all of the sudden here's this new number . . . and I have heard in the music industry "this is someone good to tour with because

they have x number of followers" or "we're interested in signing you because you've got x number of Facebook fans".... How does that translate into people in the room? ... There's this sort of conversion that doesn't necessarily happen or you can't draw a straight line between this artist has 5000 Facebook followers yet still is only drawing 30 people in this city.... In some ways I've begun to think of it as two different careers. You kind of have your online career where it's like how do you communicate with those fans and what do you do for them and how do you cultivate that interaction? And then there's also do you give a good live show and when are you coming to this city?

As musicians take on this "online career" trying to figure out "how do you communicate with those fans", they have a number of options depending, in part, on whom they imagine their Twitter audiences to be (Marwick & boyd, 2011a), what they imagine those audiences want, and what aspects of themselves they are willing to display to them. The musicians viewed it to varying degrees as a medium in which they can broadcast messages to their audiences, listen to them, be real for them, and engage them in interaction. They have varying levels of comfort with each of these and balance them in different proportions.

BROADCASTING

When Twitter is used as a broadcast medium, it fosters connection by distributing information about shows and releases to large groups of people. This sidesteps the problem of presenting an authentic self that extends beyond the musicians' already-established musical identities. "The group that I'm with now has more of a connection with fans than any other group that I've been in", said Nathan Harold, who plays bass in fun.'s touring band:

A lot of that is due to Twitter. That's really, really big with our fans and therefore with us. It's a really easy way to kind of just let the people know what you're doing, where you're going to be.... You can just send out one tweet and it goes out to hundreds of thousands of people.

Social media have particular power as broadcast media because they do not require additional effort from audience members; musicians can meet them where they are. Said Greta Salpeter of Gold Motel, "people are so busy that if you can't find a way to sneak into their daily routine, they'll miss your show".

D. A. Wallach, the frontman for Chester French (notable for, among other things, having been the first band on Facebook), was an early and avid tweeter who has over a million Twitter followers, but he uses many other social media. He asks his audience questions and strategically answers a percentage of theirs, but for him, Twitter, like blogging, is "more strictly like a broadcast". On Tumblr, which obscures follower counts, he experiences "more of a genuinely democratic discourse". David Lowery, of Camper Van Beethoven and Cracker, also described Twitter as a primarily one-to-many medium. "Twitter works a little more like an entertainment, sort of", he said, "because it's sort of broadcast, kind of like the old days". When we spoke, Lowery was seeking to use Twitter's broadcast power to mobilise fan labour for marketing (Baym & Burnett, 2009; Wikström, 2009):

> You actually enlist certain sort of self-appointed fans, basically to do much of the actual work that bands relied on record labels, and publicists, and other professionals to do. You sort of have this informal—like "get the news out on this", "tell your friends about this", "here's the link to buy tickets, pass it around". . . . That was a lot of the work of managers, publicists, record labels, et cetera, did in the past. And now you have this other kind of exchange going on where you're sort of enlisting people that could do work for you. . . . I've been sitting here . . . with the guy who's doing a couple of my Twitter accounts. And we're discussing what we're trying to get people to do. Rather than just these witty—we've been kind of doing this sort of witty back and forth thing, right, like well, now maybe actually we can enlist people to actually do something for us.

When musicians use Twitter as a broadcast medium, they cleave closely to the pre-Internet model that focusses on centralised information distribution, replacing the labels, management, and press with themselves. But this orientation towards Twitter can limit relational development on the site and run counter to site norms (Marwick, 2010).

LISTENING

Unlike mass media, social media make it as easy for audience members to contact musicians as it is for musicians to reach audiences. This makes Twitter excellent for listening to audiences, a mode of social media engagement that remains under-theorised (Crawford, 2009). Musicians listen to audiences through @replies directed at them, by searching for band and artist names, and by following audience members. "People can talk back to you", said Lowery, "and we do monitor it, and we do look at it". "For the band", said S-Endz, a rapper in the desi band Swami, Facebook is "our main forum for fan feedback", but "Twitter's probably close second to that".

Not all musicians find adequate richness in the messages they receive through social media. Gary Waleik of recently reunited late 1980s–early 1990s

Boston band Big Dipper regrets that such listening has moved from handwritten fan letters to social media:

> Everyone's font is going to look more or less the same…whereas if you write a letter, like I say, it's your handwriting, it's your stationery. Maybe those things aren't really that important, but it seems like when you take even some of the personal touch away from that, the message loses something.

Another drawback is that Twitter calls for new listening skills. Crawford (2009) wrote about what is needed when messages are plentiful, disconnected, and one has no control over who speaks when:

> The act of listening to several (or several hundred) Twitter users requires a kind of dexterity. It demands a capacity to inhabit a stream of multilayered information, often leaping from news updates to a message from a friend experiencing a stressful situation, to information about what a stranger had for lunch, all in the space of seconds. Some will require attention; many can be glimpsed and tuned out. (p. 529)

A fundamental difference between Twitter and its most-used counterpart, Facebook, at the time of these interviews was that Facebook friending was symmetrical, meaning that artists who friended their audience members instead of (or in addition to) running a fan page for themselves, had to follow their audience members. This made listening harder on Facebook than Twitter, where they can listen only when addressed. Twitter allows artists to listen selectively, offering what Stephen Mason of Jars of Clay called "a more controlled conversation". "The reciprocality of Facebook can be kind of annoying sometimes", said S-Endz. "So Twitter I think balances that out because the follower system is just a lot more friendly for that kind of thing". Most of the musicians I spoke with did not follow most or all of their audience back. Some, such as singer-songwriter Kristin Hersh, read and responded to @replies but, at the time of our interview, followed only two accounts:

> I'm not quite cut out for it. I like the note passing aspect of Twitter, I think people can be very entertaining . . . but I don't really have my head wrapped around the noise of it. I'm not a chatty person and I'm not a chatty listener either.

Others, like Mark Kelly of the first direct fan-funded band Marillion, followed some audience members back on Twitter, but "not very many".

BEING REAL

Twitter's official suggestions for musicians as well as the emergent norms of its users create a "presumption of personal authenticity and connection" (Marwick

& boyd, 2011a, p. 129) that poses particular problems for musicians. "This relationship with your audience, warned Kirsner (2009, p. 5), "cannot be faked". For some, this comes easily on Twitter, and does not feel like a conflict with commodification, creativity, or personal boundaries. For others, interacting with audiences in the way Twitter seems to demand feels inherently fake. Mason loves Twitter, saying it has been "amazing" in the way it provides "more access than, I think, historically any fan has ever been given to a band". In comparison to other sites, he describes Twitter as "the most vibrant and interactive". French-American singer-songwriter Sydney Wayser described herself as "very open about talking to my fans" on Twitter and Facebook:

> I feel like I wouldn't be here if they weren't interested in my music so for me to kind of snub them and not give them any time of day I feel like is counterproductive. I want insight, I want to know what they think and I want to build a relationship with them.

For some musicians, Twitter may even feel perfect. Lawson described the musician Rosanne Cash, who tweets many times a day, as "just like me, in that . . . her personality was on hold, waiting for Twitter to arrive, you know that she was like, 'at last, this Internet thing has been invented that fits me'".

Roger O'Donnell, former keyboardist with The Cure, described MySpace as "a bit of a wasteland" and Facebook artist pages as difficult to use, but described himself as "quite busy on Twitter and less so on everything else". O'Donnell enjoys the freedom to "be himself" on Twitter, but is not sure that kind of authenticity serves his audience:

> The trouble with Twitter is I think of it as like being at the pub, and a subject comes up, and you're standing there with three or four mates, and you're just like, 'oh, blah, blah, blah'. And then you realize there's hundreds of people reading it, and also it comes up in Google. But I haven't started filtering it. I'm still just being myself, and I don't think I'll change. So if you don't like it, if you don't like me, then that's who I am, so I've always been very honest. I've never been fake to people. So Twitter is very raw and very honest.

For O'Donnell, realness is found in speaking about the topics that he finds engaging with the people who raise them, rather than in identifying a whole audience and speaking always and only to that whole. For many artists, authenticity is tied to the topics they discuss. O'Donnell's contemporary, the more reserved singer-songwriter Lloyd Cole, would never describe himself as "raw and very honest" on social media, but he did describe Twitter as an environment where he shared aspects of himself beyond music, in this case his golf fandom: "I do tend to go on rants about FIFA and golf, which I think is potentially of no interest whatsoever to some of my fan community. . . . I don't talk about golf on Facebook."

Sports fandom is one way to perform a middle ground that will be understood as authentic self-expression but allow for musicians' privacy. McKeown used it strategically in the face of felt pressure to provide "fresh personal content":

> MySpace was less about status updates and more about just making music available in your player and kind of like collecting friends. But then Twitter and Facebook—the microblogging aspect of them kind of demanded fresh personal content, and I have certainly felt the pressure to keep up with that and that is often at odds for me with the amount of things that I'm willing to talk about with the three or four thousand people who follow me.

While some musicians embraced Twitter as a site of self-expression, and others strategically managed topics in ways that navigated between self-expression and privacy, several remained at a loss for how to communicate on the site. Schutt said:

> I'm an inveterate letter writer, I've never stopped writing letters, which is part of the reason why I hate Twitter and Facebook. . . . I'm just analog in that way. . . . I don't know where I stand. I view it as a necessary evil. That's me. Now, I have friends who are, you know, incredible with this stuff. They're on that thing all the time, retweeting, you know, hashtagging this, et cetera. And great, more power to them. I think that's awesome. If they can do that and still be great musicians, then they're better than I am. But I have a problem. I just can't do it. I don't know what to say. Everything that I say sounds dumb or sounds like a brag and I don't like to brag. And I don't really know what my fans get out of it. I'm a fan of millions of artists and I don't really want to read their Twitter stream.

Musicians often view the kind of frequent, phatic communication seen on Twitter as too trivial. Zappavigna (2011, p. 803) argued that this common criticism misses the point that "Twitter offers a medium for expressing a personal evaluation to a large body of listeners with which one can affiliate ambiently". However, if one is not experiencing that ambient affiliation, the deluge of small moments may be too much. The ska musician Chris Murray described his introduction to Twitter:

> I had limited exposure to it, but my understanding was people were posting like "Having lunch at this place," you know, just kind of like thinking out loud online. It was a lot of noise and I'm like, you know, I don't have time for that. I didn't want to know where someone is or what they had for lunch or somebody was rude to them on the street, or, you know.

Høyem was particularly vociferous:

> It's just this manic stream of just trivial information that all these celebrities and artists are just spewing out. And that's really what I see Twitter is about, just manic communication and information, and most of it is just bullshit, and I don't want to be part of it . . . . I get really amazed about how just that whole thing has changed about how artists relate to the public, how trivial it is, or how utterly mindless it is—just mindless information. It's just information for the sake of information.

For some, the performance of mundane sharing that ambient affinities require may counter their sense of themselves, making it impossible to be "more authentic" and have "more creative control" in the way Twitter and its enthusiasts say they should.

INTERACTING

Thus far I have presented Twitter as a medium through which musicians may broadcast, listen, and perform identities for their audiences. They also use it to have reciprocal interactions, to respond to one another's messages, and to create opportunities for interaction in person and in other media. Twitter's @reply design makes it very simple for artists to respond when fans talk to them. Most musicians I interviewed read @replies. This leads to the challenge of figuring out whether and how much to respond to members of their audience. Said Keating:

> Initially that was one of the things I liked about Twitter—that I can actually respond to more volume of messages via Twitter than I could with e-mail. . . . Now people they expect you to reply to them. They expect you to respond to their tweets. It's not like "oh my God, she actually wrote back." It's like "of course you wrote back."

Several musicians I spoke with used Twitter to facilitate in-person interactions, especially while touring. Keating met followers who had come to her concerts after the show. Political folk-rock singer-songwriter Billy Bragg used Twitter to find and join local protesters while on tour. Electronica star Richie Hawtin found eating companions:

> Before a show I might post and ask where people are. Like if I'm playing Korea, I might ask "where is everyone" and someone will say "they're eating salted squid next to the venue." So I'll go there and try to meet some people. I travel so much that if I didn't reach out and make connections with people there it would all be a blur.

This increased availability, expectation of response, and willingness to engage have offered musicians new, rewarding audience connections, but have also made them more vulnerable. "I think Twitter has changed a few things, and Facebook's changed a few things", said avid tweeter Brian Travers of UK reggae band UB40, who have sold more than 70 million records. "You can kind of be subject to some kind of crazy people and that could get to you. In the past, there was a kind of—I suppose a kind of barrier between you and the public". As Travers explained, such encounters are not new, but now happen "a little bit more because of the ease of contact—with Twitter especially".

Musicians also connect Twitter to other online sites of audience interaction. Some online linking is done automatically with services that cross-post messages to multiple sites. For some, this is what makes it feasible for them to use Twitter. Michael Timmins of alt-country band Cowboy Junkies said:

> If I post a blog on my site, it automatically goes up on our Facebook page, so that keeps the Facebook page active, and I just push a button, and it goes into our Twitter feed. MySpace we have to copy and paste and bring it over there, and also they've kind of fallen behind in terms of keeping up with the technology.

"I can make a status change and it can update MySpace, Facebook and Twitter and itself in this one action", said Murray, "doing it all myself, it gets a little bit arduous in cases where when I have a new show, just to post event details".

In other cases, artists tweet in ways that send their audience members to other sites, including sites where they can buy the music. Steve Lawson is an enthusiastic user of Bandcamp, a publishing platform for bands that allows them to present, stream, and sell their music direct to fans at prices they set themselves. "The combination of Twitter and Bandcamp is really quite potent", said Lawson:

> I can be chatting about things and go, 'oh by the way a new album on BandCamp' and all of a sudden I guess over the next two weeks I have people buying what I recorded last week. It's that lack of friction between the place where the narrative is being told, which is Twitter, and the way of people getting contact with that narrative, or the immediate contact to what it is we're talking about. That relationship really is really potent.

"I try to occasionally make pushes to get people from Twitter to sign up on my mailing list", said Keating, "because I always want to own my fan base".

CONCLUSION: REMEDIATION AND THE UNFORESEEN

These interviews reveal Twitter as part of multiplex relationships (e.g., Haythornthwaite, 2005), understood through a process of remediation (Bolter & Grusin, 1999) in which old and concurrent media are used to understand new ones. "The internet is particularly versatile and all-encompassing in its remediation of all that has come before", wrote Fornäs, Klein, Ladendorf, Sundén, and

Sveningsson (2002), "this way of using methods and forms from letters, books, telephones, records, radio, film, TV, and other media types is often motivated by a wish to overcome the inherent limitations of these other forms" (pp. 13–14). The musicians I have quoted understand Twitter in light of MySpace, Facebook, Tumblr, and Blip.fm, as well as older media such as letters and news, and live performance situations. They also understand Twitter in terms of an historic recording industry that no longer works as it once did. For musicians—indeed, for us all—Twitter is embedded within a complex system of multiple communication sites, each of which has its own history and forms its own backdrop for making sense of Twitter.

The interviews also demonstrate that "new forms of mediation and distribution do not simply replace existing practices", as Negus (1996, p. 97) wrote, they "set up new (often completely unforeseen) musical relationships and activities". In this case, the direct access provided through social media like Twitter has created new demand for a kind of "authentic" self-expression and interpersonal relationship in the service of commodification and branding that was not required in the earlier music industry. For some, the preferences and skills honed in other media and with other audiences translate easily to meet these demands; for others they do not. Creating this kind of "authentic" self is highly situated through performances with audiences real and imagined in a particular technology. It requires specific kinds of practice which may not be in line with anything a musician—or anyone else—would ever do elsewhere. Presenting an "authentic" self on Twitter may thus be oxymoronic for some whose real selves would only discuss music with their fans (like Høyem), or who equate authenticity with a kind of intimacy they reserve for friends and family (like McKeown). For most, presenting the kind of self that Twitter desires is neither simple nor transparent; it is real work as they struggle to make sense of the medium, to learn how it works for others and for themselves, and to strategise ways of using it with which they are comfortable.

This chapter has looked at the case of musicians, arguing in part that they are a special case for whom the tensions between authenticity, self-branding, and creativity are enhanced. Yet, the challenges they face affect all social media users. Understanding Twitter—or other social media—requires that we step back to observe how the medium is positioned relative to its alternatives, how it fits with those alternatives—both currently and over time—and how it meshes with the everyday practices of those who use (or do not use) it. Twitter bears performance demands that are constructed through its forms and its emergent norms of use. It is anything but easy.

REFERENCES

Banet-Weiser, S. (2012). *Authentic™: The politics of ambivalence in a brand culture*. New York, NY: New York University Press.

Baym, N. K., & Burnett, R. (2009). Amateur experts: International fan labour in Swedish independent music. *International Journal of Cultural Studies, 12*(5), 433–449.

Bolter, J. D., & Grusin, R. (1999). *Remediation: Understanding new media*. Cambridge, MA: MIT Press.

Crawford, K. (2009). Following you: Disciplines of listening in social media. *Continuum, 23*(4), 525–535.

Fonarow, W. (2006). *Empire of dirt*. Middletown, CT: Wesleyan University Press.

Fornäs, J., Klein, K., Ladendorf, M., Sundén, J., & Sveningsson, M. (Eds.). (2002). *Digital borderlands: Cultural studies of identity and interactivity on the Internet*. New York, NY: Peter Lang.

Goldman, D. (2010, 3 Feb.). Music's lost decade: Sales cut in half. *CNNMoney.com*. Retrieved from http://money.cnn.com/2010/02/02/news/companies/napster_music_industry/?hpt=P1

Haythornthwaite, C. (2005). Social networks and Internet connectivity effects. *Information, Communication, & Society, 8*(2), 125–147.

Hesmondhalgh, D. (2007). *The cultural industries* (2nd ed.). Los Angeles, CA: Sage.

IFPI. (2009). *Recorded music sales 2008*. The International Federation of the Phonographic Industry, London. Retrieved from http://www.ifpi.org/content/library/Recorded-Music-Sales-2008.pdf

Kirsner, S. (2009). *Fans, friends and followers*. Boston, MA: Scott Kirsner/CinemaTech Books.

Litt, E. (2012). Knock, knock. Who's there? The imagined audience. *Journal of Broadcasting and Electronic Media, 56*(3), 330–345. doi: 10.1080/08838151.2012.705195

Marwick, A. (2010). *Status update: Celebrity, publicity and self-branding in Web 2.0* (Unpublished doctoral dissertation). New York University, New York, NY.

Marwick, A., & boyd, d. (2011a). I tweet honestly, I tweet passionately: Twitter users, context collapse, and the imagined audience. *New Media & Society, 13*(1), 114–133.

Marwick, A., & boyd, d. (2011b). To see and be seen: Celebrity practice on Twitter. *Convergence, 17*(2), 139–158. doi: 10.1177/1354856510394539

Negus, K. (1996). *Popular music in theory*. Middletown, CT: Wesleyan University Press.

Senft, T. (2008). *Camgirls: Celebrity and community in the age of social networks*. New York, NY: Peter Lang.

Twitter. (n.d.). *Twitter for musicians and artists*. Retrieved from https://dev.twitter.com/media/music

Wikström, P. (2009). *The music industry: Music in the cloud*. Cambridge, UK: Polity Press.

Zappavigna, M. (2011). Ambient affiliation: A linguistic perspective on Twitter. *New Media & Society, 13*(5), 788–806.

APPENDIX: MUSICIANS INTERVIEWED

Ahmed Best (Cosmic Ghetto/STOMP!), United States

Johan Angergård (Club 8/Legends/Acid House Kings), Sweden

Anonymous Drummer, United States

Billy Bragg, United Kingdom

Stuart Braithwaite (Mogwai), United Kingdom

Rick Bull (Deepchild), Australia/Germany

Lloyd Cole, United Kingdom

Honeychild Coleman (Apollo Heights/Pollen), United States

Jonas Fårm (Starlet), Sweden

Jon Ginoli (Pansy Division), United States

Nathan Harold (fun.), United States

Richie Hawtin (Plastikman), Canada

Kristin Hersh (Throwing Muses/50 Foot Wave), United States

Sivert Høyem, Norway

Zoë Keating, United States

Mark Kelly (Marillion), United Kingdom

Gustaf Kjellvander (The Fine Arts Showcase), Sweden

Steve Lawson, United Kingdom

Rickard Lindgren (Hell on Wheels), Sweden

David Lowery (Camper Van Beethoven/Cracker), United States

Erin McKeown, United States

Stephen Mason (Jars of Clay), United States

Chris Murray, United States

Roger O'Donnell (ex-The Cure), United Kingdom

S-Endz (Swami), United Kingdom

Greta Salpeter (Gold Motel), United States

Kate Schutt, United States

Jonathan Segel (Camper Van Beethoven), United States

Thomas Seltzer (Turbonegro), Norway

Jill Sobule, United States

Sindre Solen (Obliteration/Nekromantheon), Norway

Michael Timmins (Cowboy Junkies), Canada

Brian Travers (UB40), United Kingdom

Nacho Vegas, Spain

D. A. Wallach (Chester French), United States

Gary Waleik (Big Dipper), United States

Sydney Wayser, United States

Tweeting about the Telly
Live TV, Audiences, and Social Media

18
CHAPTER   Stephen Harrington

enjoy being part of the #crowd—how Twitter lifts
the veil and makes TV's mass audience visible

*While the audience may have fragmented, the mass audience still exists
for those events that bind us together in space and time. Exploiting the
continuing power of this immediacy will be the future of commercial
television.*

—Herd, 2012, p. 313

A significant amount of attention in the media industry over the last decade
has been directed towards the idea of 'convergence'. Centrally, this has included
technological convergence, where what was once a series of discrete forms of
media content for discrete devices (e.g., a CD player can only play music, or a
game console is able only to play games) has been replaced by media content
which is accessible across a range of technologies (e.g., online videos which
may be viewed on the computer, game console, smartphone, or smart TV).

Television content has been a major part of this trend. Just over a decade ago, the only way to see your favourite television series was to watch it when your local broadcaster chose to schedule it (if at all, and perhaps months after it was available in other territories), or via a VHS recording. Now, of course, things are significantly different. Leaving aside the illegal (e.g., BitTorrent) and 'illegal but largely unpunished' (e.g., *YouTube*) recordings of television shows that viewers can easily access, many broadcasters around the world are providing viewers—perhaps to partially counter the illicit options now on offer—with access to a digital library of television content in their own time (through catch-up services such as *Hulu*, *iPlayer*, or *iView*). In changing the way content is distributed, these services have therefore radically changed how, where, and through what technologies television can be experienced, leading some to suggest that we have entered into a "post-broadcast" (Turner & Tay, 2009), or "post-network" (Lotz, 2009) era.

Revolutionary as this format shift from broadcasting to data-streaming is, such proclamations may be quite premature. Convergence, and the era of 'Web 2.0', have not seen a complete inversion of the pre-Internet power dynamics between audiences and 'big media'. It is certain that technological advances have opened up a wealth of possibilities for access to existing content—as well as the tools for amateur, DIY, or "produser" (Bruns, 2008) production and distribution—but history shows that new technologies rarely result in the *displacement* of long-standing audience practices, but are typically blended into existing routines and activities instead. Gray and Lotz (2012, p. 3), for that reason, rightly argued that television "is neither 'beating' nor 'losing' to new media in some sort of cosmic clash of technology", and, in fact, the 'old' and the 'new' have formed (and continue to develop) a complex, symbiotic relationship. Thus, it may be fruitful to move past the tired question of whether or not the forces of convergence *interrupt* or *replace* traditional television viewing patterns, and look instead at where (and how) they *enhance* them: in particular, where the television audience experience meets the new channels of "mass conversation" (Spurgeon, 2008) provided by social media, and the impact of their combined use.

In this chapter, I therefore consider the new television audience formations, activities, and experiences that are enabled by Twitter (and, by implication, other social media sites and applications). I will focus on the growing popularity of the service as a centralised, global platform that facilitates and extends conversations about television, and the many potential implications of that massively enhanced connectivity. I will also note how the television industry has responded to these developments by seeking out new ways of engaging in, and promoting, this conversation, and conclude by considering how all of this may change the way we approach audience research. In short, what are the consequences of audiences being given—through social media—an ability to converse with viewers well beyond their own lounge rooms, in real time? How might this new arrangement of viewer connectivity (however incrementally) change the way audiences watch and interact with television? And, how might the industry, and those who study it, respond to this particularly prominent example of what Burgess (2011, p. 314) calls audiences' "new visibility and publicness"?

TELEVISION AND THE SOCIAL

On 14 June 2010, approximately 20,000 football fans gathered at Darling Harbour, on the fringe of Sydney's city centre. What made this gathering unusual is that this was 2:00 a.m., the outside temperature was near freezing, and the local team was nearly half a world away. Those fans were there simply to watch TV: as their national team—the Socceroos—faced Germany in the first round of that year's FIFA World Cup (see Chambers, 2010). Although also arguably a function of the social dimension of sport fandom, there is perhaps no better demonstration of the social power of television than the fact that 20,000 Sydneysiders decided to forego the proximity, warmth, and other *accoutrements* of their own lounge rooms (where they could have watched the game for free anyway), and opted instead to watch the very same feed with a crowd mostly comprised of total strangers. It is a manifestation of the fact that television can be enhanced when experienced alongside others. It shows us that, contrary to the popular rhetoric that foregrounds isolation and antisocial behaviour, television is (and always has been) a highly *social* medium that can bring family members or even total strangers together around a shared point of interest (Lemish, 1982; Morley, 1986), and that viewers will often take steps to seek out a greater sense of community among their fellow audience members (see Fiske, 1992).

Television has occupied a large place in the daily lives of most people in the Western world for nearly 60 years It "has achieved a comprehensiveness of appeal and reach never before surpassed nor likely to be in the future" (Livingstone, 2004, p. 76), and arguably remains "the principal channel of communication in the public space between the state and the home" (Curran, 1997, p. 193) in the early 21st century. Much attention in cultural and media studies since the 1980s has attempted to demonstrate that, like other popular media, television is not just a static 'message' or product with "miraculous powers" (Morley, 1993, p. 13), but a cultural resource which audiences can interpret in a number of different ways (e.g., Fiske, 1987; Morley, 1980), "poach" for active re-use and development in social relationships (Jenkins, 1992), or "remix" and share with the wider world for reasons of pleasure or politics (Lessig, 2008). As such, it is no surprise that television is a major focus of audience activity in the digital realm as well.

TWITTER AS A VIRTUAL LOUNGE ROOM

The Internet is our friend, not our enemy. . . . People want to be attached to each other. (Leslie Moonves, Chief Executive of CBS, as quoted in Stelter, 2010)

Undoubtedly, the single most important aspect of Twitter's relationship with television concerns the opportunities the platform affords users for connecting with other viewers in real time, and engaging in a live, effectively unmediated, *communal* discussion of television programs. There already is a lot of evidence which demonstrates that these opportunities are being seized by an increasing number of viewers. One body of evidence is the range of surveys which have begun to quantify the use of social media as a 'second screen' during television viewing, including one recent study which suggested that over 60% now use social media while watching TV, with that figure growing rapidly (Ericsson, 2012). It should be cautiously noted, however, that in that particular study, users were asked about their use of social media while simultaneously watching television, and did not necessarily relate to their use of social media to *discuss what they are watching* on television. However, a second and more visible sign of the growing phenomenon is the high frequency with which hashtags or keywords that relate to television programs appear in the site's official list of 'Trending Topics' (cf. Deller, 2011, p. 225). And, finally, there is the independent research that has begun to more closely examine those hashtags and keywords, which has highlighted their uptake. For example, some 2000 tweets per minute used the #royalwedding hashtag on 29 April 2011, as millions of people around the world watched Kate Middleton marry Prince William (Bruns, 2011), while the 2012 Eurovision Song Contest saw 688,255 "#eurovision" tweets from 271,826 unique users over the course of the Europe-wide semi-final and final broadcasts (Highfield, Harrington, & Bruns, 2013).

By utilising these hashtags, and thus connecting with a group of Twitter users that extends far beyond (but is still inclusive of) one's existing followees, audience members in this online community coalesce for the specific purpose of discussing the same piece of broadcast content. Each user can engage in a running commentary or conversation on what they are seeing (be it sarcastic praise, educated guesses about what will happen next, shock about a particular plot twist, or mundane thoughts), as it happens on their television screen. Of course, audiences have always engaged in this kind of activity, but the discussion was generally very limited by physical space, or the limitations of one-to-one communication technologies (e.g., the telephone). However, the new, 'many-to-many' opportunities of social media change the landscape quite dramatically by overcoming those pre-existing limitations, creating what Harrington, Highfield, and Bruns (2012) called a "virtual loungeroom": an online space where an audience can commune and centrally share the television experience.

Given my earlier observations about the social pleasures of television, it is not hard to see why Twitter has grown quite rapidly around viewing habits. But there are still limitations to these mutual benefits, in that viewers will not necessarily settle on one single hashtag (audiences may instead be splintered across a number of them), and not every TV show has the power to attract a 'critical mass' of viewers from the demographics most likely to participate in online discussion. Indeed, the largest and most fervent hashtags for television programs tend to relate to live sports and other transnational events (such as the Oscars), and, to a lesser extent, first-run, prime-time comedy or drama. More mundane programming such as daytime soaps, or well-worn reruns of *Everybody Loves Raymond*, barely make a mark at all.

ADAPTING TO CHANGE

> Television networks can no longer assume that the production of quality content will be a distinguishing or determining factor in what programming audiences choose, as news networks often assumed in the oligopolistic era of limited viewing choices. Now, they must deliberately craft intensive *relationships* with viewers, and formulate connections that will encourage routine and repeated viewing. (Jones, 2012, p. 152)

Plagued by uncertainty and racked with worry regarding the "post-network" era of shrinking and fragmenting audiences, television networks—which have realised in recent years that they can no longer rely purely on audience share, but need to cultivate audience relationships as well (see Jenkins, 2006)—have been quick to respond to the new opportunities served up by their viewers' simultaneous use of social media. These responses can be loosely divided into three different activities.

The first concerns their overt attempts to promote and centralise the social media conversation by publicising 'formal' hashtags for a range of different programs. They do so by reminding viewers ahead of time (often through their own official accounts) about what hashtag to use, or by displaying them on screen during the broadcast and prompting viewers to 'join the conversation'. This allows the conversation to quickly centralise, rather than to converge slowly as hashtags are established and evolve (although not always without resistance from audiences). In some cases (though particularly for pre-recorded programs), specific 'conversation starters' are published to liven up the online debate, and perhaps to encourage more people to get involved and share their opinions. Here, the networks are trying to create a more *dialogical* relationship with viewers, which itself helps to create a sense of 'live-ness', even in a show which may have been recorded many months prior to being screened.

The second major response by networks has been to use Twitter not just as a backchannel, but as a return channel: where tweets from viewers are then featured in the show itself. One prominent example from Australia is the Australian Broadcasting Corporation's political discussion program *Q&A*, where a panel of politicians or other leaders discuss and debate questions asked either by the in-studio audience, or via the show's website. During the show, screened live every Monday evening, viewers are encouraged to tweet their responses using the #qanda hashtag, with a very small selection of the "best" tweets getting published on screen throughout the show. Such an arrangement not only makes the most of the discussion that was already occurring on Twitter, but brings more people into that conversation by explicitly promoting it, and presenting the obvious lure of (potentially) having one's comments broadcast on live national television.

RE-SYNCHRONISING

Q&A therefore brings us to the third and final dimension of television's co-development with social media: its increasing attention towards specific formats and shows which will not just catalyse the Twitter backchannel, but which are built around social media to such an extent that live viewership is almost a prerequisite for the full viewing 'experience'. Television-related Twitter conversation exists only fleetingly, and requires a 'live' viewing audience to bring it into being, thus the capacity to participate in that real-time second screen conversation is eliminated by the time-shifting or on-demand playback services discussed at the start of this chapter. By implication, the 'added value' of engaging with social media content alongside the broadcast may help to reinstate the televisual "flow" which Williams (1975) famously argued was central to the television-viewing experience. Just as many expected digital media to break down the power of the networks, social media—by virtue of their ability to create a 'live' experience, even for first-run, pre-recorded television—may have returned a small but significant amount of power back to those who decide what gets shown, and when.

Similar arguments have been made by Wood and Baughman (2012), who examined the long-term use of Twitter by more 'hardcore' fans of the television program *Glee*. Rather than just examining the way that people *talked about Glee* via social media in a 'live' sense, they studied the highly sophisticated use of Twitter to maintain an active and ongoing fan community for that television show. Such use included the role-playing activity that some fans engage in by setting up and maintaining user accounts for the fictional characters on the show. Even then, however, they also argued (cf. Harrington et al., 2012) that such activity can help in the re-formation of the 'live' viewing audience, because these fans will tweet using such fake accounts as the show is being broadcast, and other fans will want to follow them as this is happening, and again participate in this real-time Twitter conversation.

Wood and Baughman (2012) went on to argue, however, that the productive fans' 'unpaid labour' helps to market the show, and gives renewed value to the advertising and other "continuity" (Hartley, 1989, p. 147) embedded in the show's broadcast, and is therefore exploitative. Such a conclusion is problematic because it immediately drags a study of quite sophisticated fan activity back into the highly simplistic 'surveillance' and/or 'exploited labour' paradigm which has become fashionable within the (critical) study of social media and/ or 'Web 2.0' in recent years (see Gauntlett, 2011, ch. 8, for a useful discussion of this perspective). I would instead suggest that the biggest potential consequence of a reforming live television audience does not concern the ethics of contemporary media production dynamics (or the supposed lack thereof), but the broader social consequences. If (and this is still a very big *if*) social media can give renewed emphasis to live television—what one writer recently called "the last remaining civic common in an atomized world" (Carr, 2012)—then one has to wonder whether social media, which have been criticised for creating social disunity, may thereby be defragmenting the public sphere. If live TV can perform a "secular ritual of community building", by linking communities, or an entire nation, together in a single moment of interest (Hartley, 2000, p. 157), then it seems one ought to be at least a little positive, rather than pessimistic, about the consequences for the public sphere of the intersection of Twitter and television.

CONCLUSION: DIGITAL ETHNOGRAPHY?

Ever since notions of the 'active' audience became firmly entrenched in media studies several decades ago (Morley, 1993; Turner, 1990), television has been well understood as a medium that readily catalyses audience discussion, interaction, fandom, and other social activity. Over time, that social activity has evolved, alongside changing cultural and political conditions, and the new affordances created by rapid developments in technology. Social media—and, in particular, Twitter—are therefore not so much a rival for television, but often a complementary technology which has opened up a new space for audience-audience, and audience-text interactivity. The industry has already begun to adapt to this new (if not radically altered) environment, but the potential impact of Twitter (particularly in its potential reformation of the live viewing audience) here extends well beyond audience and industry alone. I therefore want to finish this chapter by briefly considering such an impact specifically in regard to the academic study of media audiences.

In *Television, Audiences and Cultural Studies*, Morley (1992) argued that television-audience research needed to do a better job of accounting for the "complexities" of television viewership (p. 173), and should therefore move towards an ethnographic approach which can provide the rich detail that interviews and focus groups cannot offer. More than 20 years since, Morley's goal has not been achieved, for two main reasons: first, television is embedded into our everyday lives in such a complex and multifaceted way that detailed observation and description would be extremely labour-intensive, severely limiting the potential extent of its application, and thus the reliability of the data it generates. Second, the 'observer effect' means that the sheer act of conspicuously observing a person or a group of people in a research capacity will inevitably affect the activities that they undertake, as "television watching is a touchy subject, precisely because of its association with a lack of education, with idleness and unemployment, and its identification as an 'addiction' of women and children" (Seiter, 1990, p. 64).

In order to achieve the 'ethnographic' goal, and understand the "everyday, affective practices through which [television] is experienced" (Burgess, 2011, p. 314), then, we need systems that allow us to understand the meaning-making processes that audiences engage in, even while they are not necessarily conscious of their being 'observed' by a researcher. Catherine Salmon, in the foreword to *A Billion Wicked Thoughts* (Ogas & Gaddam, 2011), made a similar argument, noting that surveys and questionnaires often force participants to disclose things they may be reluctant to reveal, and therefore "unobtrusive measures that don't require people to *actively participate* in the process of data collection" (p. x) can be far more effective in understanding people's true feelings and attitudes.

Twitter therefore represents a phenomenal opportunity. It provides us with exactly such a means of inconspicuously observing the activities of television audiences. We are left with "material traces that are left of the practice of sense-making" (McKee, 2003, p. 15), which are created 'organically', as the research participants are generally unaware of their own participation in a research study. There does, of course, exist a performative dimension to the act of tweeting, but that is not significantly different from the performance of 'audiencehood' in other research contexts (such as a focus group), and is likely to have far less of an impact than the social and cultural connotations that are broadly attached to academics undertaking (capital R) 'Research' (see Seiter, 1990). Perhaps most excitingly, we now have the capacity to efficiently collect enormous amounts of these data, at relatively minimal cost and effort, about how people react to TV *in real time* (once again, highlighting the importance of 'live-ness' within Twitter), rather than many weeks or months after a particular program goes to air, as is usually the case. The data can be analysed both quantitatively and qualitatively, and provide rich detail about who is saying what, where, when, and to whom.

In spite of the obvious possibilities to overcome the time and financial constraints that so many academics work under, the long-term challenge will be to move beyond the methodological innovations, and the *possibilities* of using data gathered via Twitter, and to get to a position where this is a firmly entrenched and widely accepted (and *respected*) form of audience research. Achieving that goal will first require us to further develop the analytical processes to more comprehensively understand the large amounts of data we gather, and to reconcile (or, indeed, 'converge') those new techniques with the existing theories through which we have conceptualised the way that television is experienced by audiences in their everyday lives.

REFERENCES

Bruns, A. (2008). *Blogs, Wikipedia, Second Life, and beyond: From production to produsage.* New York, NY: Peter Lang.

Bruns, A. (2011). Twitter and the royal wedding, Pt. 1: Something processed. *Mapping Online Publics.* Retrieved from http://mappingonlinepublics.net/2011/08/12/twitter-and-the-royal-wedding-pt-1-something-processed/

Burgess, J. (2011). User-created content and everyday cultural practice: Lessons from YouTube. In J. Bennett & N. Strange (Eds.), *Television as digital media* (pp. 311–331). Durham, NC: Duke University Press.

Carr, D. (2012). TV debates that sell more than just drama. *The New York Times*. Retrieved from http://www.nytimes.com/2012/10/15/business/media/televised-debates-that-sell-more-than-just-drama.html?_r=3&

Chambers, G. (2010, 14 June). No joy for frozen Socceroos fans at FIFA Fan Fest. *The Daily Telegraph*. Retrieved from http://www.dailytelegraph.com.au/news/socceroos-fans-cop-blast-from-germans/story-e6freuy9-1225879296184

Curran, J. (1997). Television journalism: Theory and practice. In P. Holland (Ed.), *The television handbook* (pp. 193–201). London, UK: Routledge.

Deller, R. (2011). Twittering on: Audience research and participation using Twitter. *Participations: Journal of Audience and Reception Studies, 8*(1). Retrieved from http://www.participations.org/Volume 8/Issue 1/deller.htm

Ericsson. (2012). TV and video: An analysis of evolving consumer habits. Retrieved from http://www.ericsson.com/res/docs/2012/consumerlab/tv_video_consumerlab_report.pdf

Fiske, J. (1987). *Television culture*. London, UK: Routledge.

Fiske, J. (1992). Audiencing: A cultural studies approach to watching television. *Poetics, 21*(4), 345–359.

Gauntlett, D. (2011). *Making is connecting: The social meaning of creativity, from DIY and knitting to YouTube and Web 2.0*. Cambridge, UK: Polity Press.

Gray, J., & Lotz, A. D. (2012). *Television studies*. Cambridge, UK: Polity Press.

Harrington, S., Highfield, T., & Bruns, A. (2012). More than a backchannel: Twitter and television. In J. M. Noguera (Ed.), *Audience interactivity and participation* (pp. 13–17). Brussels, Belgium: COST Action ISO906 Transforming Audiences, Transforming Societies.

Hartley, J. (1989). Continuous pleasures in marginal places: TV, continuity and the construction of communities. In J. Tulloch & G. Turner (Eds.), *Australian television: Programs, pleasures and politics* (pp. 139–157). Sydney, Australia: Allen & Unwin.

Hartley, J. (2000). The television live event: From the 'Wandering Booby' to the 'death of history.' In G. Turner & S. Cunningham (Eds.), *The Australian TV book* (pp. 155–169). Sydney, Australia: Allen & Unwin.

Herd, N. (2012). *Networking: Commercial television in Australia*. Sydney, Australia: Currency House.

Highfield, T., Harrington, S., & Bruns, A. (2013). Twitter as a technology for audiencing and fandom: The #Eurovision phenomenon. *Information, Communication and Society, 16*(3), 315–339.

Jenkins, H. (1992). *Textual poachers: Television fans and participatory culture*. New York, NY: Routledge.

Jenkins, H. (2006). *Convergence culture: Where old and new media collide*. New York, NY: New York University Press.

Jones, J. P. (2012). The 'new' news as no 'news': U.S. cable news channels as branded political entertainment television. *Media International Australia*, (144), 146–155.

Lemish, D. (1982). The rules of viewing television in public places. *Journal of Broadcasting*, 26(4), 757–781.

Lessig, L. (2008). *Remix: Making art and commerce thrive in the hybrid economy*. New York, NY: Penguin.

Livingstone, S. (2004). The challenge of changing audiences: Or, what is the audience researcher to do in the age of the Internet? *European Journal of Communication*, 19(1), 75–86.

Lotz, A. D. (Ed.). (2009). *Beyond prime time: Television programming in the post-network era*. London, UK: Routledge.

McKee, A. (2003). *Textual analysis: A beginner's guide*. London, UK: Sage.

Morley, D. (1980). *The 'nationwide' audience*. London, UK: British Film Institute.

Morley, D. (1986). *Family television: Cultural power and domestic leisure*. London, UK: Comedia.

Morley, D. (1992). *Television, audiences and cultural studies*. London, UK: Routledge.

Morley, D. (1993). Active audience theory: Pendulums and pitfalls. *Journal of Communication*, 43(4), 13–19.

Ogas, O., & Gaddam, S. (2011). *A billion wicked thoughts: What the world's largest experiment reveals about human desire*. New York, NY: Dutton.

Seiter, E. (1990). Making distinctions in TV audience research: Case study of a troubling interview. *Cultural Studies*, 4(1), 61–84.

Spurgeon, C. (2008). *Advertising and new media*. New York, NY: Routledge.

Stelter, B. (2010). Water-cooler effect: Internet can be TV's friend. *The New York Times*. Retrieved from http://www.nytimes.com/2010/02/24/business/media/24cooler.html

Turner, G. (1990). *British cultural studies: An introduction*. London, UK: Unwin Hyman.

Turner, G., & Tay, J. (Eds.). (2009). *Television studies after TV: Understanding television in the post-broadcast era*. London, UK: Routledge.

Williams, R. (1975). *Television: Technology and Cultural Form*. New York, NY: Schocken Books.

Wood, M. M., & Baughman, L. (2012). Glee fandom and Twitter: Something new, or more of the same old thing? *Communication Studies*, 63(3), 328–344.

Following the Yellow Jersey
Tweeting the Tour de France

19
CHAPTER   Tim Highfield

as #letour snakes through the French countryside, its Twitter followers are close behind

Sporting events feature among the most popular topics covered on Twitter, both in terms of volume and frequency of updates (Twitter, 2010, 2011), with spectators using social media as a backchannel to post their own commentary while watching live events. Participating in these sports-oriented discussions is not limited to the audience, though—athletes and broadcasters are also active on social media, providing additional comments and perspectives before, after, and even while competing. This chapter looks at the 2012 Tour de France cycling race and its coverage on Twitter (as both sporting and media events) during the three weeks of competition between 30 June and 22 July 2012. Watched by millions of spectators around the world as well as along the streets of France, the Tour is also covered extensively online, and social media such as Twitter can potentially foster the development of a global, participatory audience simultaneously following the race.

Online communication can change the relationship between sports fans and athletes; Hutchins (2011), for example, argued that, as sportspeople adopt

Twitter in greater numbers for such purposes as self-promotion and self-expression, this has the effect of building "a sense of 'common experience' between athletes and their followers" (p. 242). By using the same communication platforms and discussing the same topics as their fans, sportspeople and other public figures can encourage a further connection and familiarity between themselves and their audience.

Professional cyclists are not new adoptees of social media. The US rider Lance Armstrong (@lancearmstrong) had over 1.2 million followers in August 2009 (Kassing & Sanderson, 2010), and over 3.7 million followers three years later in October 2012. During the 2009 Giro d'Italia, Kassing and Sanderson (2010) studied tweets from eight competing cyclists from the US and Australia to determine the purposes for which they use Twitter, such as sharing personal commentary and opinion (including on the events of the race) and communicating with their followers; this embrace of social media could further "transform the way athletes communicate with fans and how fans in turn respond to their athletic heroes" (p. 124), creating new interactions and informality between sportspeople and their audience.

In addition to being a sporting competition, the Tour de France is also a major international media event. While Twitter users respond to each day's stage and results in their tweets, their social media coverage might also treat the race in similar ways to other television broadcasts, such as including @mentions of media personalities in relevant tweets. Deller (2011) emphasised the importance of the live broadcast to tweeting about television, since users involved in a public conversation around a common programme are required to watch simultaneously to provide a consistent context for tweets and to avoid the possibility of reading spoilers from other viewers. This communal experience of watching and commenting on television broadcasts, including live sports, can see Twitter acting as a 'virtual loungeroom', a means of connecting a show's audience and providing a public backchannel for its responses to onscreen events (Harrington, Highfield, & Bruns, 2012).

This chapter, then, provides a preliminary examination of the 2012 Tour de France and its Twitter coverage as both a sporting event and shared television experience. The analysis is guided by the following questions:

- What is the shape of the Twitter audience for the Tour, and how are users connected?
- How is Twitter used to comment on the Tour de France by its audience? Does it act as a backchannel, with users tweeting in isolation, or

does a more interactive discussion take place between those watching at home and those participating in the race?

METHOD

This project draws on an extensive collection of Twitter datasets gathered during the 2012 Tour de France. Using the open-source tool yourTwapperKeeper (for more information, see Bruns, 2012), 180 hashtag and keyword archives were created in order to collect a wider range of tweets concerning the Tour than are featured within a single hashtagged discussion. yourTwapperKeeper collects data by querying the Twitter API for each search term specified and capturing any corresponding tweets. An archive was set up for each unique term, such as common hashtags for the race, including #tdf and #letour, broadcaster-specific hashtags, and the Twitter usernames for riders, teams, commentators, and analysts. By tracking usernames as keywords, yourTwapperKeeper captures tweets containing these names as @mentions and retweets, and tweets posted by the users in question. In total, archives were created for Twitter accounts representing 120 of the 198 competitors who started the race.

Following the completion of the 2012 race, these archives were examined using a collection of Gawk scripts developed for the analysis of large Twitter datasets, aided by network visualisation in Gephi (Bruns & Burgess, 2011). The wide scope of the data collection provides an opportunity to examine several Tour-oriented discussions, as well as the uses of Twitter during the race by individuals and groups associated with the competition—and how the Tour's audience on Twitter responds to, and interacts with, these different participants. Although a full overview of the collected data is beyond the scope of this chapter, the following sections examine selected archives to identify the shape of the Tour de France's Twitter audience, and investigate how Twitter was actively used as an extension of watching the race. First, the overall audience is identified from tweets captured which included the #tdf hashtag. For the examination of specific uses of Twitter, however, the #sbstdf hashtag is studied. The hashtag is specific to the Australian multicultural public service broadcaster Special Broadcasting Service (SBS), and provides a smaller, English-language dataset than found with the #tdf tweets.

#TDF AND THE TOUR DE FRANCE'S GLOBAL AUDIENCE

The Tour de France is an international competition, with cyclists from 31 countries competing in the 2012 edition, and the race itself broadcast to 190 countries ("Le Tour: On Screens around the World!", 2012). While French riders formed the largest national group (44 entrants), six other countries—Spain, The Netherlands, Italy, Belgium, Germany, and Australia—were also represented by more than ten competitors. The race itself was also not confined solely to France's borders; the opening stages were held in Belgium, while the eighth stage finished in Switzerland. It would then be expected that the Twitter audience for the Tour de France covers a wide geographical area, with particular attention from nations with several riders competing in the race.

Examining tweets containing the #tdf hashtag provides an initial overview of the Tour's global audience. Although this was not the only hashtag used to denote Tour de France-related tweets, and there is no requirement to include this, or any other hashtag, in relevant tweets, #tdf (and its variants, such as #tdf12 or #tdf2012) was widely employed during the race: 559,569 #tdf tweets from 145,328 Twitter users were captured between 29 June and 23 July 2012.

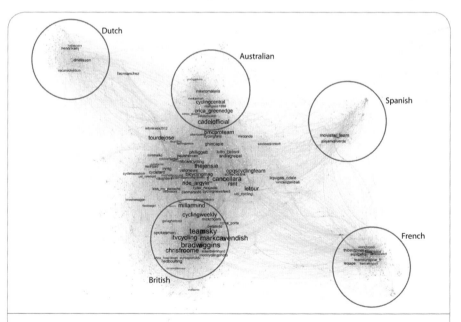

Figure 19.1: Network Map of Twitter Users Featuring #tdf in Tweets, 29 June–23 July 2012. Node Size Is Based on In-Degree—the Larger the Node, the More Connections Received

Each tweet was processed to identify any user accounts featured in the text as either an @mention or retweet. Initially, this created a network of 117,385 nodes (representing individual Twitter accounts) connected by 244,651 edges (each edge represents a link from one user to another account in the form of either an @mention or retweet). The network was then filtered to include only those nodes with a degree range of twenty or more (giving and/or receiving in combination at least twenty connections to other nodes), reducing the network to 3,083 nodes and 39,494 edges. The filtering process highlights the accounts mentioning the most users overall in their #tdf tweets, and the accounts receiving these mentions. The resulting visualisation, seen in Figure 19.1, depicts the extent to which the #tdf hashtag connects Twitter users through mentions of other accounts.

Immediately apparent are several clusters of Twitter users at the extremities of Figure 19.1; while still connected to the rest of the network, these clusters feature strong interlinking between users unique to those groups, and so are presented at a distance from the central mass of nodes. These clusters roughly correspond to national and linguistic groups: the group in the top left of the visualisation forms a Dutch-speaking corner, while a French cluster can be found in the bottom right of Figure 19.1. Between these two groups are further clusters around Australian and Spanish accounts, respectively, while the lower section of the main group is centred on British Twitter users. These are not exclusive clusters; Spanish accounts, for example, appear elsewhere in Figure 19.1, not just in the Spanish cluster, and the Dutch section features accounts based in both The Netherlands and Belgium.

The presence of these clusters suggests some separation among the Twitter audience along language, and also national lines. However, the groups share common characteristics, if not common users: unsurprisingly, given the fact that many broadcasters provide their own commentators and analysts to cover the Tour, these clusters prominently feature the accounts of local media personalities, publications, and television networks. Also noticeable within the groups are the accounts of cyclists and teams from these countries. However, the proximity of riders and teams to these clusters is not directly dependent on nationality. Although there is a Dutch-speaking cluster in Figure 19.1, the accounts of Belgian riders are distributed throughout the network, as they were members of international teams representing several nationalities. Dutch cyclists competing in the Tour, on the other hand, are more concentrated within the Dutch cluster, as the majority raced for Netherlands-based teams, such as Rabobank (@rabocycling), Argos-Shimano (@1t4i), and Vacansoleil-

DCM (@vacansoleildcm). These team affiliations also link different groups in Figure 19.1. The Spanish rider Luis Léon Sánchez (@lleonsanchez), for example, appears not within the Spanish cluster, but as a bridge between the Dutch cluster and the main network, as he was riding for the Rabobank team.

Similarly, the accounts for the BMC Pro Team (@bmcproteam) and its cyclists, such as the German Marcus Burghardt (@mburghardt83) and American George Hincapie (@ghincapie), are linked to the Australian cluster since the team's leader was the defending champion, Australian rider Cadel Evans (@cadelofficial). While the BMC accounts were mentioned by Twitter users from around the world, they were closely associated with the fortunes of Evans during the race, cited in particular by Australians hoping that Evans would repeat his 2011 success. The bridging role of cyclists, teams, and, to a lesser extent, media accounts, can be seen throughout Figure 19.1; riders competing for teams with an international roster connect different clusters, and success during the race itself can also attract attention from across the network. For example, the Swiss rider Fabian Cancellara (@f_cancellara), riding for RadioShack Nissan Trek (@rsnt) and the only cyclist other than race winner Bradley Wiggins (@bradwiggins) to wear the leader's yellow jersey, appears central to the network, since his account was linked in tweets from each of the clusters in Figure 19.1. The rider's own Twitter activity will also influence how connected they are, as fans, teams, and other riders may reply to or retweet comments made during the race weeks.

Tweeting at specific cyclists, asking questions of them and giving support and advice during the race, may take place without necessarily expecting a response—while some rider-fan interaction may result, the act of tweeting at a cyclist may also be seen as a Twitter-based representation of cheering from the roadside, or shouting at the television, in a more public manner that creates a direct connection to the subject of the comments. Similar links can also be made between the television audience and the commentators, presenters, and analysts contributing to the broadcast of the Tour. Tweets may comment on events within the race itself, but may also address phenomena related to the media coverage of the Tour that do not have any direct bearing on the final result, discussing both the sporting event and its mediatised presentation. To explore this further, the next section examines the different uses of Twitter by the audience for a specific broadcaster's coverage of the Tour: the Australian Twitter users including the #sbstdf hashtag in their tweets.

THE TROPES OF #SBSTDF

The Tour de France is broadcast live and free-to-air in Australia by SBS, with coverage starting each race day from 10:00 p.m. local time nationwide—since the programming starts at the same time locally, the two-hour winter time difference between Western Australia and the eastern states means that viewers in Perth will see less of the race than their counterparts in Sydney or Brisbane. Figure 19.2 outlines the usual broadcasting timetable for each stage on SBS.

SBS's presentation of the 2012 race was watched in total by more than six million viewers over the 21 days of competition (Nance, 2012), with individual live stages averaging between 240,000 and 380,000 viewers (Dale, 2012). These viewing figures show that although the Tour is not one of the highest-rated shows on Australian television, it does attract a returning, niche audience despite the late nights involved in watching the live broadcast. The television coverage is supplemented by online content through the SBS Cycling Central Social Hub (http://cyclingcentral.social.sbs.com.au/): video footage, links, tweets, and Facebook comments from official accounts (presenters, cyclists) and fans are presented in a central location on the SBS website, and the #sbstdf hashtag is one part of this extended media coverage of the race. In total, 39,115 #sbstdf tweets from 3,185 users were captured during the period 29 June to 23 July 2012 (covering the race, rest days, and the days immediately before and after the race). This total activity also includes variations of the #sbstdf hashtag, such as #sbstdf12.

While still discussing a sporting event, the SBS-specific hashtag allows users to also comment on the race as a mediatised event, wherein the cycling is

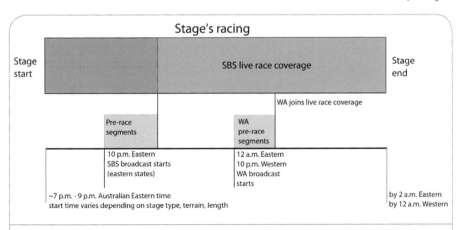

Figure 19.2: Standard Timetable for SBS's Television Broadcast of Each Stage in the Eastern States and Western Australia

just one part of a wider entertainment package. For Australian viewers, watching the Tour de France is often a ritualised experience—staying up to midnight or 2:00 a.m., depending on location, to see the end of each stage over three weeks is a sign of dedication (Mathieson, 2012)—which is further amplified by viewers' identification of common themes and repeated phenomena over the 21 days of racing (both in tweets and in other communication). SBS's presentation of each stage lends to this ritualisation by following a standard format, as shown in Figure 19.2. In the eastern states (time difference GMT +10), several segments are shown before crossing to live coverage, including a recap of the previous day's racing; a preview of the stage ahead; and a short, Tour-themed cooking programme. In Western Australia (GMT +8), this final segment is usually excised from the broadcast, as the 10:00 p.m. local start means that viewers join the live coverage over an hour after their eastern states counterparts (the exact time depends on the length of pre-race programming for viewers in these time zones). For the first part of the stage, commentary is provided by Matthew Keenan (@mwkeenan), while the final sections are covered by Phil Liggett (@philliggett) and Paul Sherwen (@paulsherwen). Following the end of the stage, analysis and comments are presented by Mike Tomalaris (@miketomalaris) and a guest expert.

The presence of these SBS personalities on Twitter—and their use of the medium for additional remarks, sharing links, and responding to follower queries—gives the #sbstdf audience a direct connection to the onscreen team. The consistent, annual appearance of Tomalaris, Keenan, Liggett, and Sherwen in the Tour coverage, and in other cycling broadcasts, aids this sense of familiarity; viewers are aware, for example, of favourite phrases used by Liggett in his commentary, such as describing the group of riders leading their respective teams as the "heads of state". Finally, the video footage of the Tour itself, highlighting castles and other local points of interest, artwork in fields, spectators in costumes, and cows near the roadside, provides the audience with familiar views each year as the cyclists race past.

For the SBS audience, these different aspects of the Tour de France translate into recurring components of the broadcast, or tropes, to be light-heartedly embraced in their repetition. Rather than a space just for serious cycling commentary (although that is still present), the #sbstdf hashtag also serves as a backchannel where viewers can share the humour of the Tour as well as the sporting contest. The extent of this use of Twitter is seen in the distribution of hashtags within the captured tweets. Although, in addition to #sbstdf itself (38,960 occurrences), the most commonly used hashtags were race-specific (such as #tdf and #tdf12, with 5,644 and 6,706 mentions, respectively), other popular hashtags reflect more humorous purposes. These include #LVDT (895 occurrences), referring to the blog of @lesvachesdutour, which highlights footage of cows during the race; and #sherliggettisms (456; 1,066 including variants), used to denote repeated, strange, offbeat, or incorrect comments by Sherwen and Liggett in their commentary. These broadcast-specific hashtags reflect different tropes of the coverage which all lead into the #drink hashtag (530); treating the Tour de France as a drinking game, the utterance of stock phrases or the appearance of visual stereotypes of the Tour (such as castles and sunflowers) are quickly tweeted as a prompt to take another drink. This does not mean that Twitter users are actually participating in a drinking game; rather, the #drink hashtag has become a staple response to clichéd behaviour, particularly within media events. Tweets concerning the SBS broadcast of the Tour, then, while still attracting a niche audience given its time slot and subject, can be seen to continue Twitter conventions from other media contexts.

The #sbstdf hashtag also features other Twitter conventions developed within unrelated situations, and which again dwell less on the actual events of the race than its tropes. While presenters Tomalaris and Keenan were active Twitter users during the Tour, their contributions were also shadowed by parody accounts in their name. Tomalaris inspired two fake accounts—@FakeMTomalaris and @FakeTomalaris—while other accounts satirised Keenan (@FakeMattKeenan) and pre-race television chef Gabriel Gaté (@fakegabrielgate). The creation of these accounts highlights the audience's familiarity with the race broadcasts, and this awareness is developed further through interactions and retweets between viewers, the fake accounts, and other humour-oriented users, especially when continuing the themes represented by other popular hashtags. Before Stage 9, for example, @FakeMTomalaris tweeted a bingo card of common phrases used by Liggett and Sherwen for which viewers should listen out, treating the broadcast in a similar manner as the drinking game trope.

The creation of fake accounts parodying public figures is not a uniquely Australian or sports-related activity, and the practice is carried out within different contexts on Twitter. Popular accounts which imitate celebrities include the parodies of Queen Elizabeth II (@queen_uk) and the actress Tilda Swinton (@NotTildaSwinton). Accounts for fictional characters and groups also spoof social media practices, such as @DeathStarPR's use of Twitter for brand management and promotion of the Galactic Empire from *Star Wars*. Satirising politicians and media figures is a newly established part of the "mediated spectacle of mainstream politics" in Australia (Wilson, 2011, p. 458); playing with the conventions of political coverage and the character of the figures parodied, Wilson (2011) found that these fake accounts are usually ongoing performances, where public attention (such as increased followers, retweets, or replies from the satire's subjects) may be the main reward. Although the scope of the Tour parodies may be less open-ended, the intentions and goals may be similar to their political counterparts—indeed, the fake accounts would often mention the real Keenan and Tomalaris accounts, and their SBS colleagues, inviting reactions even if they ultimately went unrequited.

The humorous overtones found in #sbstdf tweets also demonstrate a mixture of these various Twitter conventions and other popular Internet culture. One of the most mentioned cyclists during the 2012 race was the German rider Jens Voigt (@thejensie). Despite not being Australian, nor riding for an Australian team, Voigt was a popular figure due to his attitude to riding; for example, he once stated in an interview that his mind's response when his legs are in pain while riding is to say, "shut up legs" (Vaughan & AAP, 2012). This comment helped to promote Voigt's reputation as a tough character in the world of professional cycling. Inevitably, Voigt became the sport's equivalent of actor Chuck Norris, the subject of the Internet meme 'Chuck Norris Facts', which shared "amusing (fictional) anecdotes ostensibly about the venerable action star, but more accurately about iconic traits of hegemonic masculinity" (Dutton, Consalvo, & Harper, 2011, p. 301). During the 2011 Tour de France, a Twitter account for 'Jens Voigt Facts' (http://jensvoigtfacts.com/; @JensVoigtFacts) covered similar ground to the Chuck Norris meme; while the Twitter account was not active during the 2012 Tour, the style was appropriated by other users. Indeed, the most retweeted comment during the race with the #sbstdf hashtag (306 retweets) followed this format:

> .@thejensie has a polar bear stretched out on the floor of his den. It's not dead, it's just too scared to move. #sbstdf

Some Twitter users mixed the tropes of the SBS broadcast with Jens Voigt jokes in their tweets, creating their own new conventions. The start of the live broadcast in Western Australia, for example, would be announced by Matthew Keenan by welcoming viewers from that state. For the #sbstdf discussion, the Twitter audience regularly provided their own take on this greeting:

> Welcome to viewers in Western Australia. You just missed Jens Voigt jumping over 25 tour buses on his bike. It was SPECTACULAR. #sbstdf

> Welcome to viewers in WA. You just missed Jens Voigt riding so hard that a chopper got knocked off course by his wake turbulence. #sbstdf

The audience's familiarity with the cyclists, their personalities, and the conventions of the Tour coverage went even further, though; when it was revealed that the German cyclist Tony Martin was riding with a broken wrist, a parody account was created not for the rider himself, but for his wrist (@TonyMartinWrist). This act already demonstrates awareness of several tropes of the Tour, and the light-hearted relationship between the SBS audience and the cyclists and commentators involved in the coverage; however, there remained scope to combine conventions to an even greater extent, and this was realised in additional tweets:

> Welcome to viewers in WA. You just missed a drunken brawl between @thejensie's legs & @TonyMartinWrist over who should shut up more. #sbstdf
>
> RT @TonyMartinWrist: Shut up, Jens. #sbstdf

CONCLUSION

The tweets captured containing the #sbstdf hashtag demonstrate that the Tour de France, for its Australian viewers at least, is not just a sporting event appealing to a niche audience. Instead, the race and its result are only one component of the wider discussions taking place on Twitter. While there is certainly interest in the outcome of the Tour and the fortunes of the competitors, watching the SBS coverage can also be seen as a ritualised activity for some viewers. Tweeting about the Tour turns the rituals into a shared experience, encouraging interactions between the audience, as well as commentators and cyclists, in response to the events of the race as well as the tropes of the broadcast itself. The promotion of humorous content, especially tweets drawing on the established conventions of SBS's coverage (from stock phrases and recurring segments to the scenery and the commercials during the broadcast), make the #sbstdf discussions similar to other hashtags surrounding televised events. For instance, tropes based on recurring aspects of broadcast media spectacles are invoked by the Twitter audience for the Eurovision Song Contest, also shown in Australia by SBS (see Highfield, Harrington, & Bruns, 2013).

This chapter has focussed on the Australian context, where sports commentary is combined with irreverent remarks in a light-hearted relationship between fans, riders, and presenters. For viewers in other countries, though, the use of Twitter for commenting on the Tour de France may take very different formats and intentions. The Tour is watched around the world, attracting international attention on Twitter, and further research is required to examine how different audiences respond to the same event. In addition, analysis can expand on the examples of Kassing and Sanderson (2010) to investigate how individual cyclists, teams, and commentators use Twitter over the course of the race—alongside any satirical counterparts.

ACKNOWLEDGMENT

An early version of this chapter was presented at the 4th European Communication Conference, Istanbul, Turkey, in October 2012. The author would like to recognise the contributions of Axel Bruns and Stephen Harrington to this earlier paper, and to thank Tony Highfield for his assistance with the data collection.

REFERENCES

Bruns, A. (2012). How long is a tweet? Mapping dynamic conversation networks on Twitter using Gawk and Gephi. *Information, Communication & Society, 15*(9), 1323–1351. doi:10.1080/1369118X.2011.635214

Bruns, A., & Burgess, J. (2011, 22 June). Gawk scripts for Twitter processing. *Mapping Online Publics*. Retrieved from http://mappingonlinepublics.net/resources/

Dale, D. (2012,15 July). The ratings race: Week 28. *The National Times: The Tribal Mind (blog)*. Retrieved from http://www.nationaltimes.com.au/opinion/society-and-culture/blogs/the-tribal-mind/the-ratings-race-week-28-20120709-21qhc.html

Deller, R. (2011). Twittering on: Audience research and participation using Twitter. *Participations, 8*(1). Retrieved from http://www.participations.org/Volume 8/Issue 1/deller.htm

Dutton, N., Consalvo, M., & Harper, T. (2011). Digital pitchforks and virtual torches: Fan responses to the Mass Effect news debacle. *Convergence: The International Journal of Research into New Media Technologies, 17*(3), 287–305. doi:10.1177/1354856511407802

Harrington, S., Highfield, T., & Bruns, A. (2012). More than a backchannel: Twitter and television. In J. M. Noguera (Ed.), *Audience interactivity and participation: Interview/essays with academics* (pp. 13–17). Brussels, Belgium: COST Action Transforming Audiences, Transforming Societies. Retrieved from http://www.cost-transforming-audiences.eu/system/files/essays-and-interview-essays-18-06-12.pdf

Highfield, T., Harrington, S., & Bruns, A. (2013). Twitter as a technology for audiencing and fandom: The #Eurovision phenomenon. *Information, Communication & Society, 16*(3), 315-339. doi:10.1080/1369118X.2012.756053

Hutchins, B. (2011). The acceleration of media sport culture: Twitter, telepresence and online messaging. *Information, Communication & Society, 14*(2), 237–257. doi:10.1080/1369118X.2010.508534

Kassing, J. W., & Sanderson, J. (2010). Fan-athlete interaction and Twitter. Tweeting through the Giro: A case study. *International Journal of Sport Communication, 3*(1), 113–128.

Le Tour: On screens around the world! (2012, 27 June). *Grand Départ 2012—Tour de France 2012.* Retrieved from http://www.letour.fr/le-tour/2012/us/grand-depart.html

Mathieson, C. (2012, 19 July). The race that stops a nation's sleep. *The Age.* Retrieved from http://www.theage.com.au/entertainment/tv-and-radio/the-race-that-stops-a-nations-sleep-20120718-228yo.html

Nance, C. (2012, 24 July). SBS's Tour de France records highest online audience (Press release). *AccessPR.* Retrieved from http://www.accesspr.com.au/news/998/56/SBS-s-Tour-de-France-records-highest-online-audience

Twitter. (2010). Twitter 2010: Year in review—2010 trends on Twitter. Retrieved from http://yearinreview.twitter.com/2010/trends/

Twitter. (2011). Twitter's 2011 year in review: Tweets per second. Retrieved from http://yearinreview.twitter.com/en/tps.html

Vaughan, R., & AAP. (2012, 12 July). Evergreen Voigt still showing fight in Tour at age 40. *Cycling Central.* Retrieved from http://www.sbs.com.au/cyclingcentral/news/37746/evergreen-voigt-still-showing-fight-in-tour-at-age-40

Wilson, J. (2011). Playing with politics: Political fans and Twitter faking in post-broadcast democracy. *Convergence: The International Journal of Research Into New Media Technologies, 17*(4), 445–461. doi:10.1177/1354856511414348

Twitter and Sports
Football Fandom in Emerging and Established Markets

20
CHAPTER   Axel Bruns, Katrin Weller, and Stephen Harrington

 football clubs and fans in #epl, #bundesliga, and #aleague use Twitter to engage, with varying success

PROFESSIONAL SPORTS AND FANDOM

Twitter and other social media have become increasingly important tools for maintaining the relationships between fans and their idols across a range of activities, from politics and the arts to celebrity and sports culture. Twitter, Inc. itself has initiated several strategic approaches, especially to entertainment and sporting organisations; late in 2012, for example, a Twitter, Inc. delegation toured Australia in order to develop formal relationships with a number of key sporting bodies covering popular sports such as Australian Rules Football, A-League football (soccer), and V8 touring car racing, as well as to strengthen its connections with key Australian broadcasters and news organisations (Jackson & Christensen, 2012). Similarly, there has been a concerted effort between Twitter Germany and the German Bundesliga clubs and football association to coordinate the presence of German football on Twitter ahead of the 2012–2013 season: the Twitter accounts of almost all first-division teams now bear the official Twitter verification mark, and a system of 'official' hashtags for tweeting about individual games (combining the abbreviations of the two teams, e.g. #H96FCB) has also been instituted (Twitter auf Deutsch, 2012).

Such attempts to formalise, professionalise, and commercialise Twitter-based activities around certain sports are aimed, in the first place, at enticing sportspeople, clubs, and sporting bodies to participate in the platform more actively, from Twitter, Inc.'s perspective presumably in the hope that this will also serve to attract a greater number of fans to sign on to Twitter. As in the examples above, however, they often come well after committed fans have already discovered the platform for themselves, and have developed their own presences, conventions (such as hashtags), and dedicated accounts (in tribute to clubs and sportspeople). This may place clubs and fans, professionals and their followers, on a collision course. In turn, this both mirrors the conflicts between professional sports and traditional fandom which have already played out in a variety of other contexts (e.g., over TV broadcasting arrangements) over past decades; and the conflict between Twitter and its users which has arisen several times as Twitter has sought to formalise user-created conventions for using the platform (such as hashtags or retweets) in its further development of the underlying technology (cf. Halavais, Chapter 3 in this volume).

This chapter examines how these tensions between professional sporting bodies and their fans play out on Twitter in the context of three national football leagues at various stages of their development. Football (soccer) has grown into an enormous market: in 2009–2010, English Premier League (EPL) clubs generated nearly £2.7 billion in collective revenue; German Bundesliga clubs reached about €1.6 billion; and the entire European football market grew to €16.3 billion (Deloitte, 2011). Football clubs may now consider the Internet as a marketing tool much as other companies do (Kriemadis, Terzouidis, & Kartakoulis, 2010; McCarthy, Pioch, Rowley, & Ashworth, 2011), and Twitter now plays a part in the marketing mix.

But at the same time, sports fans, with their particular culture of fandom, cannot be compared with the customers of 'normal' brands; German football fans, for example, take a rebellious and subversive stance towards the commercialisation of 'their' teams (Merkel, 2012). Only recently has research examined how football fans make use of different online channels, for example to establish social identity (Gibbons & Dixon, 2010) and communities (Krøvel, 2012), to create "a virtual stage for their subcultural practice and performance" (Merkel,

2012, p. 369), or to "organise against the commercial power of the large football clubs" (McLean & Wainwright, 2009, p. 54). Twitter research has mainly studied sports tweets for the purpose of developing automatic techniques for event or named entity recognition (Choudhury & Breslin, 2011; Nichols, Mahmud, & Drews, 2012)—but has not yet provided many insights into the relationship between clubs and their fans. A closer examination of the interactions between fans and clubs on Twitter is necessary to both identify the opportunities for sports marketing on Twitter, and to stake out the limits which apply to the marketisation and commercialisation of voluntary fan activities on the platform. Observations from this research may also be transferable to other areas in which Twitter's attempts to commercialise its services come into conflict with its users' interests, as well as to sports marketing initiatives across other media channels.

This chapter examines the interactions between first-division football clubs and their fans in Australia, Germany, and England. We tracked fan interactions with the official accounts of the teams participating in the EPL, Germany's 1. Bundesliga, and the Australian A-League throughout the 2011–2012 seasons of each competition by capturing the tweets *from* and *to* (in the form of @replies or retweets) these accounts. Indeed, the process of identifying the accounts themselves already revealed significant differences in how the various leagues and individual clubs approached Twitter as a medium for communicating with their fans, and how well different clubs have established their Twitter presence. This comparative analysis provides a rich perspective on the different approaches to Twitter use in football fandom which are evident across such diverse markets.

Both the EPL and the Bundesliga are extremely well-established football leagues, but have different levels of global prominence. The EPL, featuring many of the best players from around the world, attracts a significant international following. The Bundesliga is also well-known and successful on an international level, but remains focussed more strongly on its domestic market, building on a very loyal local fan base. The Australian A-League, in contrast, sits very much towards the other end of the spectrum. Having only started in 2005–2006, it is still an emerging competition, and a fledgling football market, with the long-term viability of several lower-placed teams in the league remaining doubtful. In just seven seasons of the league (to 2012), three club franchises have folded. That said, its fan base is growing, especially amongst young fans and their parents, as it is seen as a less violent form of sport than the other football codes (Rugby Union, Rugby League, and Australian Rules Football) played in Australia, and its popularity has been boosted by the national team's qualification for successive World Cups in 2006 and 2010.

Such differences in the domestic and international outlook, from precarious existence to international dominance, also result in significantly different motivations for using Twitter as a means to reach out to and engage with football fans (while the fans may also have different motivations for using Twitter for sports-related communication). These differences, in turn, emerge clearly in the patterns of Twitter use which we trace here. Overall, we find that A-League clubs generally manage to enlist their fans in promoting the code by increasing its visibility on Twitter; that Bundesliga teams have taken a slower and less consistent approach to Twitter to date, sometimes clashing directly with already established fan conventions for using Twitter; and that activities around EPL clubs diverge considerably depending on the relative domestic and international standing of individual teams.

FOOTBALL CLUBS ON TWITTER

We begin by tracing the evolution of the different clubs' presences on Twitter during the 2011–2012 season, starting with the most recent of the three leagues, the Australian A-League. At the beginning of the season, in mid-2011, all ten A-League clubs had established their official Twitter accounts; one club, Adelaide United, changed its Twitter handle from @AUFC_Official to @adelutd_fc during the season. By the end of June, these accounts had managed to attract between 650 (for the regional club Newcastle Jets) and 4,800 (for the metropolitan Melbourne Victory) followers; over the course of the season, such follower figures more than doubled for most of the clubs concerned (Figure 20.1).

Perhaps helping to boost the visibility of these official club accounts—and, of course, the visibility of the league as a whole—Football Federation Australia (FFA, the sport's governing body) officially designated match-specific hashtags for the 2011–2012 season. This allows fans to more easily find and follow online conversations regarding matches, and to follow real-time updates if access to the pay-TV matchday broadcasts is unavailable. This initiative by the FFA can also be seen as a trial run for a similar framework introduced in Germany in the following year, as discussed above.

The situation in the two European leagues is noticeably more complicated. In the Bundesliga, Twitter accounts for a substantial number of clubs had already been established, but it remained difficult to ascertain whether these accounts were officially sanctioned by the clubs, or had been set up by individual fans or supporter groups. This was most notable for leading club Bayern München, whose @BayMuenchen account had been set up by the club's PR

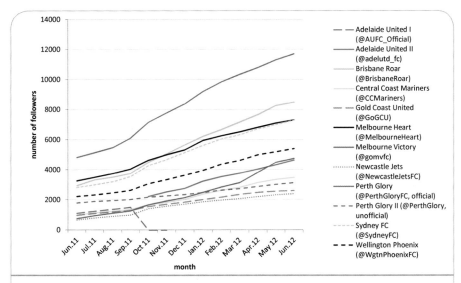

Figure 20.1: Follower Numbers of A-League Clubs during Season 2011–2012. (June 2011 to June 2012, Numbers for Figures 1–4 Usually Collected on the 3rd Thursday of Each Month.)

department, but sent only 39 tweets during the entire season, while a separate @fcbayern_news account actively tweeted news reports from the club's official Website and YouTube channel, but did so without official sanction from the club. (An official @fcbayern account has been instituted ahead of the 2012–2013 season, replacing @BayMuenchen.) Similar patterns apply for a number of other clubs as well—half of the 18 first-division clubs in the Bundesliga created new Twitter accounts during the season, renamed existing accounts to more obvious handles, or even took over originally fan-created accounts; some fan accounts which used variations on an official club name as their handle were suspended by Twitter during the season, possibly at the behest of the club (the story of one such suspension is told, by the fan who had operated the @s04 account, in Nettooor, 2011).

This considerable flux in account names and approaches to tweeting about their activities points to the German clubs' relatively late entry to Twitter (well after fans had already created their own infrastructure for tracking the latest Bundesliga news), and is in line with the generally comparatively slow adoption of Twitter in Germany (Meyer, 2012). It is only towards the end of the 2011–2012 season that the majority of clubs—with the exception of record Bundesliga title-holder Bayern München—had established an active presence on Twitter.

In spite of the confusion which these changes in accounts and account names will have caused over the course of the 2011–2012 season, the follower numbers for the accounts of Bundesliga clubs also grew during this time, as Figure 20.2 shows. Given their considerably greater domestic fan base and international exposure, at least for the leading Bundesliga clubs these numbers are substantially larger than those for A-League clubs, of course.

Finally, the situation for EPL clubs appears somewhat more stable than that for Bundesliga teams. Here, too, some further adjustments to the Twitter presence of various clubs are evident, but most clubs had already set up official accounts (if not necessarily 'verified by Twitter') by the start of the 2011–2012 season; indeed, many accounts contain the term *official* in the account name— as in @OfficialQPR or @SpursOfficial. The most significant absence from the EPL Twitter line-up, however, is that of one of its most prominent teams: Manchester United. An @MUFootballClub account has existed on Twitter since July 2011, and claims in its profile description that it is "The official Twitter page of Manchester United Football Club", but—at the time of writing—has yet to send a single tweet to its 13,000 followers. As with Bayern München, this absence of an official account provides a space for fan-generated alternatives to establish themselves; in the case of Manchester United, for example, a

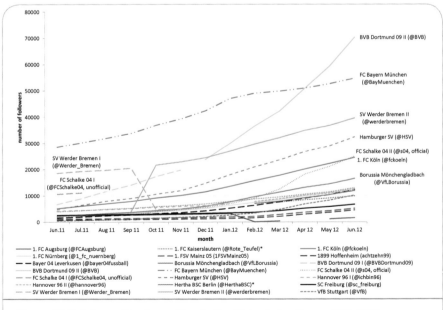

Figure 20.2: Follower Numbers of Bundesliga Clubs during Season 2011–2012.
* @HerthaBSC and @RoteTeufel Were Only Tracked from February 2012

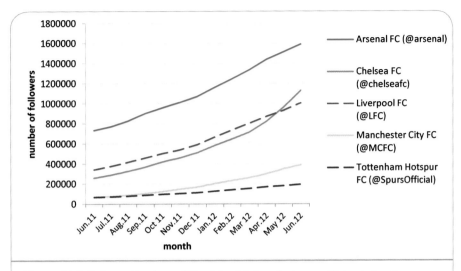

Figure 20.3: Follower Numbers of Top 5 English Premier League Clubs during Season 2011–2012

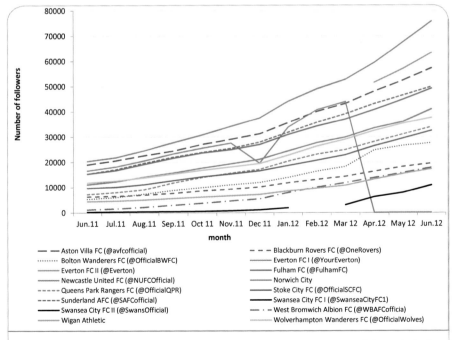

Figure 20.4: Follower Numbers of English Premier League Clubs (Top 5 Clubs Excluded) during Season 2011–2012

@manutd_fc account has sent over 6,000 tweets and has attracted over 180,000 followers, but appears simply to automatically tweet relevant headlines from the Arabic section of the *manutd.com* website.

This lack of Twitter presence is surprising for one of the biggest brands in world sports, and curiously mirrors the absence of the leading German club, Bayern München. On Twitter, three of the top five most-followed EPL players are current or former Manchester United players (Twitter UK Blog, 2012), indicating that there is no lack of public interest. At the same time, Manchester United players have been involved in scandals over injudicious tweets, and manager Sir Alex Ferguson has been quoted as saying "Twitter; I do not understand it . . . I don't know why anybody can be bothered with it" (Ladyman, 2012). Perhaps this explains the club's social media reluctance.

Follower patterns for EPL clubs also point to an even greater bifurcation in fan attention between the leading teams and the rest of the league than was already visible for the Bundesliga clubs in Figure 20.2. Consequently, we have split our graph of follower numbers into two figures. Figure 20.3 shows the top five clubs; Figure 20.4 covers the remaining EPL clubs, to facilitate comparison with the Bundesliga follower numbers. Overall, EPL clubs enjoy the largest, and still rapidly growing, Twitter audience of the three leagues. This clearly is a function of the status of the EPL as a globally marketed and broadcast league. But mirroring criticisms about the level of competition in the league itself (only five teams have won the premiership in 21 years of competition, with Manchester United winning 13 times), follower numbers are similarly dominated by three of the traditional 'Big 4': Arsenal, Liverpool, and Chelsea, with Manchester City and Tottenham Hotspur the only other genuine competitors.

This divergence in the three leagues' online presences is in keeping with the different contexts in which they operate. In their need to become competitive against the three other, more widely televised 'football' codes in a country with a strong Twitter adoption rate (*ABC News*, 2010), it makes sense for A-League clubs to have been early adopters on Twitter; through their efforts, they appear to have successfully engaged and enlisted fans in the campaign to build a sustainable base for football in Australia. Both sides are interested in growing the code, and both sides are—for now—pushing in the same direction.

The German Bundesliga, on the other hand, is a well-established competition that is widely broadcast on free-to-air and pay-TV, while Twitter adoption in Germany remains comparatively low (Meyer, 2012) and clubs have been slow to explore this space as a further channel for sports communication. Here, enterprising fans were left to create their own fan spaces on the platform, which now conflict, to some extent, with a more coordinated, professional approach to Twitter. Attracting official club accounts may be seen by Twitter, Inc. as a useful marketing tool, but as with other top-down Twitter initiatives, such formalisation has the potential to break more organically grown, user-initiated structures (cf. Halavais, Chapter 3 in this volume).

The EPL, finally, presents a more complicated picture, as it constitutes—more so than the Bundesliga or A-League—an intersection of domestic and international interests. For the handful of clubs which have a major worldwide following, Twitter provides a useful channel to connect with these international fans, whose domestic media may not cover the EPL in detail. For the rest of the league, their communicative orientation is more akin to their Bundesliga counterparts, except that the English clubs' Twitter presences are already well-established, without overriding grown fan ecosystems.

APPROACHES TO FAN INTERACTION

In addition to such broad distinctions across the three leagues, more specific differences in how accounts are maintained in day-to-day club activity and interaction with fans also provide important insights into how the various clubs position their Twitter accounts within the context of their overall public relations efforts. Of particular interest in this context is whether clubs restrict themselves simply to posting the latest news and information (especially in the form of URLs pointing to further information), or whether they also directly @reply to comments and questions from their followers, and even retweet other users' messages.

Figure 20.5 shows these patterns for the clubs which participated in the 2011–2012 A-League season (however, @adelutd_fc was only created in October 2011). Perhaps unsurprisingly, the accounts belonging to 2011 and 2012 champions Brisbane Roar, and 2011 and 2012 runners-up Central Coast Mariners and Perth Glory were especially active over the course of the season; major metropolitan clubs Sydney FC and Melbourne Heart also posted well above 2,000 tweets over this time frame. Notably, however, there are also substantial differences in tweeting styles: while the three most active clubs largely posted original tweets, some fifty per cent of the tweets by @MelbourneHeart and @PerthGloryFC consisted of @replies to or retweets of other Twitter users. This can be seen as a conscious attempt to generate and maintain a Twitter 'buzz' around these accounts, thereby positioning fans as part of an 'inner circle' connecting them to other fans, and encouraging them to attend live matches or otherwise continue

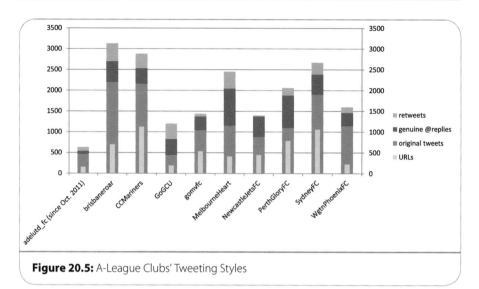

Figure 20.5: A-League Clubs' Tweeting Styles

to support the club (additional, in-depth analysis of tweets can reveal which messages generated the greatest levels of club and/or fan engagement). Overall, however, with the exception of the newly-created Adelaide United account and the @GoGCU account of the financially troubled Gold Coast United, A-League tweeting styles are relatively uniform across all clubs.

The situation in the German Bundesliga is considerably more complex. Figure 20.6 provides an overview of the major club accounts both before and after the various changes during the season: it includes multiple accounts for several teams, due to the various account suspensions and renamings, as well as the general confusion over their status as 'official'.

Overall, Bundesliga clubs used Twitter considerably less than their Australian counterparts: while most Australian clubs came close to posting at least 1,500 tweets over the course of the season, the majority of Bundesliga clubs failed to reach even 1,000 tweets, even if tweets posted from various alternative accounts are combined. In total, for example, the two accounts of 2011 and 2012 champions Borussia Dortmund (@BVBDortmund09 and @BVB) still posted fewer than 600 tweets throughout the season—less than one fifth of the more than 3,000 tweets posted by their Australian counterparts Brisbane Roar. Additionally, the vast majority of the tweets posted by German club accounts were original tweets—the clubs used Twitter almost exclusively as a means to disseminate information, not to engage with fans through @replies or even to retweet their messages. For many clubs, Twitter is not a 'social' platform at all, as the vast

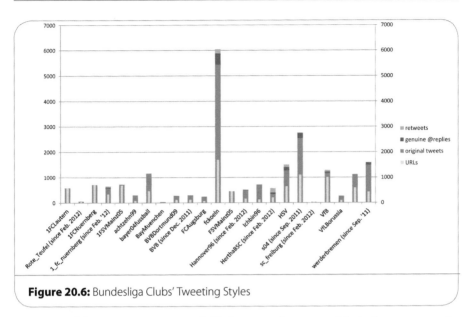

Figure 20.6: Bundesliga Clubs' Tweeting Styles

majority of their tweets contained URLs—another sign of their focus on purely top-down information dissemination.

The major exception to this rule is the @fckoeln account of FC Köln. Though still largely focussed on original tweets, Köln did respond to (and even retweet) its fans more often than any other Bundesliga club, and it participated on Twitter at a rate which far surpassed any other team in the Bundesliga or A-League. It also instituted special Twitter activities: for example, @fckoeln hosted interview sessions with club officials on Twitter. Using the hashtag #fragfc ("ask FC"), fans could talk to manager Stale Solbakken, for instance, with responses posted through the @fckoeln account. Later in the season, @fckoeln also published real-time updates during matches. Such activity may be in keeping with FC Köln's struggle to remain in the first division, as an attempt to maintain the loyalty of the fan base and ensure their turnout in the stadium; it may also simply reflect a very different understanding of Twitter as a medium for sports communication.

Finally, Twitter activity patterns in the English Premier League vary considerably across clubs (Figure 20.7); at some 3,000 tweets over the course of the season, the average level of Twitter activity exceeds that of A-League clubs (around 2,000 tweets), however, and several clubs managed well over 4,000 tweets during the season. Notably, as with Köln in the Bundesliga, the most active on Twitter are generally not amongst the leading EPL clubs: with the exception of eventual Premiers Manchester City (@MCFC), teams such as @NorwichCityFC, Sunderland (@SAFCofficial), and Wolverhampton (@OfficialWolves) performed

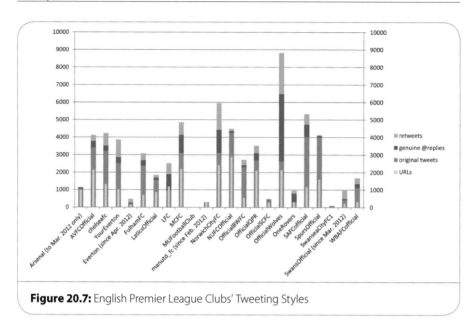

Figure 20.7: English Premier League Clubs' Tweeting Styles

comparatively poorly during the season, with Wolverhampton suffering relegation. In the EPL as in the Bundesliga, then, eventually relegated teams proved most active on Twitter, while in the A-League, the 2011–2012 champions led the Twitter activity table.

Overall, however, in their tweeting styles EPL clubs resemble Bundesliga more than A-League clubs, as far as their interactions with fans are concerned: while there are some notable exceptions—by far the largest component of Wolverhampton's tweets are @replies—the majority of EPL clubs' tweets are original tweets, followed by retweets. Again, a substantial number of tweets also contain URLs, pointing to a communicative preference for information dissemination rather than fan engagement.

FAN RESPONSES

Such attempts by the clubs to reach out to their fans and followers tell only one side of the story, of course: fan reactions to the clubs' Twitter activities must also be considered—indeed, a three-way relationship between the size of established fan bases, sporting performance on the field, and (social) media performance by the club is likely to determine club-fan interactions on Twitter. For the A-League, it is evident that the number of @mentions of the clubs' accounts, and the number of retweets of the clubs' messages, do not match these accounts'

activities particularly closely (Figure 20.8): while the @brisbaneroar account
of the repeat champions leads both rankings, and other highly active accounts
(@SydneyFC and @MelbourneHeart) also received a substantial amount of fan
attention, the highly active @CCMariners account of 2011 championship runners-up Central Coast Mariners does not generate the level of fan responses
which would be expected; by contrast, the far more limited activity of Melbourne
Victory's @gomvfc account is sufficient to propel the club into second place in
the fan activity rankings.

It is notable in this context that the accounts receiving the most @mentions
and retweets in Australia are those of the clubs based in Australia's three major
metropoles: Sydney, Melbourne, and Brisbane. By contrast, the accounts of
regional clubs such as Central Coast Mariners (based in Gosford, north of
Sydney), Newcastle Jets, and even of clubs based in smaller state capitals, such
as Adelaide United and Perth Glory FC, receive comparatively less fan interaction—even where, in the case of @CCMariners or @PerthGloryFC, the clubs
played in A-League Grand Finals in 2011 and 2012, respectively. What audience interaction patterns seem to indicate, then, is a mixture of the relative success of the clubs during the year (2011/12 champions Brisbane Roar still lead,
despite the fact that Brisbane is the smallest of the three major metropoles in
Australia), *and* of general (Twitter) demographics in the country.

Additionally, it is evident that audience interaction with the accounts is
mainly through @mentions rather than through retweets of the accounts'

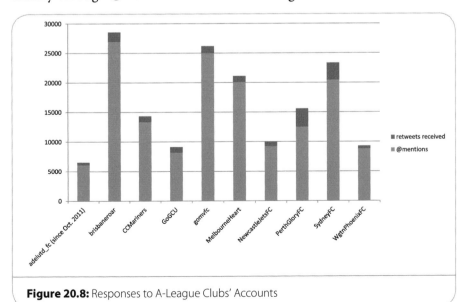

Figure 20.8: Responses to A-League Clubs' Accounts

messages. This points strongly to the fact that there is a substantial amount of fan activity *around* the clubs on Twitter (@brisbaneroar received some 28,600 @mentions and retweets over the course of the season), into which fans attempt to include the official club account by mentioning it, but that the clubs' official Twitter activities themselves—their own tweets—are not yet central to such expressions of fandom.

The situation in the Bundesliga is considerably different once again. Here, the majority of clubs struggled to reach more than 1,000 @mentions and retweets during the season—well below even the least visible A-League accounts (Figure 20.9). Only a handful of clubs stand out significantly from the rest, and also receive a notable amount of retweets. This is remarkable for the @BVB and @so4 accounts, which were only set up during the season itself, while @fckoeln's highly active outreach efforts explain its placing. Indeed, Dortmund's less memorable Twitter handle @BVBDortmund09 received some 3,000 @mentions and retweets, while the switch to the more straightforward @BVB in December saw that number increase by a factor of five. Similarly, Schalke 04's embrace of @so4, and the suspension of the @FCSchalke04 fan account, also appear to have focussed fan energy on the official account. By contrast, the unofficial and eventually suspended @BayerLeverkusen account generated almost as much fan engagement as the official, unintuitively named @bayer04fussball account—memorable Twitter handles *and* active use of club accounts clearly mattered in Germany.

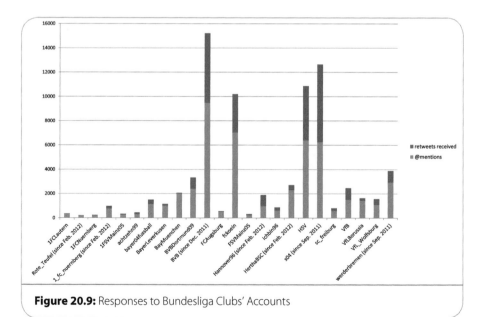

Figure 20.9: Responses to Bundesliga Clubs' Accounts

Overall, however, the lack of @mentions for the vast majority of club accounts points to the conclusion that fan activity around football clubs in Germany does not, in the main, happen on Twitter so far (especially if compared to the massive fan interest as expressed in audience turnouts, fan club activities, and media coverage); it remains to be seen whether Twitter Germany's campaign to get clubs tweeting during the 2012–2013 season can change this situation.

Fan interactions with English Premier League clubs, finally, reflect the much broader audience for this competition, but also a very uneven distribution of attention (Figure 20.10). Here, the average number of @mentions and retweets received by most club accounts is well above 30,000 over the course of the season, with several clubs surpassing that mark by a substantial margin: 2012 champions Manchester City (@MCFC), Liverpool (@LFC), Arsenal (@arsenal), and eventual 2012 UEFA Champions League winners Chelsea (@chelseafc) each attracted several hundred thousand tweets from fans—indeed, Chelsea received some 675,000 @mentions and nearly 300,000 retweets during this time. As noted above, the obvious exception on this list of leading EPL clubs is Manchester United, which did not operate an official Twitter account; the two unofficial accounts for the club received just over 20,000 tweets.

The four leading EPL clubs on Twitter also received substantially more retweets than the minor clubs; this points to the existence of a two-tier structure within the English Premier League itself, reflecting a distinction between those EPL clubs which are globally recognised, and which regularly participate in major international competitions, and those whose focus remains mainly on the domestic league. While the minor teams' domestic Twitter fan base still

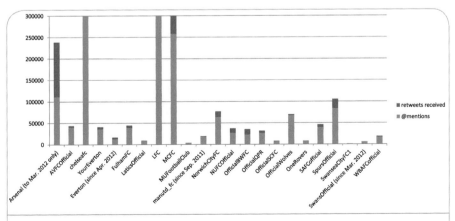

Figure 20.10: Responses to EPL Clubs' Accounts (Vertical Axis Truncated at 300,000 Tweets for Readability)

exceeds that of comparable Bundesliga clubs by an order of magnitude—pointing again to the slower, more limited take-up of Twitter in Germany, as well as to potential language barriers between German accounts and international fans—it necessarily pales against the substantial international following which successful clubs such as Chelsea or Manchester City have attracted.

CONCLUSION: TWITTER AND SPORTS FANDOM

The comparison between the three national contexts is telling, and demonstrates how a range of factors influence the ways Twitter is used to engage with fans. In Australia, Twitter has been used in a concerted manner, with many levels of the sport attempting quite overtly to leverage the social network's possibilities for the promotion of the A-League to a larger audience. This is driven by necessity, as the sport works to establish itself in a marketplace already crowded with other 'football' codes. The long-term stability (and financial viability) of the A-League depends on effective marketing; engaging with fans on Twitter is just one component of that effort. That outreach effort, however, has been successful: fan mentions of A-League clubs' accounts substantially surpass those of most Bundesliga clubs, and (per capita of the population) even compare well against those EPL clubs which compete mainly at a domestic level.

The situation in the European leagues—where clubs (and the leagues themselves) have existed for several decades and have managed to build support for generations, and where this code of football is the most popular form of sport by some margin—is quite different. English Premier League clubs divide into two categories: internationally recognised brands, whose visibility on Twitter is assured by the substantial activities of a global fan base; and domestic-grade teams, whose Twitter accounts speak mainly to a more localised fan base. Here, active outreach via Twitter can make a difference: much like @fckoeln in the Bundesliga, Wolverhampton, the most active Twitter user in the EPL, also received the second most @mentions of all the clubs within this second tier of EPL teams, even in spite of its poor sporting performance. Overall, however, it is likely that for the EPL clubs, the platform is only one part of a wider marketing mix: an addition to mainstream activities, but far from central to their efforts.

Finally, in Germany, the 2011–2012 season saw a relatively sluggish and seemingly *ad hoc* use of Twitter: at best, most clubs made some exploratory steps into the social media arena. With many clubs using the service merely in a broadcast mode, their interest in using Twitter as a new avenue for fan interaction was clearly limited during the season. Here, then, it is Twitter, Inc.

itself which appears most interested in expanding the clubs' presence on the platform, as part of its efforts to gain a stronger foothold in the German social media market; the success of its Bundesliga promotions during the 2012–2013 season remains to be seen.

It is worth emphasising that our analysis in this chapter clearly represents a sports *marketing* perspective. Even where the clubs themselves do little to encourage tweeting about football, it is likely that there will still be considerable fan activity on Twitter around clubs, players, competitions, and matches (for EPL and Bundesliga especially also involving international audiences). Such wider sports *fandom* may be studied most effectively by tracing activities around the keywords and hashtags associated with major competitions (e.g., #Brazil2014), players (e.g., Rooney, Robben), or clubs (e.g., Manchester United, Bayern München), rather than by assessing the performance of the official Twitter accounts of the clubs. What the present chapter illuminates, then, are the specific activities of, and fan responses to, the clubs' official accounts on Twitter. The considerably more complex story of how domestic and global football fandom unfolds in more general terms, and of how such fandom intersects with the sports marketing efforts of clubs, leagues, and Twitter, Inc. itself, has yet to be told.

REFERENCES

ABC News. (2010, 1 Feb.). Tweets ahead: Aussies embrace social media. Retrieved from http://www.abc.net.au/news/2010-02-01/tweets-ahead-aussies-embrace-social-media/317364

Choudhury, S., & Breslin, J. (2011). Extracting semantic entities and events from sports tweets. In *Proceedings of Making Sense of Microposts Workshop (#MSM2011)* (pp. 22–32), Crete, Greece. CEUR Workshop Proceedings, Vol. 718.

Deloitte. (2011). Annual review of football finance 2011: Foreword. Retrieved from http://www.deloitte.com/assets/Dcom-UnitedKingdom/Local%20Assets/Documents/Industries/Sports%20Business%20Group/uk_sbg_arff11_foreword.pdf

Gibbons, T., & Dixon, K. (2010). Surf's up! A call to take English soccer fan interactions on the Internet more seriously. *Soccer and Society, 11*(5), 599–613.

Jackson, S., & Christensen, N. (2012, 26 Nov.). Twitter flies in to meet leaders. *The Australian*. Retrieved from http://www.theaustralian.com.au/media/twitter-flies-in-to-meet-leaders/story-e6frg996-1226523789033

Kriemadis, T., Terzouidis, C., & Kartakoulis, N. (2010). Internet marketing in football clubs: A comparison between English and Greek websites. *Soccer and Society, 11*(3), 291–307.

Krøvel, R. (2012, 7 May). New media and identity among fans of a Norwegian football club. *First Monday, 17*(5). Retrieved from http://firstmonday.org/htbin/cgiwrap/bin/ojs/index.php/fm/article/viewArticle/2882/3208

Ladyman, I. (2012, 16 Aug.). Man United star Rio hit with £45,000 fine from FA over Cole 'choc-ice' tweet. *Daily Mail*. Retrieved from http://www.dailymail.co.uk/sport/football/article-2189769/Rio-Ferdinand-fined-Ashley-Cole-choc-ice-tweet.html

McCarthy, J., Pioch, E., Rowley, J., & Ashworth, C. (2011). Social network sites and relationship marketing communications: Challenges for UK football clubs. In *Proceedings of MindTrek 2011* (pp. 145–152)., 28–30 Sep. 2011. Tampere, Finland. New York, NY: ACM.

McLean, R., & Wainwright, D. W. (2009). Social networks, football fans, fantasy and reality: How corporate and media interests are invading our lifeworld. *Journal of Information, Communication and Ethics in Society, 7*(1), 54–71.

Merkel, U. (2012). Football fans and clubs in Germany: Conflicts, crises and compromises. *Soccer and Society, 13*(3), 359–376.

Meyer, D. (2012, 27 Mar.). What does Twitter want with Germany? *GigaOM*. Retrieved from http://gigaom.com/2012/03/27/what-does-twitter-want-with-germany/

Nettooor. (2011, 30 Aug.). Account @S04 suspended #fail Schalke. Retrieved from http://www.nettooor.be/?p=1327

Nichols, J., Mahmud, J., & Drews, C. (2012). Summarizing sporting events using Twitter. In *Proceedings of 17th International Conference on Intelligent User Interfaces (IUI '12)* (pp. 189–198), Lisbon, Portugal. New York, NY: ACM.

Twitter auf Deutsch. (2012, 24 Aug.). Die Fußball-Saison 2012/2013 auf Twitter. Retrieved from http://blog.de.twitter.com/2012/08/die-fuball-saison-20122013-auf-twitter.html

Twitter UK Blog. (2012, 27 Sep.). Discover the Premier League. Retrieved from http://blog.uk.twitter.com/2012/09/twitter-is-great-place-for-football.html

Public Enterprise-Related Communication and Its Impact on Social Media Issue Management

21
CHAPTER   Stefan Stieglitz and Nina Krüger

 what decision makers in enterprises can learn from @adidas, @toyota and @qantasairways

In recent years, companies have realised that providing a website is not sufficient to satisfy customer's online needs. Further effort is required to better exploit the potential benefits that arise from social media (Kietzmann, Hermkens, McCarthy, & Silvestre, 2011), such as improved stakeholder management, stakeholder integration, open innovation, and crowdsourcing activities. Communication data in public social media can be understood as a rich source of information that can be utilised by enterprises. Additionally, enterprises are also able to interact directly and publicly with their target groups. However, social media platforms like Twitter are still regarded by most companies as black boxes when it comes to interactions with stakeholders (Rui, Liu, & Whinston, 2010): enterprises face the challenge, for example, of having to identify relevant pieces of communication, of having to react appropriately to messages from customers, or of being suddenly affected by negative feedback, or even by social media "shitstorms" (social crises). Already, several studies have explored the relevance of public social media for enterprises (e.g., Parveen, 2012; Reinhold & Alt, 2012). It has been observed that customers (among other stakeholders) have an interest in gathering information about brands and products on the Internet, and that public information (e.g., recommendations or complaints) may influence their buying decisions. Additionally, it has been shown that companies aim to identify feedback and complaints provided by customers (Raisinghani, 2012).

Overall, from the perspective of enterprises, regular, day-to-day communication (DC) can be differentiated from issue-related communication (IC). In this sense, DC is understood as communication which takes place among stakeholders and which is not dominated by a timely, limited discussion about a certain event or issue. Following the issue management model, certain topics may evolve into crises that are subsequently able to harm the enterprise (for example, by influencing people not to buy a company's products). One such issue, which was discussed in social media and evolved into a full-blown crisis, was experienced by Domino's Pizza: two employees posted a video on YouTube which showed shocking actions performed on customer's meals. Rapid diffusion of this issue on Twitter resulted in a volume of some 20,000 tweets, generated in a time span of eight days (Park, Cha, Kim, & Jeong, 2012).

To date, little research has been done to better understand these two types of communication (DC and IC) and their distinct characteristics. As a result, there also remains a lack of knowledge about how to measure crisis communication (CC) in social media. Furthermore, methods and best practices for gathering and analysing the necessary data are missing. This is one aspect which prevents enterprises from developing appropriate social media strategies (Culnan, McHugh, & Zubillaga, 2010).

We address this field by providing insights into three different cases which cover enterprise-related communication on Twitter in times of crisis, as well as during day-to-day communication phases. The comparison of three cases with differing backgrounds helps to identify and to better understand dynamics in Twitter, and therefore also might support decision makers (see Nitins & Burgess, Chapter 22 in this volume, for another use case and a complementary, user-centric perspective on brand communication and an in-depth discussion of the relation between brands and their users).

The remainder of the chapter proceeds as follows: in the next section we provide an overview of related scholarly work about the topic of enterprise-related communication in social media. Following this section, we give a short overview of social media-related issue management. Next, we introduce the three case studies by first explaining the methodology and then describing the general characteristics of each case. We then draw conclusions by comparing all three cases. The chapter ends with a conclusion and an outlook for further research.

RELATED WORK

Research in the field of enterprise-related communication in public social media remains very sparse, especially regarding microblogging platforms such as Twitter. This is surprising because of the potentially strong impact which social media communication might have on enterprises. Rui, Liu, and Whinston (2010) demonstrated Twitter's suitability as a platform for stakeholder communication, and suggested that it can be understood as a worthwhile source of information for both researchers and practitioners. Insights gathered from social media data might lead to a better understanding of the black box of (online) word-of-mouth. In times of crises (often first mentioned in traditional media), there is a certain danger that stakeholders catch up such issues and start to communicate negatively about the enterprise or its brands (e.g., criticising the information strategy or the abilities of a product) in social media spaces. Social media-based communication in this context might have an even stronger effect on enterprises than traditional press releases, because, as studies show, peer communication has a larger impact on future buying decisions (Rui, Liu, & Whinston, 2010). Additionally, those issues which were not published or which were only briefly mentioned in traditional media, may, based on social media communication, evolve and develop into a corporate crisis (Stieglitz & Krüger, 2011). Stakeholders, in this context, are understood as "any group or individual who can affect or is affected by the achievement of the organization's objectives" (Freeman, 1984, p. 46): customers, investors, shareholders, suppliers, employees, government, and media.

Recently, Park et al. (2012) analysed the Domino's Pizza crisis; in their study, the authors came to the conclusion that there exist certain patterns for how information spreads through the Twitter network, resulting in the emergence of an issue. For example, tweets that contain URLs spread faster through the network, on average (Park et al., 2012). This has also been shown for the political context by Stieglitz and Dang-Xuan (2012). Additionally, they also showed that tweets featuring words that indicate either positive or negative sentiment tend to receive more retweets than neutral posts. Furthermore, it has to be taken into account that certain individuals are much more active in social media, and therefore might become important opinion leaders and multipliers of information. For example, Heil and Piskorski (2009) investigated a sample of 300,000 randomly chosen tweets and showed that Twitter communication is characterised by a strongly unequal distribution (top 10 percent of prolific Twitter users publish more than 90 percent of tweets). Numerous other studies have identified Twitter as a platform that is used intensively to spread information about enterprises, products, and political parties (e.g., boyd, Golder, & Lotan, 2010; Stieglitz & Dang-Xuan, 2012; Stieglitz & Krüger, 2011; Tumasjan, Sprenger, Sandner, & Welpe 2010).

As outlined by Larson and Watson (2011), companies are interested in making their stakeholders aware of new products and campaigns, while customers want companies to be aware of their complaints about products and services as well as their suggestions for new ones. Further, customers want to communicate directly with other customers and share both positive and negative experiences they had with products or brands. Following Berthon, Pitt, and Watson (1996), there is a clear distinction between this bidirectional interaction and the traditional PR/marketing and customer service, which is based on the concept of a unidirectional "customer notification".

SOCIAL MEDIA ISSUE MANAGEMENT

Coombs (2007) defined a corporate crisis as "a sudden and unexpected event that threatens to disrupt an organisation's operations and poses both a financial and a reputational threat" (p. 164), and pointed out that this definition implies the impact of an issue on both the company and its stakeholders. So far, there is a lack of research into the patterns of corporate crises in general, and into how such crises unfold in the social media sphere in particular. In fact, the characteristics of corporate communication in public social media discussions in general are also under-researched. In both areas, further research can improve our understanding of patterns in public communication, and might aid enterprises in improving their adoption of social media. The gathering of detailed empirical data about patterns and dynamics in enterprise-related discussions in social media is crucial to this endeavour.

Of course, enterprises are aware that traditional public communication (as it takes place in newspapers, radio, or television) might affect their business, and that issues may evolve to a corporate crisis when they actually or potentially concern the enterprise. Therefore, larger enterprises have established well-directed issue management processes in order to monitor or even influence public opinion about their products, services, and reputation. This is not a trivial task, since aspects which have to be considered include the heteroge-

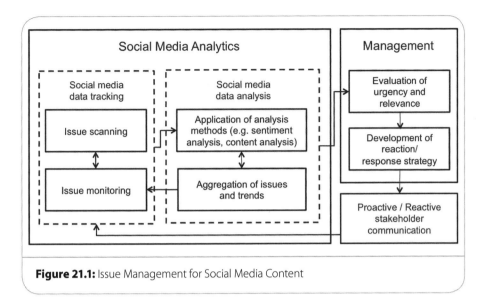

Figure 21.1: Issue Management for Social Media Content

neous expectations of different stakeholders, the fact that pieces of communication can be interpreted in various ways, and the problem that the reactions of an enterprise might also contain further potential to extend the crisis, or might lead to polarisation among the target group.

However, issue management in social media is more complex even than in traditional media. First of all, considerably more content has to be taken into account every day, and relevant content will usually be unstructured and hard to identify. As Figure 21.1 shows, enterprises need to scan (search for new relevant issues) and monitor (observe already known, potentially critical issues) the public social media sphere continuously. Following this, the gathered data have to be prepared (e.g., remove spam or filter relevant subsets) and analysed by applying automatic or manual methods of content analysis (e.g., Einspänner, Dang-Anh, & Thimm, Chapter 8 in this volume; Stieglitz & Dang-Xuan, 2012; Stieglitz & Krüger, 2011). At this point, the appropriateness of certain methods depends on the specific goals of the analysis (e.g., identification of lead users, investigation of dynamics in sentiments of communication about a product). In a next step, the information about issues and trends has to be aggregated in order to develop a response strategy. One crucial step for companies is then to evaluate the urgency and relevance of the discovered issues, in order to develop a response strategy, and to decide on whether to act proactively or reactively (Stieglitz & Krüger, 2011).

Determining the appropriate reaction to issues in social media is difficult, since, for the most part, best practices have not yet been established.

Additionally, it has to be considered that crisis situations usually have a unique character which makes it difficult to elaborate a structured management process. However, as the model implicates, companies which establish a comprehensive social media analytics process will be able to better understand the stakeholder's concerns in times of crisis, and might therefore develop better response strategies. Furthermore, social media analytics might also extend the management's response time by discovering critical issues in an early state.

CASE STUDIES OF ENTERPRISE-RELATED TWITTER COMMUNICATION

In order to learn more about the characteristics of enterprise-related communication on Twitter, we conducted three explorative case studies which examined adidas, Toyota, and Qantas. Two of these cases have already been published and extensively discussed (Krüger, Stieglitz, & Potthoff, 2012; Stieglitz & Krüger, 2011). However, to date, none of the data sets has been compared to other cases. Since it has already been described in other publications, we will outline the method of data collection and data analysis only briefly here, and instead focus on comparing the cases and their potential implications for issue management.

METHODOLOGY

From 2011 to 2012, we collected Twitter communications about several enterprises by tracking the tweets which contain keywords relating to these companies. We conducted a basic analysis on all data sets in order to identify crisis situations, which we found for Qantas and Toyota. By contrast, adidas has not been affected by any obvious crises during the period of investigation, providing us with an opportunity to compare crisis situations in social media with general day-to-day communication (DC).

The tweets were tracked by applying a software prototype which uses the Twitter API to collect tweets containing specific hashtags or keywords (for a description of the prototype, see Stieglitz & Kaufhold, 2011). Furthermore, metadata such as the timestamp, account name, language, hashtags, and URLs for each tweet are also available. Based on these data it is possible to conduct temporal, user-centric, or group-centric analysis (see Chapter 6 by Bruns & Stieglitz in this volume).

CHARACTERISTICS OF THE CASE STUDIES

adidas

The case study includes 289,513 tweets, which were collected over 60 consecutive days. Overall, the amount of tweets per day did not change substantially within the time frame observed. This result could be expected, as there were no serious public discussions about adidas during this time span. It turned out that adidas followed a clear social media strategy, and established hashtags and maintains several user accounts for different purposes (e.g., news, discussion) and different target groups (e.g., in different countries). In general, a large amount of attention is paid to adidas on Twitter even during times which are not characterised by specific brand issues.

Toyota

Based on the keyword combination of "Toyota" and "recall" (referring to a global recall of various Toyota models, due to faulty brakes), we collected 37,323 tweets over a time frame of 26 weeks. To receive these data we first tracked all tweets containing the keyword "Toyota" (730,000 tweets), and then extracted those tweets which additionally contain the keyword "recall". Within this time frame, certain peaks and troughs of activity could be identified. In most cases, peaks seemed to be triggered by new press releases about the brake problems on Toyota cars. It can also be shown, however, that even during such peaks, users do not increase the frequency of postings significantly; rather, the peaks in Twitter activity are caused by more users entering the discussion. Further, it turned out that overall communication about this issue was clearly dominated by a small group of user accounts. Overall, 10 lead users (0.07% of all users who published at least one tweet containing the two keywords) published more than 17.5% of all the tweets in the data set.

Qantas

This dataset consists of some 240,000 tweets which were collected based on the keyword "Qantas", during two major brand crises affecting the Australian airline Qantas—the volcanic ash cloud caused by the eruption of Chilean volcano Puyehue in June 2011, and the global grounding of Qantas flights ordered by management in the course of an industrial dispute in October–November 2011. The results of our examination of Twitter content during the two crises show that there are vivid discussions about enterprise-related issues in the network. Results of the case study show that communication of the first, brand-related crisis was based more strongly on news sharing, while the other was focussed more strongly on discussion. We also found that this distinction is most pronounced for those user groups which play a leading role within the overall communication process, as measured based on the volume of their contributions. Therefore, the accounts of lead users, including the relevant companies themselves, play pivotal roles within overall discussion.

DISCUSSION

In order to develop a better understanding of enterprise-related communication on Twitter, we considered data from three different case studies. From the perspective of decision makers in enterprises, one important aspect to establish is the potential relevance and impact which Twitter communication might have. As a simple approach to contribute to this, we compared the dynamics of communication in all three cases, and observed the intensity of peaks in crisis situations in contrast to regular day-to-day communication.

Figure 21.2 displays the dynamics of the tweet volume during time periods analysed for each case. To provide a better overview and visualisation, we have transformed the individual periods of each case to one scale (time: x-axis). In the context of each of the three different cases *100* is a marker for the specific amount of days, the crisis-related communication has been investigated (adidas: 100 = 60 days; Toyota: 100 = 182 days; Qantas: 100 = 150 days). Similarly, at the y-axis we use a marker (1–8) to provide information about the tweetvolume of each of the three different cases (adidas: 1 = 3,113 tweets; Toyota: 1 = 1,000 tweets; Qantas: 1= 3,169 tweets).

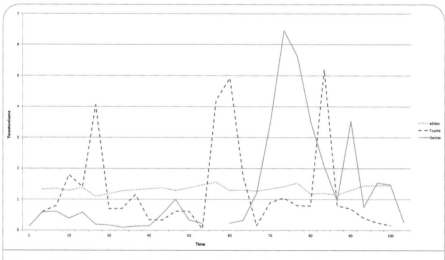

Figure 21.2: Dynamics of Tweet Volume in the Three Case Studies

In one sense, the figure neither reflects the correct relation between the total tweet volumes of the adidas, Toyota, and Qantas cases, nor did the evaluated crises take place at the same time. However, it does make obvious that crisis situations (such as those in the Toyota and Qantas cases) seem to cause a strong relative growth of Twitter activity beyond the normal, day-to-day baseline of activity.

A deeper investigation of the Toyota data set reveals that in times of crisis, significantly more user accounts start to publish comments (Stieglitz & Krüger, 2011). Therefore, it can be said that Twitter communication is strongly affected by crisis situations, since a significantly greater amount of stakeholders join the discussion during such times. It is also obvious that communication on Twitter is strongly influenced by articles in traditional media. Therefore, in the case of Toyota, it might have been a promising strategy to concentrate social media analytics strategies on monitoring the impact of Toyota's press releases, in order to improve and shape further PR activities on that information.

In contrast to Toyota, the Qantas dataset includes two different crises affecting the same company. The first crisis was caused by a natural disaster, whereas the second one was initiated by the management of the airline. We found that the patterns of communication differ between these two crises, both with regard to the behaviour of the lead users in the discussion and across the overall data set. Communication in the first crisis is primarily focussed on spreading new information (with a high number of retweets and URLs), while communication around the second crisis is characterised by a high number of original tweets and @replies: this indicates that the second crisis (caused by Qantas management) resulted predominantly in more discussion among users.

Enterprises therefore need to develop strategies for how to react to such situations. It also seems to be necessary to identify different types of crises which might affect enterprises. From our findings, it appears that some crisis situations are more interactive than others; influencing factors might be the foreseeability of the event, or the degree to which users think that their opinions might change the company's behaviour. It can be assumed that users expect companies to participate especially in those situations which are characterised by a high degree of interaction and discussion; therefore, issue management needs to identify the type of crisis in order to develop appropriate response strategies. Another important point would be to establish early-warning systems within the social media activities by improving issue scanning and issue monitoring, in order to identify upcoming types of crisis as soon as possible. Furthermore,

Qantas management would have needed to develop different strategies to deal with the Twitter discussions in each case, given the different issues at stake.

As mentioned above, communication about adidas was not affected by any crisis. Figure 21.2 shows that this resulted in a relatively stable amount of tweets per day. In all three cases, it could be observed that the enterprise account itself did not play a major role, and generally published a low amount of tweets. Based on a genre analysis on the adidas data set, we found that this enterprise account mostly publishes content on their own, without really starting a discussion or answering other users (Krüger et al., 2012). For example, adidas primarily provides information on its brand account about issues directly related to the brand (such as new products), or about topics which are indirectly related to the brand (such as sports and sporting events). Similarly, users seem to communicate about adidas in order to make other participants of their network aware of their preference for the brand (named as signalling), rather than to get into contact with the brand itself. In this sense, the day-to-day communication observed here features only a low level of bidirectional interaction. Our data indicate that, even though adidas follows a clear strategy for its Twitter activity, the resultant effect seems to be rather low. Based on this case, various implications can be drawn for social media analytics and management: e.g., adidas would be able to monitor Twitter communication within the investigated time frame and discover that there are no crises concerning the brand, and might therefore concentrate more fully on identifying lead users in its Twitter fan base and building strong connections with them. Based on this, adidas might be successful to encourage more intensive interaction between stakeholder and company, and at the same time increase customer loyalty.

As our study shows, issue management can be supported by social media analytics in various ways. Supported by keyword tracking, it is possible to identify enterprise-related communication on Twitter and establish a continuous monitoring which allows a better understanding of the dynamics of certain issues. A deeper understanding of various case studies may therefore help to develop more advanced methods for the early detection of brand crises. When a crisis actually takes place, however, it is difficult to develop a general response strategy, since communication about an issue is usually case-specific and therefore requires the development of individual response strategies. Nevertheless, by gathering more information about certain types of crises and their effects on stakeholder communication, it might become possible to develop frameworks or general guidelines for the development of effective response strategies.

CONCLUSION

Our study shows that the issue management of enterprises has to consider the change of communicative behaviour as a greater amount of public communication transitions to social media. The considerable number of messages on Twitter and the high numbers of participants demonstrate the need for continuous issue scanning and issue monitoring as a starting point for the development of communication strategies which might include proactive and/or reactive participation in discussions at an early stage. However, it has to be noted that our results are based on only three case studies, and that their scope is therefore limited. Based on the cases which we have been able to observe, communication on Twitter exhibits a high level of complexity. Not only does day-to-day communication seem to be significantly different from crisis-related communication, but there are obviously different types of communication patterns which unfold in different types of crises. What is needed next is to develop a framework which considers structural elements (such as the usage of URLs, retweets, etc.), but also addresses content-related aspects (such as foreseeability, the perceived potential to influence management decisions, etc.).

REFERENCES

Berthon, P., Pitt, L. F., & Watson, R. T. (1996). The World Wide Web as an advertising medium: Toward an understanding of conversion efficiency. *Journal of Advertising Research, 36*(1), 43–54. doi:10.1017/S0021849996960067

boyd, d., Golder, S., & Lotan, G. (2010). Tweet, tweet, retweet: Conversational aspects of retweeting on Twitter. In *Proceedings of the 43rd Hawaii International Conference on System Sciences (HICSS-43)* (pp. 1–10). doi:10.1109/HICSS.2010.412

Coombs, W. T. (2007). Protecting organization reputations during a crisis: The development and application of situational crisis communication theory. *Corporate Reputation Review, 10*(3), 163–176. doi:10.1057/palgrave.crr.1550049

Culnan, M. J., McHugh, P. J., & Zubillaga, J. I. (2010). How large U.S. companies can use Twitter and other social media to gain business value. *MIS Quarterly Executive, 9*(4), 243–259.

Freeman, R. E. (1984). *Strategic management: A stakeholder approach.* Cambridge, UK: Cambridge University Press.

Heil, B., & Piskorski, M. (2009). New Twitter research: Men follow men and nobody tweets. Retrieved from http://blogs.harvardbusiness.org/cs/2009/06/new_twitter_research_men_follo.html

Kietzmann, J. H., Hermkens, K., McCarthy, I. P., & Silvestre, B. S. (2011). Social media? Get serious! Understanding the functional building blocks of social media. *Business Horizons, 54*(3), 241–251. doi: 10.1016/j.bushor.2011.01.005

Krüger, N., Stieglitz, S., & Potthoff, T. (2012). Brand communication in Twitter: A case study on adidas. In *Proceedings of the Pacific Asia Conference on Information Systems (PACIS) 2012*, Paper 161. Retrieved from http://aisel.aisnet.org/pacis2012/161/

Larson, K., &Watson, R. T. (2011). The value of social media: Toward measuring social media strategies. In *Proceedings of the International Conference on Information Systems (ICIS) 2011* (pp. 1–18).

Nielson, J. (2006). Participation inequality: Encouraging more users to contribute. Retrieved from http://www.useit.com/alertbox/participation_inequality.html

O'Riordan, L., & Fairbrass, J. (2008). Corporate social responsibility (CSR): Models and theories in stakeholder dialogue. *Journal of Business Ethics, 83*(4), 745–758. doi: 10.1007/s10551-008-9662-y

Park, J., Cha, M., Kim, H., & Jeong, J. (2012). Managing bad news in social media: A case study on Domino's pizza crisis. In *Proceedings of the International AAAI Conference on Weblogs and Social Media (ICWSM) 2012*. Retrieved from https://www.aaai.org/ocs/index.php/ICWSM/ICWSM12/paper/view/4672

Parveen, F. (2012). Impact of social media usage on organizations. In *Proceedings of the Pacific Asia Conference on Information Systems (PACIS) 2012*, Paper 192. Retrieved from http://aisel.aisnet.org/pacis2012/192/

Raisinghani, M. (2012). Social media and e-commerce: A strategic perspective. *Proceedings of the 18th Americas Conference on Information Systems (AMCIS) 2012*, Paper 68. Retrieved from http://aisel.aisnet.org/amcis2012/proceedings/Posters/68/

Reinhold, O., & Alt, R. (2012). Social customer relationship management: State of the art and learnings from current projects. In *Proceedings of BLED 2012*, Paper 26. Retrieved from http://aisel.aisnet.org/bled2012/26

Rui, H., Liu, Y., & Whinston, A. B. (2010). Chatter matters: How Twitter can open the black box of online word-of-mouth. In *Proceedings of the International Conference on Information Systems (ICIS) 2010* (pp. 499–505). Retrieved from http://aisel.aisnet.org/icis2010_submissions/204/

Stieglitz, S., & Dang-Xuan, L. (2012). Social media and political communication: A social media analytics framework. *Social Network Analysis and Mining.* doi:10.1007/s13278-012-0079-3

Stieglitz, S., & Kaufhold, C. (2011). Automatic full text analysis in public social media: Adoption of a software prototype to investigate political communication. In *Proceedings of the 2nd International Conference on Ambient Systems, Networks and Technologies (ANT-2011) / The 8th International Conference on Mobile Web Information Systems (MobiWIS 2011). Procedia Computer Science, 5*, 776–781.

Stieglitz, S., & Krüger, N. (2011). Analysis of sentiments in corporate Twitter communication: A case study on an issue of Toyota. In *Proceedings of the 22nd Australasian Conference on Information Systems (ACIS)*, Paper 29. Retrieved from http://aisel.aisnet.org/acis2011/29/

Tumasjan, A., Sprenger, T. O., Sandner, P. G., & Welpe, I. M. (2010). Election forecasts with Twitter: How 140 characters reflect the political landscape. *Social Science Computer Review.* Advance online publication. doi: 10.1177/0894439310386557

Twitter, Brands, and User Engagement

22
CHAPTER   Tanya Nitins and Jean Burgess

brand communication from play to parody and
PR #fails – and where does Twitter's business
model fit in?

In social media services, users mostly generate unverified information—
both true and false—and put forth ideas about organisations that can
differ greatly from what organisations share with the public—that is, an
organisation's own idea of what it is or what it wants to be.

—Aula, 2010, p. 45

Businesses spend millions of dollars every year carefully tailoring their brands, and even more protecting them. This process was relatively easy to manage through traditional media with their one-to-many approach, with control over the brand's aura remaining with the advertiser. Yet with the emergence of social media, the traditional brand communication process has reached something of a crisis. Traditional communication lines are rapidly breaking down, with social media disrupting the relations among brand owners, consumers, competitors, and other stakeholders to encompass more dialogic, even antagonistic, models of communication; Twitter is a particularly educative example of this shift. In this chapter, we are concentrating on the two-way communicative environment that Twitter generates, and the discussion between brands and users that can ensue (see Stieglitz & Krüger, Chapter 21 in this volume, for a complementary, data-centric analysis of brand communication that focusses on different business strategies for employing Twitter).

As soon as it became clear that Twitter was becoming an important social networking site and a public communication platform, a number of businesses and "social media marketing" professionals attempted to exploit the platform for commercial purposes—from straight public relations and advertising through to more underhanded viral marketing tactics. For many businesses, it was the "popularity of communities on the Internet [that] captured the attention of marketing professionals" (McWilliam, 2000, p. 43). With the promise of instant, free access to consumers around the world gathered together in dominant platforms like Facebook and Twitter, it was little wonder that so many brands suddenly launched themselves into the social media space. However, the reality is that many of these businesses entered into the social media environment without fully appreciating its already-established dialogic culture. In many cases, standard advertising models were simply transferred into the online environment provided by platforms like Twitter without first considering the possible repercussions on their brands.

The qualities and meanings attached to a brand—that is, the very value of the brand itself—become vulnerable to constant renegotiation in "this ebbing and flowing space that is subject to a wide variety of influences moving increasingly beyond the control of the organisation" (Ind, 2012, p. 36). The shift to incorporate the "many-to-many" communicative affordances of Twitter can have significant implications for the standard brand communication process by disrupting the traditional "top down" models of marketing and brand communication. Through Twitter, consumers are now not only able to "talk back" to companies—even very large global corporations—but to do so in public; they can share their pleasure, or displeasure, with potentially millions of other consumers without significant effort.

Yet, despite the risks posed by the new transparency of social media, businesses are potentially disadvantaged if they avoid engaging in social media. The reality is that "conversations [regarding their brands] are [already] taking place whether or not companies are participating in them" (Thoring, 2011, p. 142). According to Kinzey (2009), "organizations that fail to listen and engage with customers, special interest groups, and employees in the social media world will likely find they have relinquished control of their reputation to others". This lack of control becomes particularly apparent during moments of controversy and bad press. Messages that are quickly spread through social media services are increasingly being picked up and reported upon by the mass media (Singer, 2012). Because of this, it is becoming much more difficult for businesses to hide behind their brand during moments of crisis (Aula, 2010, p. 43). It is vitally important that businesses acknowledge the impact Twitter can have on their brand, and adjust their brand communication strategies accordingly.

TWITTER AND BRAND CONVERSATIONS

The "brand conversations" that occur on Twitter operate on a variety on levels. First, Twitter provides an open space for consumer engagement and participation. Instead of trying to control or silence these conversations through heavy-handed measures, some businesses have successfully maximised the impact of this online participation and engagement by providing them with an official space to congregate and "play" with their brand. For example, Nike recently introduced its +GPS app that encourages their fan base to monitor and share their fitness training with friends and other Nike users through social media services such as Twitter (Business Wire, 2012). Starbucks' Facebook page has over 200,000 fans and over 2 million followers on Twitter (Gembarski, 2012), with people logging on regularly into their online cafe to share experiences and reviews. According to M2 PressWIRE (2012), "Amazon's UK facility to easily 'tweet purchases' is [also] a big factor in helping the online retailer to generate brand awareness amongst wider online social circles".

However, Twitter has its own culture, with a prevalent libertarian and/or anti-establishment ideology which is generally resistant to overtly commercial uses of the platform. Twitter users frequently delight in "gotcha" moments, picking up on PR mistakes and gaffes, and then exploiting them for parody purposes. There are now so many parody accounts that Twitter has had to develop its own policy guidelines on what is considered acceptable use (Twitter, 2012). Indeed, some parody accounts have a greater online presence than the official sites they are copying—during the 2010 Gulf oil spill crisis (an oil spill at the Gulf of Mexico, caused by the Deepwater Horizon drilling rig explosion in April 2010), BP America's feed on Twitter was being "drowned out" by the satirical account @BPGlobalPR (Cohen, 2010, p. 18). The sheer popularity of some of these accounts have even forced Twitter to recently restore a New York Times parody account, despite the official news service having filed an official complaint (Asian News International, 2012). So while parody accounts in particular demonstrate the cultural and political dynamism of Twitter, there is no guarantee that developing an online presence through Twitter will prove to be beneficial to a brand.

The impact and effect of poorly judged advertising campaigns in this space can be immediate—if businesses are not monitoring and gauging the mood of the Twitter community, the results can be disastrous. The Qantas Luxury Twitter campaign, for example, has been described by some PR experts as "perhaps Australia's greatest public relations failure" (Taylor, 2011). The competition gave Twitter users the opportunity to win one of 50 luxury first-class amenity packs—all users had to do was to define what "Qantas luxury" meant to them. What the public relations department had failed to appreciate was how volatile public sentiment at that time was, launching the "luxury" campaign days after failed union negotiations had grounded the Qantas fleet and left thousands of travellers stranded around the world. The Qantas Twitter account was subsequently flooded with thousands of angry or satirical posts on the #qantasluxury hashtag, which was then reported upon and relayed by mainstream news services (ABC News, 2012), amplifying the social media response very significantly.

Qantas attempted to duplicate a traditional public relations strategy reminiscent of the "in 25 words or less" marketing campaigns often used in marketing through mainstream media. It was applied to an online social environment with little or no adaptation. In more traditional campaigns, businesses were able to filter the responses once they had been received, and only publish the ones that were the most flattering and conducive to their brand message. Qantas failed to appreciate that once you enter the online social environment, you can no longer control the message nor censor what is or is not seen by the wider online community. As Bruns (2012) noted, "choosing Twitter as the platform for their promo activities . . . Qantas didn't have access to similar forms of censorship; once unleashed, there was nothing they could do to stop the barrage of criticism".

Some of the tension and resistance to advertising from the Twitter user community also seems to be connected to the uncertainty around Twitter's own business model, and therefore its moral contract with the user. Despite its remarkable growth, Twitter has notoriously struggled to find a business model that could be retrofitted to the service; and is particularly challenged in its attempts to integrate advertising into the platform. Simply put, because Twitter never had an established advertising focus nor a commercial business model, users of the site often rebel against what they perceive to be an encroachment on their personal space (one that previously had been relatively free of commercial activity). This rebellion is particularly evident in cases where an attempt has been made to camouflage commercial messages as genuine tweets, or as referred to in other media, "cash for comments".

For example, it was recently revealed that the South Australian Tourism Commission had been paying various celebrities to "endorse" the virtues of Kangaroo Island through their personal Twitter accounts. According to ABC News (2012), various celebrities, including celebrity chef Matt Moran and singer Shannon Noll, were "paid up to $750 for one tweet about the island", depending on the size of their followers. Officials at the South Australian Tourism Commission seemed to be puzzled by the backlash that occurred from the Twitter community. Paid celebrity endorsements are standard practice in traditional marketing campaigns; however, these practices cannot be easily transferred to the specific dynamic of the Twitter space. As digital media commentator Stilgherrian stated, "Twitter is about being authentic . . . if someone comes out and says something and doesn't declare that they're being paid to do it, then they're being dishonest and unethical" (as quoted in Watson & Novak, 2012).

In each one of these examples, the primary issue was in relation to companies failing to acknowledge or even to recognise the importance of the discursive nature of the online social environment. They enter into the space and attempt to control how users engage with their brand whilst employing traditional, one-to-many brand communication strategies in a space that naturally encourages many-to-many discussion. To further illustrate the importance of these online conversations, we now draw upon a notable incident of brand crisis to examine the types of conversation that were occurring around the brand at the time, and explore the way in which the company did—or did not—engage effectively with their online consumers, and the impact this had on consumer engagement. In this case study, we collected data containing the #sony hashtag to track the conversations during the 2011 Playstation hacking incident, and used thematic analysis to catalogue the areas of most concern to the online Sony community. This analysis also helped determine the level of Sony's engagement with this online community and the ways in which the company did—or did not—use Twitter to keep the public informed during the crisis.

THE SONY PLAYSTATION NETWORK HACKING INCIDENT

The Playstation Network (PSN) was established by parent company Sony in 2006. The online gaming and media service "was at the heart of the company's efforts to differentiate itself" from rival competitors (The Wall Street Journal,

2011). By 2011, the network had approximately 77 million registered users and was growing. To register, users had to log into the system with a username and password combination and provide credit card details—regardless of whether they used the online purchasing component of the service. The high rate of registration meant that Sony's attempts to "establish a [new] business model that links gadgets to an online network of games, movies and music" was rapidly becoming a success (The Wall Street Journal, 2011). This emphasis on developing an online network, in addition to the already extensive online consumer base whose members were active in the space already through online games and discussions, would suggest a strong need for Sony to know how to effectively operate in this space. However, as the evidence suggests, Sony is still stuck in the traditional brand communication mindset when it comes to engaging with its online consumer base.

In mid-April 2011, the Playstation Network was suddenly shut down without explanation. Frustrations quickly spread through social media sites such as Twitter, as gamers around the world voiced their annoyance at not being able to access their online games. Their frustrations grew as Sony remained silent on the reasons behind the shutdown and provided no indication of when the Network would be operational again. It would be over a week until Sony admitted that the closure of the Playstation Network had been in response to a massive security breach that had compromised the personal details (possibly including credit-card information) of its 77 million registered users (Goldberg, 2011). The response on Twitter was instantaneous—the tone of the tweets using the #sony hashtag quickly changed from frustration and impatience to shocked anger. The #sony hashtag increasingly became dominated by retweets of news feeds detailing the admission by Sony, as the community began spreading the word to one another.

By gathering the tweets containing the #sony hashtag which were published during the peak period (17 April–15 May 2011), we were able to reconstruct a detailed timeline of the community's real-time response to these events:

17–19 April 2011:	Hacking of PSN commenced
20 April 2011:	Site "undergoing maintenance" statement issued
25 April 2011:	Senior Director "no ETA" (estimated time) of PSN available
26 April 2011:	First mention of personal details hacked
1 May 2011:	Sony "welcome back" program
2 May 2011:	Press related to possible stolen credit-card information
15 May 2011:	PSN back online

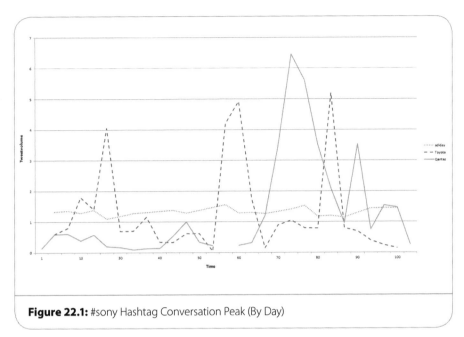

Figure 22.1: #sony Hashtag Conversation Peak (By Day)

We identified a significant peak in Twitter activity following Sony's media release about the hacking incident—particularly on Wednesday, 28 April 2011, as people began retweeting the news to other users. In addition, by breaking down the initial data set to an hour-by-hour analysis, it was possible to isolate a specific, 16-hour period when conversation using the #sony hashtag was most intense.

The tweets published during this 16-hour window still numbered over 78,000, so every 20th tweet was sampled in order to conduct manual thematic analysis on a more manageable data set of approximately 4000 tweets.

This specific time frame offers an excellent opportunity to gain direct insight into the public's reaction to a brand crisis—in particular, the emotional reactions of consumers to a brand when something goes wrong. Some of the key questions are: exactly what issues are of most concern to the hashtag community (security, lack of information, etc.)? How do they feel about the Sony brand itself during the crisis? Who do they hold accountable (Sony or the hackers)? Would this event change their perceptions towards and interactions with the Sony brand in future?

We coded the tweets into four top-level categories to separate out the tweets that would be the richest source of data on community responses to the hacking incident. These categories were: Discussion/Commentary (specifically on the Sony hacking incident), LOTE (languages other than English), News (spe-

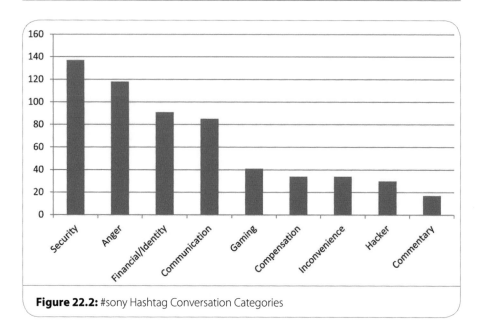

Figure 22.2: #sony Hashtag Conversation Categories

cifically about the Sony hacking incident), and Sony General/Other (designed to filter out unrelated Sony mentions, advertisements for Sony products and general spam).

By conducting manual thematic analysis on the Discussion/Commentary category of the isolated data set, we were able to chart the particular issues and topics that people were discussing most frequently. As represented in Figure 22.2, the nine thematic categories were: Security, Anger, Financial/Identity, Communication, Gaming, Compensation, Inconvenience, Hacker, and Commentary.

While common sense would suggest that the most discussed category would be in relation to the financial/identity risk associated with the hacking, the analysis of the Twitter stream reveals something else entirely. The main topics of discussion during the peak period in Twitter conversation related to people's anger and disbelief over Sony's lapse in security:

Hey #Sony, there's a new thing called encryption . . . google it sometime!

Sources at Sony seem to be confirming no/low encryption on personal data. V worrying.

Major breach of security and Sony deserve the bad press they are getting due to it. They should have been better prepared, they were not.

The second largest category of discussion contained purely emotional expressions of anger and disgust at Sony, with no specification of exactly what the main concerns were:

Sony = Pathetic.

Well crap, thanks Sony.

Fuck PSNetwork, Fuck Sony.

These comments, whilst useful in monitoring the general atmosphere and sentiment towards Sony during the incident, are not helpful in clarifying for the company the exact nature of the complaints. As the brand monitors customer attitudes, such comments could therefore be similarly used as a gauge of public sentiment.

The high number of tweets related to the lack of communication from the company during this period should be of great concern to Sony. They were divided into two subcategories—first, the initial delay from Sony in informing customers of the security breach:

I think it is pretty disgusting that Sony have waiting 7 days to tell users that their Credit Card details may have been compromised.

Why does it [take] seven days for Sony to report the PSN issue?

The second subcategory was in relation to updates—or lack thereof—from Sony once the initial breach was confirmed:

Sony! Please state CATEGORICALLY whether credit card details have been stolen or not. Do it soon while you still have a reputation.

I bet the hacker will get emails out quicker than Sony!

Any updates or is Sony still not sayin anythin useful?

From a brand management perspective, this lapse in communication was incomprehensible to consumers. The lack of regular updates and information from Sony only served to incense users further, as they struggled to determine what was fact and what was rumour on Twitter. Sony's lack of response in immediately addressing these issues only magnified the overall negative impact. Sony had not embraced the defining feature of social media—that 'top down' control no longer works in this space.

In comparison, when Toyota had to recall a number of its cars in 2009 and 2010 due to serious safety faults which had resulted in the deaths of over 50 people (CBS News, 2010), they immediately went into damage control. "As soon as the recall crisis started getting media attention, Toyota quickly put together an 'Online Newsroom' and a 'social media strategy team' to coordinate all the media releases from different organisations of the company" (Rajasekera, 2010, p. 9). While there was still anger and negative viewpoints shared through social media services, the company was able to minimise their impact by eliminating confusion and keeping the consumer base regularly informed of developments. If Sony had employed a similar approach, they might have significantly reduced short-term confusion and anger amongst their online consumers, thereby minimising long-term negative associations with the Sony brand.

CONCLUSION

This case study has highlighted not only the importance but also the risks of businesses actively engaging with their user base in online social environments. In too many cases, businesses have sought to capitalise upon the growth in popularity of social media sites such as Twitter without taking the time to fully understand the dynamics that are at play in this space. Social media sites like Twitter evolved separately from commercial enterprises, and quickly developed their own culture and rules of engagement. Many businesses—quite arrogantly—assumed that they could enter into these spaces and still maintain control over their brand and the 'consumer experience'. They transferred traditional, top-down business models and advertising campaigns into these spaces, and then were surprised at the often negative responses these ventures received.

As McWilliam (2000, p. 44) stated: "Brand managers need to understand the bases for dialogue that can lead to strong relationships, which in turn provide the foundations for online brand communities". The emphasis here is on dialogue. The mistake of companies like Sony is that they are happy to try and cash in on social media environments—but only when it is on their terms. But Sony does not really get to set the terms: social media environments like Twitter are defined by two-way communication. As this case study illustrates, you cannot enter this space and not engage with users; particularly in the case of a consumer technology brand like Sony, whose community is highly active and literate in digital media.

It is an emerging truth of social media that, in a shift away from purely symbolic brand power, "the brand will ultimately be judged on the quality of the experience it offers through its community" (McWilliam, 2000, p. 51). Once users have developed an opinion about an organisation, "they share it with others and the subjective truth turns into a collective truth about what an organisation is and what it should be" (Aula, 2010, p. 46). If Sony had simply engaged with its online consumer base on Twitter and kept them informed of the situation with regular updates of new developments, they could have significantly reduced the negative impact on their brand. By refusing to participate in this online conversation, companies risk more than some bad press: they risk alienating a loud and powerful consumer base with a global reach.

REFERENCES

ABC News. (2012, 24 Apr.). Tourism boss defends cash for tweets. Retrieved from http://www.abc.net.au/news/2012-04-24/tourism-sa-twitter-celebrities/3968920

Asian News International. (2012, 21 Nov.). Twitter restores New York Times parody account after outcry. Retrieved from http://www.aninews.in/newsdetail4/story85562/twitter-restores-new-york-times-parody-account-after-outcry.html

Aula, P. (2010). Social media, reputation risk and ambient publicity management. *Strategy & Leadership, 38*(6), 43–49.

Bruns, A. (2012). How not to use Twitter: Lessons from Qantas and Westpac. The Conversation. Retrieved from http://theconversation.edu.au/how-not-to-use-twitter-lessons-from-qantas-and-westpac-5342

Business Wire. (2012, 21 June). Nike + running app for Android allows runners to track, share and compare their runs. Retrieved from http://www.businesswire.com/news/home/20120621006611/en/Nike-Running-App-Android-Runners-Track-Share

CBS News. (2010, 2 Mar.). U.S.: 52 deaths reportedly tied to Toyota. Retrieved from http://www.cbsnews.com/2100-500395_162-6258603.html

Cohen, N. (2010, 8 June). Twitter satirists are having a field day with BP: Parodies get spotlight while company's efforts fade into background. *International Herald Tribune*, p. 18.

Gembarski, R. (2012). How Starbucks built an engaging brand on social media. Retrieved from http://www.brandingpersonality.com

Goldberg, A. (2011, 28 Apr.). Sony's fiasco spells disaster for ailing company. *McClatchy-Tribune Business News*. Retrieved from https://www.tmcnet.com/usubmit/-news-feature-sonys-fiasco-spells-disaster-ailing-company-/2011/04/30/5477786.htm

Ind, N. (2012). *Brand together: How co-creation generates innovation and re-energises brands*. London, UK: Kogan Page.

Kinzey, R. (2009, 27 July). Managing your reputation in the social media world. *The Business Journal*. Retrieved from http://www.bizjournals.com/triad/stories/2009/07/27/smallb2.html?page=all

M2 PressWIRE. (2012, 3 Feb.). Amazon UK leads retailers in positive post-Christmas feedback on Twitter. Retrieved from http://www.realwire.com/releases/Amazon-UK-Leads-Retailers-in-Positive-Post-Christmas-Feedback-on-Twitter

McWilliam, G. (2000). Building stronger brands through online communities. *Sloan Management Review, 41*(3), 43–54.

Rajasekera, J. (2010). Crisis management in social media and digital age: Recall problem and challenges to Toyota. GSIM Working Papers—International University of Japan, (Working Paper No. IM-2010-02) (pp. 1–16). Retrieved from http://nirr.lib.niigata-u.ac.jp/bitstream/10623/31308/1/EMS_2010_06.pdf

Singer, P. (2012, 22 Oct.). Media's RTs help Twitter parodies take flight. *Gannett News Service.* Retrieved from http://www.delawareonline.com/article/20121023/BUSINESS08/310230027

Taylor, R. (2011, 23 Nov.). 'Epic PR fail' at Qantas Airlines. *The Gazette.*

Thoring, A. (2011). Corporate tweeting: Analysing the use of Twitter as a marketing tool by UK trade publishers. *Publishing Researchers Quarterly, 27*(2), 141–158.

Twitter. (2012). Parody, commentary, and fan accounts policy. Retrieved from https://support.twitter.com/articles/106373-parody-commentary-and-fan-accounts-policy

The Wall Street Journal. (2011, 6 May). As Sony counts hacking costs, analysts see billion-dollar repair bill. Retrieved from http://online.wsj.com/article/SB10001424052748703859304576307664174667924.html#

Watson, C., & Novak, L. (2012, 25 Apr.). Cash for your tweet. *The Advertiser.*

Political Discourses on Twitter

Networking Topics, Objects, and People

23
CHAPTER   Axel Maireder and Julian Ausserhofer

 political discourses enter Twitter, develop a life of
their own and become part of the
#networkedpublicsphere

On 2 August 2011, the deputy governor of the Austrian province of Carinthia, Uwe Scheuch, was sentenced to six months imprisonment for corruption. The court was convinced that he had offered Austrian citizenship to a Russian investor in exchange for a party donation. The conviction was the top news story in the Austrian media for days, and triggered strong reactions from Scheuch's opponents and supporters, the latter claiming he was innocent and the victim of a political conspiracy. Outside the mass media, the conviction was also heavily debated on Twitter. Twitter users discussed the impact of the event on Austria's political system and culture, commented on the story's development, got upset about the reactions of politicians, and cracked jokes about Scheuch's upcoming imprisonment. They linked to news stories, documents, critical blog posts, and satirical videos. They also heavily referred to each other, retweeted one another's messages, responded to arguments, and approached each other for a reaction. On Twitter, Scheuch's conviction was not just a news story, but a public conversation engaging hundreds of politically interested Austrians.

The opportunities and challenges of the Internet for citizens to access and participate in political discourses are major strands of discussion within the academic debate on the nature of contemporary democracy (see Farrell, 2012, for a review). The open, transparent, and low-threshold exchange of information and ideas Twitter allows shows great promise for a reconfiguration of the structure of political discourses towards a broadening of public debate by facilitating social connectivity. Based on extensive empirical research into practices and patterns of political tweeting in Austria, we will describe those discourses from three perspectives:

1. Networking topics, in terms of the inclusion of information, interpretation, and views into a debate;
2. Networking media objects, driven by hyperlinking practices and resulting in a reconfiguration of Web spheres; and
3. Networking actors, driven by @mentioning practices, resulting in new patterns of interaction between political actors and citizens that reshape the participation structure of the public sphere.

Connecting those perspectives can be fruitful for understanding the processes of the creation and negotiation of political meaning through Twitter, and the way Twitter usage may shape citizens' approaches to political information and participation.

NETWORKING TOPICS—
SOCIAL REALITY TESTING THE NEWS

Twitter is an awareness system that allows for an immediate, fast, and widespread dissemination of information (e.g., Kwak, Lee, Park, & Moon, 2010). The platform offers diverse means to share news from various sources, resulting in a stream of information, opinions, and emotions (Papacharissi & de Fatima Oliveira, 2012) that presents a multifaceted experience of ambient news (Hermida, 2010). Within political discourses, various political actors as well as individuals use Twitter to spread information on political events and to state their opinions (Small, 2011). The Twitter stream potentially provides multiple viewpoints on political debates (Yardi & boyd, 2010), and holds unique opportunities to structure those debates by the use of common hashtags (Bruns, 2012).

Hashtag-driven political discourses are largely connected to events reported by mass media, at least in terms of topics taken up and quantity of messages sent (Bruns & Burgess, 2011; Larsson & Moe, 2011). Despite this connection, the Twitter agenda is likely to differ from the media agenda, because "events and themes are filtered through the community's own established interests and news frames, resulting in a distribution of attention that is different from that of the mainstream media or of general public debate" (Bruns & Burgess, 2011, p. 45).

This holds true for political tweeting in Austria. In a study conducted on the tweets of the 374 most active users in discussions on Austrian domestic politics, Ausserhofer and Maireder (2013) found that the mass media and Twitter agendas differed considerably, in terms of attention given to certain topics. While long-lasting and complex issues like the financial crisis or the wage negotiations of the metal industry were subject to detailed media reporting, hardly anyone mentioned those topics on Twitter. At the same time, the multinational treaty for intellectual property rights, ACTA, was heavily debated by the tech-savvy Twitter community, but almost ignored by news media for a long time. Short-lived and eventful topics like the heavy protests against a prom of Vienna's right-wing fraternities, or political scandals, were reported both by news media and on Twitter. On Twitter, however, political news were not only reported on, but also interpreted and actively connected to other topics by the users at the same time. This contrasts with the traditional Two-Step-Flow of Communication model (Katz & Lazarsfeld, 1955), which researchers have used for decades to describe the interrelation of interpersonal and mass communication. While perceiving the news from the mass media and discussing it within the personal social network are somewhat separated activities in this model, they are not on Twitter. The two steps of the communication flow dissolve as reports by news media and interpretation by the personal social networks become part of the same news stream, and any single message may include both information and commentary on an event.

Following up on these findings, Maireder (2012) focussed on Twitter discourses connected to three outstanding political events in Austria, and showed that Twitter users extensively share political views and interpretations, besides the news itself. The three case studies were (a) the conviction of Uwe Scheuch for corruption, introduced earlier; (b) the proposal of Austria's minister of education, Karl-Heinz Töchterle, to reintroduce university tuition fees; and (c) the announcement of the assignment of a former official of the Social Democratic Party of Austria (SPÖ), Niko Pelinka, to a high-level post at Austria's public service television station ORF. All of these cases were widely covered by the news media, as well as discussed on Twitter. In a content analysis of news reports from Austrian Press Agency's comprehensive database (na = 188, nb = 293, nc = 394) and tweets connected to the event collected by combined keyword queries (na = 1492, nb = 612, nc = 1955), the stories' development, in terms of topics addressed, was examined, comparing news media and the Twitter discourses. Each news item and tweet was assigned to one or more aspects of the story it was about. Each case had its own categories, of course, but the categories were consolidated into three types: Information on the actual political development (reports on the initial events or statements by actors involved, for example); context, meaning information related to other incidents connected to the story (political events in the past or current events); and general commentary.

Figure 23.1 illustrates the frequency of these types of Twitter activity on the three cases: In the Scheuch and Pelinka cases, people tweeted a lot of general commentary from the beginning, while in the Pelinka case, general commentary was the major type of content throughout the time. Some peaks in the discussion can be traced to specific events in the stories' developments, an interview broadcast or a parliamentary speech, for example. The figure also shows that from the minute the news on Uwe Scheuch's conviction, Niko Pelinka's

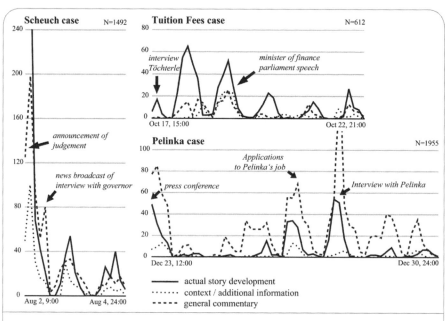

Figure 23.1: Number of Tweets Posted in the Different Categories of Story Aspects During the Three Hours Following the Event

appointment at the ORF, and Karl-Heinz Töchterle's push for tuition fees broke, Twitter users began interpreting the events. While the majority of tweets in the first hour after the initial event included short information reporting the incident itself, most of the tweets were not informational only. Users reported Scheuch's conviction by briefly stating that he had been sentenced to six months imprisonment, but often accompanied this information with a short emotional or interpretative personal remark, signified by expressions like "Yeah!", "It was about time", or an emoticon. The tweets announcing Niko Pelinka's promotion largely included expressions of disbelief or anger, and those on the minister of education's statement mostly expressed either support for his proposal or opposition to it.

In the hours and days after each of the initial events, the news media continuously reported on new developments within the political arena. All major online news sites had articles on the discussion of Scheuch's case by legal experts and politicians; on the reactions to the tuition fee proposal by parties, universities, and the student union; as well as on the official statements by the journalists' union and others on the controversial appointment in the ORF. Alongside some background information on the central actors and political history, the news media concentrated, to a large extent, on reporting the actual events that took place in the arena of professional political actors.

After the initial spreading of the news as such, the Twitter discourses developed differently than the mass-media reporting. Twitter users infrequently passed on information on the discussion in the arena of professional politics as reported by the news media, but rather, provided alternative background information and interpretation. In the Pelinka case, for example, users reflected on the long history of nepotism in Austria, and brought comparable cases within the ORF and the SPÖ to mind. Some users analysed Pelinka's career, asking for the reasons he was qualified for the job. Others developed theories on the role of his father (an influential journalist) and his uncle (a famous political scientist) in the events. Prominent journalists publicly announced on Twitter that they would apply for the job themselves, stating that they would be much better qualified, according to the original job description. Some users called for civil disobedience to protest the decision of the ORF director by stopping payment of the TV licence fees, which was broadly supported by other Twitter users. In the Scheuch case, users drew comparisons to other court decisions, praised or condemned the judge, and raised questions about the legal base for Scheuch's announcement to stay in power until the appellate proceeding. They reflected on the history of corruption in Austria, discussed the impact on the federal elections, and the future of Carinthia's government. Users discussed whether politicians are out of touch with reality in general, satirically envisioned how Scheuch would survive in jail, and stated which of Austria's politicians should be imprisoned next.

Communication research has long emphasised how the reception of political and societal events depends on conversations about news in people's immediate social context. It helps them to make sense of what happens in the world by connecting the news to personal experiences, embedding them into social relevance structures. They put the news to a "social reality test" and shape "public perceptions of issue salience" (Erbring, Goldenberg, & Miller, 1980, p. 41). On Twitter, such processes of social negotiation of the meaning of news happen right away, because the messages diffusing the news may already include interpretation. Twitter users often connect current events to personal experiences, opinions, and world views: they explain, classify, interpret, and reinterpret what they have received. This way, a much wider range of aspects may be included in Twitter discourses than in news reports. Events may get connected to other topics by the way they are framed by the users. Thus, Twitter may provide information and commentary far beyond the event itself, massively enriching the traditional news media reporting—or even triggering it, as in the discussion about ACTA, for instance. Observing political discourses unfold on Twitter is observing the process of the social negotiation of the meaning of news.

NETWORKING OBJECTS—MODELLING THE NETWORKED PUBLIC SPHERE

Tweets as media objects are often connected to other objects by hyperlinks. This network of objects is part of the 'material' base of the networked public sphere, and following the links between those objects—surfing the Web—is the central mode of activity to access information. In 2002, Foot and Schneider coined the term "web sphere" for a relatively stable cluster of websites defined by their structure of interlinkages that 'host' discourses on certain broadly defined topics such as domestic politics. The notion of "blogosphere" has the same meaning for clusters of blogs, and has been used in research to map the virtual places certain discourses become manifest in, and the interconnections of those places (e.g., Bruns & Adams, 2009; Etling, Kelly, Faris, & Palfrey, 2010). Research on Twitter spheres has focussed on networks that emerge from the common use of hashtags (Bruns, 2012; Bruns & Burgess, 2011). However, a lot

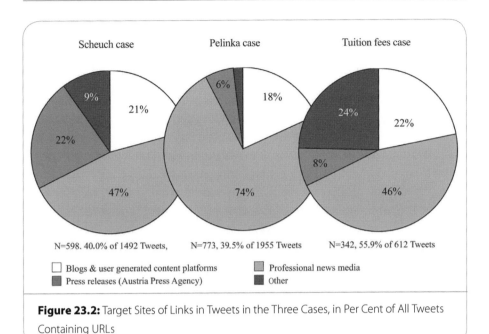

Figure 23.2: Target Sites of Links in Tweets in the Three Cases, in Per Cent of All Tweets Containing URLs

of tweets do not contain hashtags (Ausserhofer & Maireder, 2013; Small, 2011), although they are still part of a specific discourse.

Besides forming a sphere themselves, tweets provide access points to the networked public sphere in general, because they heavily link to content elsewhere on the Web (Maireder, 2011). In a media ecology where the circulation of content heavily depends on the users' active participation (Jenkins, 2006, p. 3), these links are important for the distribution of attention to specific media objects. Between 40 and 56% of all tweets analysed within the three case studies discussed earlier included links (Maireder, 2012). A large share of those links, between 46 and 74%, referred to news media reports; about 20% to blogs and other user-generated content; up to 22% to press releases on the platform of the Austrian Press Agency; and the rest to content published by political parties, NGOs, NPOs, or companies other than media (see Figure 22.2). In the Scheuch and Pelinka cases, a substantial share of tweets directly linked to press releases by politicians published on the platform of Austrian Press Agency's distribution service. In all three cases, the first hours after the initial incidents were dominated by links to news media, while links to blog posts and other user-generated content were posted later on. News articles were shared within a limited period of time after their initial publishing, usually a couple of hours, but popular blog posts were shared and reshared for days. It seems that professional news is faster, but commentary has a longer life span.

While it is obvious that articles produced by professional editors are of high interest within general political discourses (even in the context of social media), the other content is particularly interesting. In the Scheuch and tuition fees cases, more than half of the links tweeted led to content not produced by news media, and in the Pelinka case about a quarter. For example, in the latter case, four individual, private blog posts were shared several times, all of them taking Pelinka as a starting point for a general critique of the allegedly nepotistic and corrupt political culture of Austria. In the Scheuch case, a YouTube video of a 1990 song by German punk-rock band Die Ärzte, entitled "Uwe sitzt im Knast" (Uwe is in prison), was shared several times. Another piece linked to a number of times was a five-year-old press release by the Austrian Freedom Party (FPÖ) pointing to a lawsuit filed in Hungary for a different charge in which Scheuch was condemned. While a member of the FPÖ in 2011, Scheuch had been a member of another party (BZÖ) in 2006. By tweeting this press release, users emphasised the fact that the same actors that defended Scheuch in the current case had condemned him for similar reasons back in 2006. In the tuition fees case, several links led to a list of political demands that a popular student protest movement had drafted in 2009.

None of this content had a direct connection to the current cases, since all of it had been produced in other periods of time and other contexts. Nonetheless, they were included into the discourses, because they carried new meaning within the current contexts. The students' demands were reread on the background of the ministers' proposal; the old press release on Scheuch pointed to the flip-flopping of political personnel on the far right and the contradiction of political messages; and the music video helped in abstracting the case in a humorous way. Users had reframed the content to connect it to current discourses.

In communication research, the concept of framing refers to techniques used in texts to semantically emphasise "specific aspects of perceived reality" (Scheufele, 2006, p. 65). Frames "draw boundaries, set up categories, define some ideas as out and others in, and generally operate to snag related ideas in their net" (Reese, 2007, p. 150). Traditionally, the term is used to refer to the way journalists make certain schemata manifest within their texts, but Weaver (2007, p. 144) emphasised the ambiguity and the comprehensive nature of the framing concept that can be applied to many different aspects and types of messages.

For discourses manifest in networked media elements, the way the relation between the elements is constructed may be crucial to the way users perceive them (Harrison, 2002). In the cases discussed here, users pre-framed media objects to integrate them into the current discourses by referring to the cases within the text part of their tweets. This kind of framing, however, was not specific to the links mentioned above, but was observed for large parts of the links in general. An analysis of the tone of the messages showed that links to news reports were framed less interpretatively than those to press releases or blogs and other user-generated content. However, depending on the case, between 22 and 50% of the links to news media were framed by a personal interpretation, with about two thirds commenting in a sober tone, and one third sarcastically or aggressively. These numbers are even higher for other content.

Links in tweets connect the Twitter discourse to the networked public sphere in general, providing access to media objects and their relations that form its material base. In the political discourses examined here, the links in tweets referred to manifold news reports, blog posts, YouTube videos, press releases, and much more, connecting those objects to the Twitter conversations. By framing the links, users introduced certain schemata to perceive the objects linked to, reinterpreting their meaning and negotiating their position within the networked public discourses.

NETWORKING PEOPLE—CUTTING ACROSS SOCIAL BOUNDARIES

Twitter is a social network medium, because the structure of the information flow is based on networks between accounts that represent social actors. Beyond that, Twitter's @mention function is used to address or reference other users, enabling conversation throughout a network of interconnected actors that boyd, Golder, and Lotan (2010, p. 1) described as "a public interplay of voices that gives rise to an emotional sense of shared conversational context". Research on Twitter and political protest found that the platform facilitates the integration of very different actors into a common conversation (Maireder & Schwarzenegger, 2012), and holds opportunities to cut across and connect diverse social networks (Segerberg & Bennett, 2011). Political conversations on Twitter thus hold opportunities for users to enlarge their personal network, and for political actors to connect to other professionals as well as politically active citizens.

In Austria, Twitter is only used by about 1% of the population, but it is increasingly popular with professionals operating around the political centre (Ausserhofer & Maireder, 2013). Many journalists, PR professionals, politicians,

political activists, and experts have turned to Twitter for news sharing, self-presentation, and conversation among people with an interest in domestic politics.
Because those actors rather address each other than a general public (at least
compared to mass media), they may form what Davis (2010, p. 754) has called
an "online elite discourse network". Like the Swedish political Twittersphere
researched by Larsson and Moe (2011; see also Larsson & Moe, Chapter 24 in
this volume), Austria's political Twitter users rather form an information and
conversation network of people already engaged in politics than a communication platform that integrates the political centre and the periphery.

The users identified as Austria's political Twitter elite in the study on the
Austrian political Twittersphere introduced in the first section (Ausserhofer
& Maireder, 2013) intensely interact with each other. More than two thirds of
the tweets on domestic politics included at least one @mention to another user,
and about half of the @mentions referred to a user of the elite network itself.
This means that political professionals form a densely knit communication network among themselves, but at the same time, each of them also heavily interacts with dispersed users outside of the core network. In the exchange of news,
arguments, and interpretations on political events described above, they connect to each other on a day-to-day basis, and form a political discourse sphere
structurally independent from the traditional arena of politics, but, of course,
connected to it by their official affiliations and real-life interactions.

In addition to such political professionals, there are several users in the centre of the network that have no professional affiliation to the traditional political arena, as well as some political actors who would traditionally be located
at the periphery of the national political arena, for instance, backbenchers in
parliament, local politicians, or political activists. Some of these actors from
the political periphery have a prominent position within specific discourses.
They are niche authorities, for potentially different reasons: Some may address
journalists and politicians on a given issue, which may result in these groups
addressing them in return; others may have become respected experts on a
topic due to their knowledge and role as disseminators or opinion leaders in
the political Twittersphere.

A network of Twitter interactions in the Austrian political sphere is illustrated for the Pelinka case in Figure 23.3. The node size is calculated by the
number of received @mentions on the topic; the node position represents the
centrality of the account within the network (based on all @mentions the user
received). The TV journalists @ArminWolf and @DieterBornemann were the
first to tweet about Pelinka's hiring, and are among the most frequently addressed

users in the Pelinka discourse. While news anchor @ArminWolf is central within different discourses, @DieterBornemann is particularly important within the discussions of the Pelinka case. Other central nodes include journalists of different media companies, such as @MartinThuer and @florianklenk; experts like @HubertSickinger; and 'casual citizens' such as @AnChVIE. Except from the oppositional Green party's @michelreimon, a local representative in the province of Burgenland, there are hardly any politicians addressed in the discussions about Pelinka. No member of the Social Democratic Party, who could have defended Pelinka's appointment, was participating.

For the political arena in the United Kingdom, Davis (2010) has stated that the Internet has led to "a significant increase in the communicative links between those in and around the UK political centre" (p. 754), and thus more means

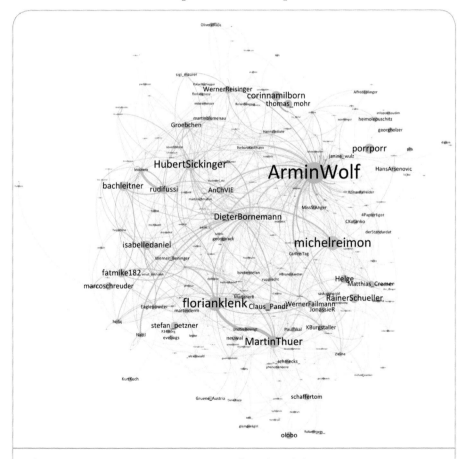

Figure 23.3: Main Twitter Interaction Network on the Pelinka Case

of exchange and deliberation. This is certainly true for the Austrian political Twittersphere, but analyses have also shown that Twitter holds opportunities for politically interested but unaffiliated users to become integral actors within the sphere of discourse of the political centre. Moreover, Twitter allows casual citizens to observe conversations of the political elite and, if they like, to participate in those conversations. Even though the elite preferably refer to each other, they do interact with other users, and from time to time, include their views into the debate by retweeting them or referring to them.

CONCLUSION

Deuze (2006) described Internet users as "bricoleurs" to emphasise the "highly personalized, continuous, and more or less autonomous assembly, disassembly, and reassembly of mediated reality" in digital culture (p. 66). The reality of political discourses Twitter users experience is shaped by the bricolage of messages and media objects they access through their individually composed streams, an assembly produced in a process of networking meaning by dispersed actors mutually referencing each other. The networking of topics, media objects, and people in the course of political discourses, as described in this text, are heavily entangled processes that reorganise the users' experiences of the political. Those users participating in the discourses find themselves within a public social negotiation of the meaning of political events—for themselves, for their social network, for the actors of the political arena, and thus, for society in general. The arguments presented here are another indicator of the gradual reallocation of the construction of political meaning from the mass-media system to a "networked public sphere" (Benkler, 2006), advanced by the socialisation of media experiences.

REFERENCES

Ausserhofer, J., & Maireder, A. (2013). National politics on Twitter: Structures and topics of a networked public sphere. *Information, Communication & Society, 16*(3), 291–314. doi:10. 1080/1369118X.2012.756050

Benkler, Y. (2006). *The wealth of networks: How social production transforms markets and freedom*. New Haven, CT: Yale University Press.

boyd, d., Golder, S., & Lotan, G. (2010). Tweet, tweet, retweet: Conversational aspects of retweeting on Twitter. In *Proceedings of the 43rd Hawaii International Conference on System Sciences* (pp. 1–10). doi:10.1109/HICSS.2010.412

Bruns, A. (2012). How long is a tweet? Mapping dynamic conversation networks on Twitter using Gawk and Gephi. *Information, Communication & Society, 15*(9), 1323–1351. doi:10. 1080/1369118X.2011.635214

Bruns, A., & Adams, D. (2009). Mapping the Australian political blogosphere. In A. Russell & N. Echchaibi (Eds.), *International blogging: Identity, politics, and networked publics* (pp. 85–110). New York, NY: Peter Lang.

Bruns, A., & Burgess, J. E. (2011). #ausvotes: How Twitter covered the 2010 Australian federal election. *Communication, Politics and Culture, 44*(2), 37–56.

Davis, A. (2010). New media and fat democracy: The paradox of online participation. *New Media & Society, 12*(5), 745–761. doi: 10.1177/1461444809341435

Deuze, M. (2006). Participation, remediation, bricolage: Considering principal components of a digital culture. *The Information Society, 22*(2), 63–75.

Erbring, L., Goldenberg, E. N., & Miller, A. H. (1980). Front-page news and real-world cues: A new look at agenda-setting by the media. *American Journal of Political Science, 24*(1), 16–49.

Etling, B., Kelly, J., Faris, R., & Palfrey, J. (2010). Mapping the Arabic blogosphere: Politics and dissent online. *New Media & Society, 12*(8), 1225–1243. doi: 10.1177/1461444810385096

Farrell, H. (2012). The consequences of the Internet for politics. *Annual Review of Political Science, 15*(1), 35–52. doi:10.1146/annurev-polisci-030810-110815

Foot, K. A., & Schneider, S. M. (2002). Online action in Campaign 2000: An exploratory analysis of the U.S. political Web sphere. *Journal of Broadcasting & Electronic Media, 46*(2), 222–244.

Harrison, C. (2002). Hypertext links: Whither thou goest, and why. *First Monday, 7*(10). Retrieved from http://firstmonday.org/htbin/cgiwrap/bin/ojs/index.php/fm/article/view/993/914

Hermida, A. (2010). Twittering the news: The emergence of ambient journalism. *Journalism Practice, 4*(3), 297–308. doi:10.1080/17512781003640703

Jenkins, H. (2006). *Convergence culture: Where old and new media collide.* New York, NY: New York University Press.

Katz, E., & Lazarsfeld, P. F. (1955). *Personal influence.* New York, NY: Free Press.

Kwak, H., Lee, C., Park, H., & Moon, S. (2010). What is Twitter, a social network or a news media? In *Proceedings of the 19th International Conference on the World Wide Web* (pp. 591–600) , Raleigh, NC. doi:10.1145/1772690.1772751

Larsson, A. O., & Moe, H. (2011). Studying political microblogging: Twitter users in the 2010 Swedish election campaign. *New Media & Society, 14*(5), 729–747. doi:10.1177/1461444811422894

Maireder, A. (2011). *Links auf Twitter. Wie verweisen deutschsprachige Tweets auf Medieninhalte?* [Links on Twitter: How do German-language tweets refer to media content?]. Retrieved from http://phaidra.univie.ac.at/o:64004

Maireder, A. (2012). *Evolution von Nachrichten in der Netzöffentlichkeit* [Evolution of news in the networked public sphere]. Retrieved from http://phaidra.univie.ac.at/o:258194

Maireder, A., & Schwarzenegger, C. (2012). A movement of connected individuals—Social media in the Austrian student protests 2009. *Information, Communication & Society, 15*(2), 171–195. doi: 10.1080/1369118X.2011.589908

Papacharissi, Z., & de Fatima Oliveira, M. (2012). Affective news and networked publics: The rhythms of news storytelling on #Egypt. *Journal of Communication, 62*(2), 266–282. doi:10.1111/j.1460-2466.2012.01630.x

Reese, S. D. (2007). The framing project: A bridging model for media research revisited. *Journal of Communication, 57*(1), 148–154. doi:10.1111/j.1460-2466.2006.00334.x

Scheufele, B. (2006). Frames, schemata, and news reporting. *Communications, 31*(1), 65–83. doi:10.1515/COMMUN.2006.005

Segerberg, A., & Bennett, W. L. (2011). Social media and the organization of collective action: Using Twitter to explore the ecologies of two climate change protests. *The Communication Review, 14*(3), 197–215. doi:10.1080/10714421.2011.597250

Small, T. A. (2011). What the hashtag? A content analysis of Canadian politics on Twitter. *Information, Communication & Society, 14*(6), 872–895. doi:10.1080/1369118X.2011.554572

Weaver, D. H. (2007). Thoughts on agenda setting, framing, and priming. *Journal of Communication, 57*(1), 142–147. doi:10.1111/j.1460-2466.2006.00333.x

Yardi, S., & boyd, d. (2010). Dynamic debates: An analysis of group polarization over time on Twitter. *Bulletin of Science, Technology & Society, 30*(5), 316–327. doi:10.1177/0270467610380011

Twitter in Politics and Elections
Insights from Scandinavia

24

CHAPTER — Anders Olof Larsson and Hallvard Moe

 during #elections in Sweden, Denmark and Norway, Twitter is used as a #megaphone

Alongside blogs and sites such as YouTube and Facebook, Twitter by now seems to have established itself as an everyday part of the arsenal of political communication in many parts of the world. Campaigners, lobbyists, companies, NGOs, as well as activists commonly use the platform to spread their messages, or to connect with and receive feedback from potential voters or clients.

Researchers have approached the political uses of Twitter in a number of different contexts. Attention has been given to the use of Twitter during uprisings in totalitarian countries (e.g., Gaffney, 2010; Lotan et al., 2011), but also in more stable, democratic contexts. Beyond attempts to predict election results using Twitter data (e.g., Tumasjan, Sprenger, Sandner, & Welpe, 2010), studies have primarily focussed on political Twitter use at the hands of politicians.

In the US, Lassen and Brown (2011) assessed factors affecting Twitter adoption among members of Congress. While finding no definitive results, Twitter users in the U.S. Congress tended to be younger, to belong to the minority party, to serve in the Senate, and to have been urged by their party leaders to tweet.

Golbeck, Grimes, and Rogers (2010) found tweets from members of the U.S. Congress to be largely "vehicles of self-promotion" (p. 1612). Beyond the U.S. context, Sæbø (2011) studied Twitter use by members of the Norwegian parliament, finding similar results to the previously mentioned U.S. study: tweets were mainly used for providing information on professional activities, to express views on current topics, and to discuss issues with fellow politicians. Only to a lesser extent was Twitter used to engage in discussion with citizens. Focussing on the 2009 European Parliament election campaign in the Netherlands, Vergeer, Hermans, and Sams (2011) found opposition politicians to be more progressive in their use of Twitter—a result that mirrors the findings reported by Lassen and Brown (2011). These results largely correspond to studies performed on more general aspects of Web 2.0 use among U.S. and European politicians, which in the main also found a slow but steady uptake of features allowing for more participation and discussion (e.g., Larsson, 2011; Lilleker et al., 2011; Schweitzer, 2011; Wattal, Schuff, Mandviwalla, & Williams, 2010).

But what about the role of Twitter in politics and elections more generally, as a tool for public communication by citizens? General user statistics for a range of countries tell us that Twitter is used only by a specific subset of the wider population. But how is this use fashioned during periods of heightened attention to politics? Who uses Twitter, and how does this use differ among user groups? An interesting question is whether or not new and larger user groups join when much is at stake, and, if so, how such users behave in relation to more frequent users.

To gauge this question, this chapter presents findings from a comparative study of political Twitter use during recent election campaigns in the three Scandinavian countries—Sweden, Denmark, and Norway. Beyond the analytical advantages offered by such case-specific comparisons (e.g., Raats & Pauwels, 2011; Ragin, 1987), there is a two-sided rationale for our selection of countries. First, it moves us beyond the Anglo-American context (as suggested by Goggin & McLelland, 2009; Moe, 2011). Second, as the Scandinavian countries all boast comparatively egalitarian practices of media use, high levels of Internet penetration and use, as well as high levels of voter turnout (e.g., Syvertsen, Enli, Mjøs, & Moe, 2014), they are interesting cases for a study of political Twitter use.

In what follows, we first elaborate on the case characteristics. Next, we explain our methodological approach, which builds on previous efforts (e.g., Bruns & Burgess, 2011; Larsson & Moe, 2012), in collecting and analysing tweets and their metadata based on the key hashtags in each case. On this basis, we provide empirical insights into how Twitter practices are fashioned during elections in established democracies.

CASES

The three Scandinavian countries have consistently reported high numbers of Internet penetration and use (e.g., Nordicom, 2010). However, their level of adoption when it comes to services like Twitter is not necessarily as high. Under Swedish conditions, a 2010 survey specified that a mere one percent of Internet users made use of Twitter during the course of a typical day (Facht & Hellingwerf, 2011). For Denmark, a survey from the same year disclosed that of all online Danes, about three per cent identified themselves as Twitter users ("Befolkningens brug af Internet 2010", 2011). Similarly, just below four percent of Norwegian Internet users reported that they used Twitter on a weekly basis (NRK/Ipsos MMI, 2011).

While the electoral system for each country under scrutiny differs slightly, voter turnout for elections is steady at high levels. Our analyses are based on the latest elections from each of the case countries. In Sweden, parliamentary, regional, as well as local elections are held in conjunction every four years. The latest election took place on 19 September 2010. For Denmark and Norway, elections at the different levels of government are held at separate times. In Denmark, the latest election was for the national parliament, and was held on 15 September 2011. As for Norway, the latest election dealt with regional and local matters, and was held on 12 September 2011. The cases, then, offer an opportunity to compare among three similar contexts, with elections in relatively close temporal proximity.

METHOD

Data were collected using the yourTwapperKeeper tool, an open-source platform designed to collect tweets and their metadata. Focussing on hashtags deemed as relevant for each election, we constructed three separate archives, the details of which are outlined in Table 24.1.

Table 24.1 shows the total number of tweets archived per country, often spanning over multiple hashtags. Hashtags are thematic keywords which are convenient for demarcating Twitter searches and archiving tweets. As hashtags were used to guide data collection in the present study, Twitter content not

Table 24.1: Summary of Archives Used for Data Collection

Case Country	Hashtag Archives	Number of Tweets Archived across Archives in Each Case	Time Frame
Sweden	#val2010 (election2010)	99,348	19 Aug. 2010–22 Sep. 2010
Denmark	#fv11 (parliamentary election2011), #valg2011 (election2011), and #valg11 (election11)	28,489	15 Aug. 2011–18 Sep. 2011
Norway	#valg2011 (election2011), #valg11 (election11), and #kommunevalg (municipality election)	29,423	12 Aug. 11– 15 Sep. 11

tagged accordingly was not included in our archiving processes. It should also be noted that, at the time, yourTwapperkeeper did not collect so-called button retweets (see Bruns, 2011; Moe & Larsson, 2012 for further discussion of the method of data collection). Building on the conceptual model introduced by Bruns and Moe (Chapter 2 in this volume), a focus on hashtagged communication allows for scrutiny of the macro layer of communication on Twitter: a hashtagged tweet potentially reaches well beyond a user's existing number of followers. Hashtags can help coordinate communication about a topic: they make messages searchable for any user. By adding a topical hashtag to a tweet, then, the user deliberately inserts their message into a wider context, lifting the tweet above the meso level of followers, and into the macro level of a wider public.

We selected suitable, popular hashtags, determined by close observation of the unfolding pre-campaign communication on Twitter, for each of the cases: #val2010 ("election 2010") for the Swedish, #fv11 (an abbreviation for "parliamentary election 2011") for the Danish, and #kommunevalg ("municipality election") for the Norwegian elections. Moreover, two hashtags covered both the Danish and the Norwegian cases: as "valg" means "election" in both languages, #valg2011 and #valg11 served as hashtags in both cases. These two archives were filtered based on the language information provided in the tweet metadata, and manually checked. Likewise, all five archives were comprehensively manually checked for irrelevant content (e.g., where the hashtags referred to unrelated topics). To include the entire campaign period, the data sets cover one month (31 days) before election day in each case. Furthermore, to grasp some of the post-election Twitter communication, archiving was continued until three days after each election, and then terminated.

The collected data were processed in a series of different analyses, employing Gawk scripts to extract information, check, sort, and filter the data sets, and descriptive statistics were established using Excel and SPSS.

RESULTS AND DISCUSSION

Figure 24.1 compares the distribution of tweets over time as day-to-day percentages of the total number. In general, the three cases follow a similar development: from a quiet start, the activity increases incrementally, with clear intensification when approaching election day. Election day and night themselves constitute notable peaks, generating over 20 per cent of the total number of tweets in all three cases. The post-election day exhibits the second highest volume in the Swedish and Danish cases, and the third in Norway. The majority of these tweets come during the first hours of the day—that is, late on election night.

Beyond these similarities, also matching previous, similar work (Burgess & Bruns, 2012), Figure 24.1 also exhibits some differences. First, the Danish hashtags reveal very low activity for the first 11 days. On 26 August, day -20 in Figure 24.1, the activity peaks. This coincides with a press conference where the incumbent Prime Minister officially announced the election. Sweden and Norway have fixed election dates. Since a fixed date facilitates a longer, more planned campaign compared to an announced election, this could explain the higher levels of early activity in these two cases.

Second, it is worth noting that the 22 July 2011 terror attack in Norway had two somewhat contradictory consequences for the election campaign and for Norwegian political debate. On the one hand, it directed attention to fundamental issues of democracy, openness, political participation, and the rule of law, resulting in more media debate about such issues—possibly also affecting political communication on Twitter. On the other hand, the election campaign was officially delayed due to the atrocities. The official start was 13 August—day -30 in Figure 24.1.

Finally, a considerable number of the peaks visible in Figure 24.1 can be understood in conjunction with different forms of mediated content. For all three elections, televised debates or individual interviews with politicians tended to result in such increases in Twitter activity. As such, while the bulk of activity takes place on election day for all three cases, we can distinguish clear bursts of intense activity, often as reactions to televised programming.

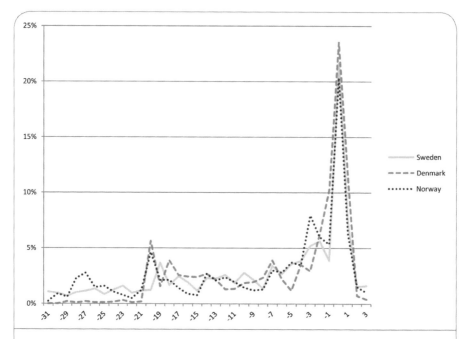

Figure 24.1: Distribution of the Total Number of Tweets in Each Case over the Campaign Period (Percentage of Total Number of Tweets), 31 Days Pre-Election, and Three Days Post-Election. (Sweden, N=99,348; Denmark, N=28,489; Norway, N=29,423)

In their 2004 study of political Web-campaigning practices in a series of countries throughout the world, Gibson (2004) claimed that "one of the major traits of parties' and politicians' exploitation of the web around the world is its 'stop-start' nature, in that it is largely structured around election cycles" (p. 102). While such comparably early research efforts primarily dealt with the online features of what can perhaps best be described as a Web 1.0 variety of online politics, the result appears to hold true also for the political audience in the alleged Web 2.0 era, as illustrated in Figure 24.1. Such patterns are also present in other, similar studies.

While Figure 24.1 provides us with an overview of Twitter activity in all three cases, it does not tell us more about the actual use patterns that emerged during the three campaigns. Table 24.2 and Figure 24.2 offer insights into these matters.

Table 24.2: Comparison of User Activity from Sweden (SE, N of tweets: 99,348, N of users: 9,285), Denmark (DK, N of tweets: 28,489, N of users: 3,185), and Norway (NO, N of tweets: 29,423, N of users: 6,981)

User Group	Number of Users			Tweets/User		
	SE	DK	NO	SE	DK	NO
Least active	8,328	2,861	6,272	<19	<15	< 9
Highly active	863	291	638	19–147	15–139	9–44
Lead users	94	33	71	>147	>139	> 44

As new communication technologies are conceived, their spread among the general public, and their respective uses, can be understood and explored theoretically in a number of different ways. One such approach, the 90-9-1 rule suggested by Nielsen (2006), was employed for our analyses. Nielsen's rule postulates that, with some variation, most net-based groups can be divided into three subgroups. The largest group (i.e., 90% of all users) are the least active, while levels of activity increase among the remaining groups: the next 9% of highly active users, and the final 1% of lead users. (Also see Chapter 6 by Bruns and Stieglitz in this volume, which further explores the use of such standardised Twitter activity metrics.)

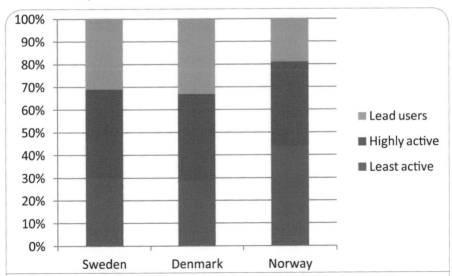

Figure 24.2: Distribution of Activity among Three User Groups as Percentages of the Total Number of Tweets in Each Case

When applied to our data, several results stand out. As shown in Table 24.2, for Sweden and Denmark, the ratio of tweets per user is considerably higher for the lead users. This group may be smaller in numbers, but makes a clear mark because of its level of activity (shown in Figure 24.2). While this is to be expected, we should also note that the majority of tweets in these cases tend to emanate from the second group of highly active users. As such, while the lead users are most active per capita, the bulk of tweets originate from more casual users. Second, while similar trends can also be shown for the Norwegian case, the data from this particular campaign tell a somewhat different story. The distribution of tweets per user appears less skewed here, indicating a more proportionate use of Twitter across the three groups visible. So, while the Swedish and Danish cases exhibit the expected 90-9-1 distribution, the 2011 Norwegian election yielded patterns of Twitter use that appear comparatively more evenly distributed across the three groups of users.

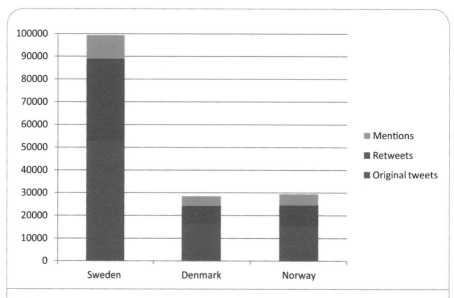

Figure 24.3: Distribution of Types of Tweets in the Three Case Countries (Sweden, N=99,348; Denmark, N=28,489; Norway, N=29,423)

One of the factors that might help us understand this difference is related to the wider societal context: as noted, the terror attack that shocked Norway just weeks before the campaign left its mark. The attack was aimed especially at the ruling Labour Party and its youth organisation, but also triggered fundamental discussions about freedom of speech, the rule of law, and democracy.

As a consequence, the upcoming election was given a new meaning, and the role of politicians in general and of Labour politicians in particular changed. Another factor relates to the election itself: whereas the Swedish and Danish elections primarily focussed on the national level—with well-known members of parliament running for re-election against comparative newcomers, the Norwegian election took place at the local and regional levels only. Although party leaders, of course, also appeared in this campaign (for instance in the televised debates), issues as well as candidates mainly related to the local level. This might lead to more diverse or dispersed patterns of communication on Twitter. A third potential factor should also be mentioned: the uptake of Twitter among the general Internet population in Norway seems to be slightly higher than in the other two countries, which means that the basis for "recruitment" to the practice of election tweeting was larger.

As Twitter has been heralded for its communicative potential in political contexts, we assessed the presence of original tweets, @mentions, and retweets. The results of this comparison between the three countries are shown in Figure 24.3.

While slight variations between the three cases are evident, the overall picture is one of stable distributions: original tweets account for between 52% and 58%, retweets range from 27% in the Danish case to 36% in the Swedish, and between 10% and 17% are @mentions. The default distribution of messages is the dominant mode of communication, therefore, and dialogic communication through @mentions is the least widespread. It follows from this that the majority of activity on the platform at hand is not geared towards discussion and deliberation; however, it must be noted that our data set focusses only on tweets which included predefined hashtags. Any message not featuring the specific hashtag was therefore not captured during the data collection phase; as we may assume that hashtags are often omitted in more interpersonal modes of communication through @replies, this must be seen as a limitation of the chosen approach.

Thus, the results presented in Figure 24.3 present a picture of mostly original tweets being transmitted. While patterns of more conversational use could be discerned, Twitter was mostly used as a "megaphone" of sorts in all three cases. This finding falls in line with previous research on online political communication, where digital media are seen to provide "normalisation"—an enhancement of already established modes of discussion and campaigning, rather than a shift regarding such practices (e.g., Larsson, 2011, 2012). In the contexts studied here, Twitter was used to strengthen already existing patterns of societal debate—not to change or revolutionise it.

CONCLUSION

Just as with the launch and spread of the Internet itself, the Web 2.0 paradigm has carried with it certain expectations regarding its potential for political activity. The present study has provided a cross-national examination based on extensive data collection, employing state-of-the-art methods for analysis. As the results showed more evolutionary than revolutionary tendencies, we align ourselves with Kalnes (2009) in suggesting that "one should be careful not to overemphasize these changes at the expense of continuity" (p. 251). While novel services such as Twitter in some cases provide the electorate as well as politicians with online spaces to meet and discuss, these interactions tend to take on rather traditional patterns. Finally, while case studies from individual contexts have provided useful insights into political uses of Twitter and other online phenomena, there is a need for clearer comparative efforts as well as longitudinal study designs, assessing how these practices evolve over time and space. To develop a more comprehensive understanding, there is also a need to address the modes of communication via social networking sites that are not captured with the kind of data-collection tools employed in the present study. Future research might find it helpful to gauge such online practices with these aspects in mind.

REFERENCES

Befolkningens brug af Internet 2010. (2011). Statistics Denmark. Retrieved from http://www.dst.dk/en/Statistik/emner/informationssamfundet.aspx

Bruns, A. (2011). How long is a tweet? Mapping dynamic conversation networks on Twitter using Gawk and Gephi. *Information, Communication & Society, 15*(9), 1323–1351. doi: 10.1080/1369118x.2011.635214

Bruns, A., & Burgess, J. (2011). #ausvotes—How Twitter covered the 2010 Australian federal election. *Communication, Politics & Culture, 44*(2), 37–56.

Burgess, J., & Bruns, A. (2012). (Not) the Twitter election. *Journalism Practice, 6*(3), 384–402.

Facht, U., & Hellingwerf, K. (2011). *Nordicom-Sveriges Internetbarometer 2010.* Göteborg, Sweden: Nordicom.

Gaffney, D. (2010, April). *#iranElection: Quantifying online activism.* Paper presented at the WebSci10: Extending the Frontiers of Society On-Line conference, Raleigh, NC.

Gibson, R. (2004). Web campaigning from a global perspective. *Asia-Pacific Review, 11*(1), 95–126.

Goggin, G., & McLelland, M. (2009). Internationalizing Internet studies: Beyond Anglophone paradigms. In G. Goggin & M. McLelland (Eds.), *Internationalizing Internet studies: Beyond Anglophone paradigms* (pp. 3–18). New York, NY: Routledge.

Golbeck, J., Grimes, J. M., & Rogers, A. (2010). Twitter use by the US Congress. *Journal of the American Society for Information Science and Technology, 61*(8), 1612–1621.

Jackson, N. A., & Lilleker, D. G. (2009). Building an architecture of participation? Political parties and Web 2.0 in Britain. *Journal of Information Technology & Politics, 6*(3), 232–250.

Kalnes, Ø. (2009). Norwegian parties and Web 2.0. *Journal of Information Technology & Politics, 6*(3), 251–266.

Larsson, A. O. (2011). 'Extended infomercials' or 'Politics 2.0'? A study of Swedish political party Web sites before, during and after the 2010 election. *First Monday, 16*(4). Retrieved from http://firstmonday.org/htbin/cgiwrap/bin/ojs/index.php/fm/rt/printerFriendly/3456/2858

Larsson, A. O. (2012). 'Rejected bits of program code'—Why notions of 'Politics 2.0' remain (mostly) unfulfilled. *Journal of Information Technology & Politics*. Advance online publication. doi: 10.1080/19331681.2012.719727

Larsson, A. O., & Moe, H. (2012). Studying political microblogging: Twitter users in the 2010 Swedish election campaign. *New Media & Society, 14*(5), 729–747. doi: 10.1177/1461444811422894

Lassen, D. S., & Brown, A. R. (2011). Twitter: The electoral connection? *Social Science Computer Review, 29*(4), 419–436.

Lilleker, D. G., Koc-Michalska, K., Schweitzer, E. J., Jacunski, M., Jackson, N., & Vedel, T. (2011). Informing, engaging, mobilizing or interacting: Searching for a European model of Web campaigning. *European Journal of Communication, 26*(3), 195–213. doi: 10.1177/0267323111416182

Lilleker, D. G., & Malagón, C. (2010). Levels of interactivity in the 2007 French presidential candidates' websites. *European Journal of Communication, 25*(1), 25–42. doi: 10.1177/0267323109354231

Lotan, G., Graeff, E., Ananny, M., Gaffney, D., Pearce, I., & boyd, d. (2011). The revolutions were tweeted: Information flows during the 2011 Tunisian and Egyptian revolutions. *International Journal of Communication, 5*, 1375–1405.

Moe, H. (2011). Mapping the Norwegian blogosphere: Methodological challenges in internationalizing Internet research. *Social Science Computer Review, 29*(3), 313–326.

Moe, H., & Larsson, A. O. (2012). Methodological and ethical challenges associated with large-scale analyses of online political communication. *Nordicom Review, 33*(1), 117–125.

Nielsen, J. (2006). Participation inequality: Encouraging more users to contribute. Retrieved from http://www.useit.com/alertbox/participation_inequality.html

Nordicom. (2010). Media trends and media statistics in the Nordic region. *Nordicom—Nordic Information Centre for Media and Communication Research*. Retrieved from http://www.nordicom.gu.se/eng.php?portal=mt

NRK/Ipsos MMI. (2011). *Nettbrukerundersøkelse. 2011*. Oslo, Norway: NRK.

O'Reilly, T. (2005). What is Web 2.0? Design patterns and business models for the next generation of software. Retrieved from http://www.oreillynet.com/lpt/a/6228

Raats, T., & Pauwels, C. (2011). In search of the Holy Grail? Comparative analysis in public broadcasting research. In K. Donders & H. Moe (Eds.), *Exporting the Public Value Test. The regulation of public broadcasters' new media services across Europe* (pp. 17–28). Göteborg, Sweden: Nordicom.

Ragin, C. (1987). *The comparative method: Moving beyond qualitative and quantitative strategies*. Berkeley, CA: University of California Press.

Sæbø, Ø. (2011, Aug.-Sep.). *Understanding Twitter™ use among parliament representatives: A genre analysis*. Paper presented at the Third International Conference on eParticipation (ePart 2011), Delft, The Netherlands.

Schweitzer, E. J. (2011). Normalization 2.0: A longitudinal analysis of German online campaigns in the national elections 2002–9. *European Journal of Communication, 26*(4), 310–327. doi: 10.1177/0267323111423378

Stromer-Galley, J. (2000). On-line interaction and why candidates avoid it. *Journal of Communication, 50*(4), 111–132.

Syvertsen, T., Enli, G., Mjøs, O.J., & Moe, H. (2014). *The Media Welfare State: Nordic Media in the Digital Age*. Ann Arbor, MI: University of Michigan Press.

Tumasjan, A., Sprenger, T. O., Sandner, P. G., & Welpe, I. M. (2010). Election forecasts with Twitter: How 140 characters reflect the political landscape. *Social Science Computer Review, 29*(4), 402–418. doi: 10.1177/0894439310386557

Vergeer, M., Hermans, L., & Sams, S. (2011). Is the voter only a tweet away? Micro blogging during the 2009 European Parliament election campaign in the Netherlands. *First Monday, 16*(8). Retrieved from http://firstmonday.org/htbin/cgiwrap/bin/ojs/index.php/fm/article/view/3540/

Wattal, S., Schuff, D., Mandviwalla, M., & Williams, C. B. (2010). Web 2.0 and politics: The 2008 U.S. presidential election and an e-politics research agenda. *MIS Quarterly, 34*(4), 669–688.

Winston, B. (1998). *Media, technology and society—A history: From the telegraph to the Internet*. New York, NY: Routledge.

The Gift of the Gab
Retweet Cartels and Gift Economies on Twitter

25
CHAPTER

Johannes Paßmann, Thomas Boeschoten, and Mirko Tobias Schäfer

 among Dutch parliamentarians, @replying is unaffected by party affiliation while RTs are structured by it

Whether Twitter is viewed as a platform for narcissistic self-representation or a catalyst of political change, the service provides for the circulation of brief messages among connected users. These users participate actively in this circulation by retweeting, favouring (or 'faving'), and replying to messages and drawing additional attention to them, stimulating even more circulation through other users' retweets and favourites of the initial message. This chapter looks at the modes of circulation of Twitter messages and will reveal user practices for retweeting. It shows that users make pragmatic choices when retweeting or faving messages, and illustrates how these choices are embedded in a socio-cultural context.

The support of other users and their willingness to share a message with their range of followers is crucial for distributing tweets successfully. In *Debt: The First 5000 Years*, anthropologist David Graeber explained that the commercial exchange of goods is different from the exchange of gifts because trading partners have the opportunity to even things out by paying their debts and parting ways (Graeber, 2011, p. 105). However, in the case of neighbourly relationships, not paying back 'debts' can actually create and consolidate relationships. On this point, Graeber referred to Laura Bohannan's (1954) anthropological novel *Return to Laughter*, where she explained how the Tiv people in rural Nigeria base their communities on a perpetual circulation of *gifts* (Graeber, 2011). Tiv customs require the receiver of the present, the presentee, to eventually return the favour—not immediately, but after a while. Can we argue—keeping in mind the protocol behind the exchange of gifts in Tiv communities in Nigeria described by Graeber—that the successful circulation of communication on Twitter relies heavily on pervasive mutual indebtedness?

The philosopher and ethnologist Marcel Hénaff argued that in the past the ceremonial, mutual exchange of gifts was limited to segmentary societies and was the common way to publicly acknowledge and show respect to a presentee. According to Hénaff, this way of demonstrating recognition has become obsolete in today's political societies, because social status is regulated by law. The gift has become a purely private matter (Hénaff, 2008, p. 237). If social media revive gift exchanging as a popular form of public appreciation—whether by retweeting or faving on Twitter or by liking on Facebook—the concept of the gift would reveal a new perspective on social interaction in social media.

In order to understand circulation via social media, Henry Jenkins, Xiaochang Li, Ana Domb Krauskopf, and Joshua Green made the same distinction between the circulation of commerce and gifts that many other scholars have made before them, and which is most famously explored by Graeber, Malinowski, and Gregory. Specifically, the authors distinguish in cultural production in social media—alluding to the novelist Lewis Hyde—between a 'commodity culture' and a 'gift economy' (Jenkins, Li, Domb Krauskopf, & Green, 2009, p. 45). This distinction enables them to 'develop a better model' (Jenkins et al., 2009, p. 46) than does the concept of viral distribution, which degrades users to "involuntary 'hosts'" (Jenkins et al., 2009, p. 8) of a virus. Different models for the dissemination of communication in online media have been proposed, such as the abovementioned viral distribution, a term coined by Chris Anderson (2004) and further elaborated by Charles Leadbeater (2008) and Clay Shirky (2008).

This chapter does not propose a superior model for the circulation of messages on Twitter, but rather tries to map the practices users actively employ for spreading their messages. We essentially assume that the hybrid infrastructure of Twitter, since it consists of a software design and user activities, will remain dynamic and subject to design and appropriation processes that significantly affect the modes of circulation. For example, retweeting used to be a user-initiated practice, a form of citing, in which 'RT @username' was manually added to the written text. Later, the retweet button was introduced, one of many changes in Twitter's software design that altered its modes of circulation (see Chapter 3 by Halavais, in this volume).

In a qualitative analysis, we map user perceptions of how to successfully use Twitter, and how users think Twitter communication works. We also elaborate on these findings with a quantitative analysis of two different examples of highly active Twitter users.

FOLLOW THE NATIVES

We will refer to two cases that empirically show how circulation is conducted on Twitter. They also show how sample messages are distributed. Because of Twitter's social network infrastructures and hierarchies, anyone attempting to explain how circulation is conducted cannot only focus on content. We also reject the notion of a stable distribution model, as we view Twitter as a socio-technological setting, where users appropriate technology and media practices, while the platform provider also constantly readjusts the platform's information management and distribution mechanisms.

In case 1, a mapping of the Dutch parliamentary Twitter sphere reveals functional interactions between professional elites. Case 2 is an analysis of German Twitter users, which reveals two loosely connected networks with quite different core interests: net politics and fun. Both networks are dominated by retweet cartels that are crucial for pushing messages beyond the attention threshold of a wide audience. Our quantitative approach was able to retrieve the actual flow of messages through a network, and can trace in detail when which topic was raised by whom and to what effect. Our qualitative research, meanwhile, was able to reveal the factors that this communication thrived on: social interaction, face-to-face communication, mutual respect, and the individual standing of a sender within the network.

POLITICAL PARTIES AS RETWEET CARTELS

Mapping the activity of Dutch politicians on Twitter shows that the party affiliation of the initial sender and those who subsequently retweet the message is crucial for the circulation. The Dutch parliament has a multi-party system based on proportional representation. From 2010 to 2012 there were 150 members of parliament representing 10 parties, roughly divided into left-wing, right-wing, and centre parties. Precisely this multi-party system is reproduced in the scene's Twitter communication.

We gathered all the tweets sent by members of the Dutch parliament between 1 February 2012 and 31 August 2012. Two data sets were prepared: one with all replies by politicians, and another consisting solely of retweets. For Figure 25.1 and Figure 25.2 we filtered both data sets in order to show only the mutual relationships between members of parliament. We used Gephi, an interactive

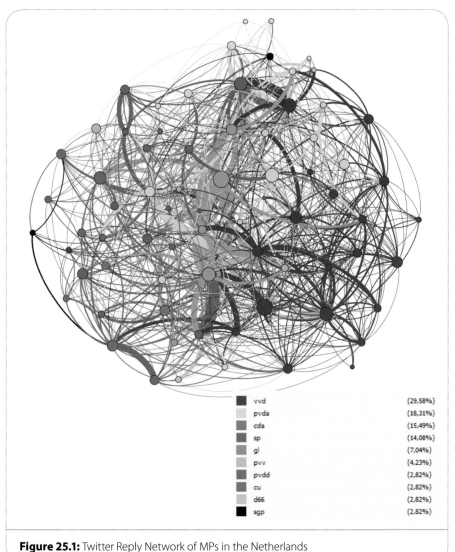

vvd		(29,58%)
pvda		(18,31%)
cda		(15,49%)
sp		(14,08%)
gl		(7,04%)
pvv		(4,23%)
pvdd		(2,82%)
cu		(2,82%)
d66		(2,82%)
sgp		(2,82%)

Figure 25.1: Twitter Reply Network of MPs in the Netherlands

visualisation platform, to visualise the data by applying the Force Atlas 2 algorithm to it (with the same settings for both data sets).

The reply network (Figure 25.1) shows that many members of parliament communicate frequently with each other, and reciprocate regardless of their party affiliation. Their communication on Twitter is essentially unaffected by party affiliation. Therefore, the graph has an almost perfect, round shape, with many users connected to a wide variety of colleagues from different parties. Some members of the same parties flock close together, forming a cluster (especially the Dutch Liberal Party, VVD), but in general, party membership hardly affects with whom they communicate via Twitter. The clusters of parties are well connected to other parties. Some of them, like the Socialist Party (SP), the Christian Democrats (CDA), and the Labour Party (PvdA), do not form clear clusters at all.

While replying is widely unaffected by party affiliation, retweeting is very much structured by it. Dutch politicians tend to prefer retweeting their own

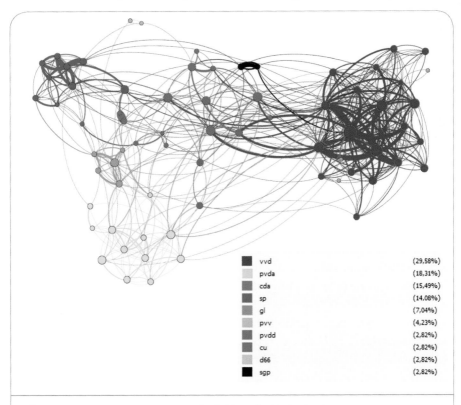

■	vvd	(29,58%)
▫	pvda	(18,31%)
■	cda	(15,49%)
■	sp	(14,08%)
▫	gl	(7,04%)
▫	pvv	(4,23%)
■	pvdd	(2,82%)
■	cu	(2,82%)
▫	d66	(2,82%)
■	sgp	(2,82%)

Figure 25.2: Twitter Retweet Network of MPs in the Netherlands

party members' messages than retweeting messages by members from opposition parties. That is why, instead of a highly intertwined network, the retweet network shows almost isolated clusters of parties. On the right, we see the VVD, closest to the parties with whom they formed a government in the previous cabinet (CDA and the Party for Freedom, PVV). On the left, we see the opposition, the left-wing parties, with the nodes forming clusters and some weak ties between the clusters.

The difference between Figure 25.1 and Figure 25.2 suggests that retweeting and replying are treated as different media practices: retweets are often seen as a form of endorsement, while replies appear to be a mode of communication among colleagues. So the retweet network resembles the political organisation, with the different parties clustering together next to their political kin. MPs' tendency to prefer their own MPs for retweeting above others is a form of homophily:

> Similarity breeds connection. This principle—the homophily principle—structures network ties of every type. . . . The result is that people's personal networks are homogeneous with regard to many sociodemographic, behavioral, and intrapersonal characteristics. Homophily limits people's social worlds in a way that has powerful implications for the information they receive, the attitudes they form, and the interactions they experience. (McPherson, Smith-Lovin, & Cook, 2001, p. 415)

Earlier research also found several forms of homophily on Twitter (see Java, Song, Finin, & Tseng, 2007; Weng, Ling, Jiang, & He, 2010; and Wu, Hofman, Mason, & Watts, 2011). This suggests that users tend to flock in homogeneous networks, in terms of values or social status. Even though MPs' behaviour demonstrates homophily, it should be noted that it is a very specific form of homophily. It refers to a specific legal form of organisation, namely the political parties representing their shared values. Politicians do not just retweet people who are similar to them or share their values. They retweet people from their own party, and this behaviour evokes Durkheim's concept of 'mechanical solidarity' (Durkheim, 1984, ch. 2). This mechanical solidarity is what is behind this specific brand of homophily, which can be called a retweet cartel. Here, the practice of retweeting takes place in the context of membership in a political organisation, whereas its gifting character apparently does not initially generate relationships. The choice to retweet their fellow party members over other politicians is an affirmation of offline affiliations, and as such, reproduces social structures existing also 'outside' of Twitter.

OPENING GIFTS IN THE GERMAN FAVSTAR SCENE

The Favstar scene is a widely popular network among German Twitter users. Favstar is a Web application that tracks retweets and favourites, called Favs. Favstar generates rankings of users and awards them for particular achievements, such as having received 50 or more Favs. Users ranking high on Favstar are sometimes ironically referred to as members of a 'Twitter Elite' by other users. When mainstream media refer to tweets that report current events in Germany, they frequently refer to accounts held by 'elite' members. Members try to write tweets that receive a maximum of retweets and Favs, in order to increase their status. Status in this group is gained both by the number of followers and the number of received Favs, retweets, and Favstar awards users accumulate. While the politicians mentioned above have the advantage of being known to a large audience through their mainstream media appearances, Favstar members frequently have to build up their audience from scratch after setting up what are often pseudonymous accounts.

Apart from the skill it takes to write witty messages, there are other practices that help users to establish an audience of followers. We experimented by searching for tweets that have received Favs from popular accounts, and then randomly awarded Favs to as many tweets as possible. The result was a sharp increase in the Favs we received. Some users who had received Favs from us returned the favour by sending out recommendations to follow our accounts. This is a well-known strategy among heavy users and the Favstar scene, but anyone who employs this strategy repeatedly risks being labelled an 'Allesfaver' (someone who faves anything).

This practice of awarding Favs evokes the 'opening of gifts' as described in Malinowski's work: at the beginning of an exchange ceremony, potential partners are lured with an opening gift. If one of the participants accepts it, he or she has to reciprocate with a 'clinching gift' that establishes a relationship with certain obligations. The actual exchange takes place after this initial opening ritual (cf. Malinowski, 1932, pp. 98, 352ff., 472f., 487f.). Awarding Favs to other Twitter users is similar to an opening gift. However, Twitter is not coercive about the clinching gift in the case of Favs and retweets. The circulation of these gifts is not necessarily mutual. While Malinowski's account of the gift exchange appears to be shaped by tradition and thrives on rather explicit social coercion, gifts in Twitter thrive on the expectation that some Twitter users will return the gesture.

After accepting the opening gift, the future relationship between two users on Twitter can evolve into an alliance, where both pragmatically retweet each other's tweets in order to have access to each other's audience. The gift in the digital realm is not pricy, which is why some successful members of Favstar give away opening gifts in large numbers. Some users award up to 200,000 Favs per year, and this strategy rewards them with many followers. The inflated number of Favs in question here sheds doubt on their value as a gift. Some Favstar scene members award up to 4,000 Favs per day. We might almost speak of a

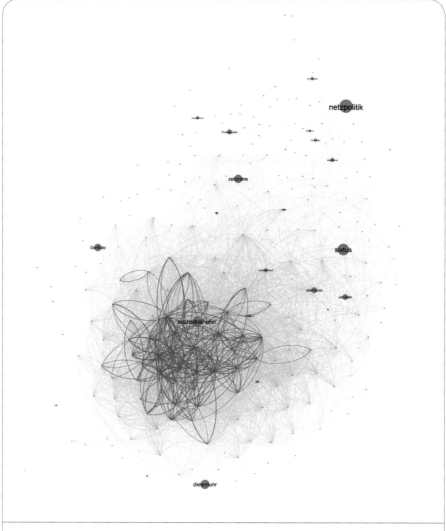

Figure 25.3: Frequent Distribution of Favs among 350 Popular German Accounts

gift simulation here, an ephemeral gesture of endorsement; the presentee is not required to reply in kind, and the donor has an abundance of Favs to distribute.

Quantitative analysis also sheds light on the practice of ritual faving. Figure 25.3 and Figure 25.4 show networks of about 350 popular Twitter accounts in the German Twitter sphere. Linking these 350 accounts to their Favstar records, we built a database consisting of the 100 most popular tweets sent by each of these accounts, and traced all the users who retweeted or faved them.[1] We used that database to create two visualisations, filtering the Favs (Figure 25.3) and retweets (Figure 25.4) that had been exchanged between the 350 accounts only. In both cases, this maps out a part of at least two German gift economies on Twitter.

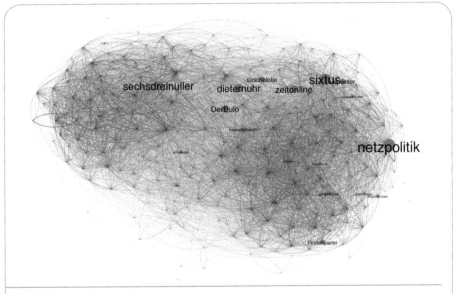

Figure 25.4: Retweets in the Favstar Scene (left) and Other Accounts, Often Affiliated to Net Politics (right)

The dark and thick lines in Figure 25.3 show at least five mutual Favs out of 100 Tweets. The thin lines show one-way Favs. The size of the nodes corresponds to the number of toptweets that each account has written. Some large nodes in the diagram have no connection to others, as they use Favs much like bookmarks. Others, gathered in the dark cluster of accounts, use the favourite function excessively. Here we see a specific scene emerging with a specific gifting practice: the Favstar scene. Almost half are involved in the mutual exchange of Favs.

In this scene, Favs are awarded much more often than retweets. As was the case with the politicians in the example above, retweets indicate a stronger commitment. Taken together, these quantitative findings and Paßmann's participant observations reveal an economy of gift exchanging that stabilises and maintains the popularity of the accounts in question through Favs.

We also analysed another gifting practice that contributes more obviously to the circulation of tweets. We mapped the retweets of the 350 popular German accounts in Figure 25.4, which enables us to show the circulation of tweets on German Twitter.

Two clusters are discernible in Figure 25.4: the left one represents the Favstar scene discussed above, while the right one consists of accounts mainly involved in net politics. The right cluster resembles use similar to what we found with Dutch politicians, because the retweet also serves to promote shared objectives and values, as well as being an effective form of information distribution.

Looking at the two clusters, we notice that the practice of retweeting is different. We found that the Favstar accounts retweet each other much more frequently than the accounts in the right cluster. We interpret this to be the result of a stronger social cohesiveness consolidated by the practice of gifting.

The gift economy is most distinctive in the cluster where the circulation of messages is perceived as a desirable end in itself, and is therefore much more present in the Favstar scene. This leads to a more homogeneous cluster of accounts that are exclusively concerned with the distinct memes, habits, and communication patterns of this same scene. The accounts related to net politics show a more heterogeneous mix of participants, ranging from activists to mainstream media accounts. The Favstar scene has been facilitated by a form of technology appropriation described by Schäfer (2011): originally created as a way of bookmarking, the Fav button is now fundamental to the gift economy of the Favstar scene. The accounts displayed in the other cluster have not developed such a salient form of technology appropriation.

Contrary to the Fav, the retweet is a demonstration of public commitment, and is therefore used less frequently, especially in the Favstar scene, but also among the politicians who are very selective about whose messages they are willing to multiply. Retweets are common among people in the Favstar scene who already have an established relationship in mutual exchange, while in the political sphere people mainly stick to retweeting members of their own party. Unlike the politicians, people in the Favstar scene are not burdened with many formal or professional obligations, and have relatively little in common outside their Twitter activities. Members of the Favstar scene rely heavily on other people's support for the circulation of their messages, while for the politicians, Twitter is only marginally important—'just another' channel to promote their agendas and a means to communicate (Schäfer, Overheul, & Boeschoten, 2012).

The quantitative description and our interpretation are supported by Paßmann's interviews. By confronting members in the Favstar retweet clusters depicted in Figure 25.4 with the findings, he tried to retrieve their personal view of their practice of awarding Favs and retweets. At the end of an almost four-hour-long conversation with @sechsdreinuller, the most retweeted account of the Favstar scene, he said: "Of course there are cartels, and of course we invest in them and use them. Why should I retweet someone who will never retweet me back or promote something that is already on the mass-media anyway?" (Personal notes from conversation with @sechsdreinuller in Frankfurt on 27 July 2012).

When asked if he has ever retweeted a tweet from a non-governmental organisation or other charitable organisation, he answered: 'I did that once because that was extremely important to me. But, you know, things like that cost me a massive amount of followers. My followers follow me for the punch lines, not for what I want them to do' (Personal notes from conversation with @sechsdreinuller in Frankfurt on 27 July 2012).

The Twitter users we have described above are aware of the fact that they depend on others to maximise the distribution of their messages and form useful alliances. While the politicians reproduce their political alliances on Twitter, the Favstar members initiate them implicitly through their gifts. Making these alliances explicit is—at least among Favstar members—objectionable. In an interview, user @goganzeli calls it a 'form of cheating'. The user @sechsdreinuller was only willing to speak about the retweet cartels after Paßmann could show that his pseudonymous account appeared in the same retweet cartel. This reveals two sides to the alliances Favstar users forge: on the one hand, mutual support is necessary for distributing messages successfully, and on the other hand, the alliance must remain latent.

CONCLUSION

We have shown that the circulation of messages on Twitter is co-shaped by consolidation of relationships through mutual gift exchanging and the reproduction of existing social relationships. A quantitative analysis of Favs and retweets revealed distinct clusters of users who prefer to circulate messages by members of the same cluster. This circulation might be based on shared values, a political affiliation, or other things people have in common outside the world of Twitter, as the example of the politicians indicated, but it could also be the result of a common practice of using retweets, Favs, replies, and other gifts to establish mutual relations that extend beyond the existing range of potential circulation. Employing Malinowski's (1932) term "gift economy" has made it possible to explain the patterns of message circulation revealed by our quantitative analysis, and to back them up with qualitative findings.

We observed that the gift in the Favstar scene resembles a revived form of public recognition. This is useful for analysing interaction on social media in general. The term "gift economy" has been repeatedly used to describe forms of 'immaterial' exchange in online networks (see, for example, Rheingold, 2000, p. 49).[2] Investigating how content spreads online, Jenkins and colleagues have revived the notion of gift economy in their book *Spreadable Media*:

> As a rule, we are misled when we focus on what media does to people rather than trying to understand what people are doing with media and why. We start from the premise that consumers only help facilitate the circulation of media content when it is personally and socially meaningful to them, when it enables them to express some aspect of their own self-perception or enables valued transactions that strengthen their social ties with others. (Jenkins et al., 2009, p. 43)

Our analysis elaborates on this argument, and provides empirical data to support the notion of gift economies as a modus operandi on social media platforms. We could show a difference between gift economies as Malinowski described them and those on social media platforms. Gifts are available and distributed in abundance; contrary to 'material' gift economies, their pecuniary value is insignificant. As such, the exchange of gifts described in our research corresponds with the notion of information gift economies. Here, sharing information is considered an inexpensive 'gift' with the added benefit that one receives information in return (Mackaay, 1990).

The opening gift provides a strong incentive to distribute content, and this is even encouraged by the interface design of social media platforms, such as the retweet and favourite buttons, though factually appropriated by the users. These buttons lower the threshold to distribute an opening gift and establish contact. The design features for ephemeral communication provided by the platform providers fuel the user interaction and communication. These features facilitate gift-giving, which initiates social interaction and the collaborative use of the networked infrastructure in order to circulate content.

NOTE

1   The sample consists of Twitter accounts that have been retweeted by the account @toptweets_de (which belongs to the Twitter corporation) at least three times between 9 September 2011 and 9 March 2012. The toptweets account uses an algorithm to define a range of accounts and a range of tweets. Messages that receive the status of toptweet as defined by Twitter are retweeted through the various language-based toptweets accounts. Here we focussed on the German edition of toptweets. Other publications also refer to @toptweets_de retweets, or mention a criterion for the range of accounts (see Neuberger, vom Hofe, & Nuernbergk, 2009). We would like to thank Martijn Weghorst for retrieving the data. He was most helpful in visualising data and commenting on the findings.

2   We want to emphasise that our understanding of the immaterial is only related to the non-haptic nature of commodities online. Like Van den Boomen et al. (2009), we recognise the material nature of digital artefacts and online practices in their economic, social and political relations and effects.

REFERENCES

Anderson, C. (2004, Oct.). The long tail. *Wired Magazine, 12*(10). Retrieved from http://www.wired.com/wired/archive/12.10/tail.html

Bohannan, L. (1954). *Return to laughter.* New York, NY: Harper.

Durkheim, E. (1984). *The division of labour in society.* Basingstoke, UK: Palgrave Macmillan.

Graeber, D. (2011). *Debt: The First 5000 Years.* New York: Melville House.

Gregory, C. A. (1982). *Gifts and commodities.* London, UK: Academic Press.

Hénaff, M. (2008). *Der Preis der Wahrheit: Gabe, Geld und Philosophie.* Frankfurt am Main, Germany: Suhrkamp.

Java, A., Song, X., Finin, T., & Tseng, B. (2007). Why we Twitter: Understanding microblogging usage and communities. *Proceedings of the Joint 9th WEBKDD and 1st SNA-KDD Workshop.* Retrieved from http://aisl.umbc.edu/resources/369.pdf

Jenkins, H., Li, X., Domb Krauskopf, A., & Green, J. (2009). *Spreadable media: Creating value in a spreadable marketplace.* Retrieved from http://educat.dsm.usb.ve/wp-content/uploads/2010/11/Spreadability.pdf

Leadbeater, C. (2008). *We think: Mass innovation, not mass production.* London , UK: Profile Books.

Mackaay, E. (1990). Economic incentives in markets for information and innovation. *Harvard Journal of Law & Public Policy, 13*(3), 867–910.

Malinowski, B. (1932). *Argonauts of the Western Pacific. An account of native enterprise and adventure in the archipelagoes of Melanesian New Guinea.* London, UK: George Routledge & Sons.

McPherson, M., Smith-Lovin, L., & Cook, J. M. (2001). Birds of a feather: Homophily in social networks. *Annual Review of Sociology, 27*(1), 415–444. Retrieved from ftp://www.soc.cornell.edu/csi/Networks/mcpherson%20smith-lovin%20cook%20ars.pdf

Neuberger, C., vom Hofe, H. J., & Nuernbergk, C. (2009). *Twitter und Journalismus: Der Einfluss des 'Social Web' auf die Nachrichten*. LfM Dokumentation Band 38. Retrieved from http://www.lfm-nrw.de/fileadmin/lfm-nrw/Publikationen-Download/LfM_Doku38_Twitter_Online.pdf

Rheingold, H. (2000). *The virtual community*. Cambridge, MA: MIT Press.

Schäfer, M. T., Overheul, N., & Boeschoten, T. (2012). Een netwerkanalyse van twitterende Nederlandse politici. In C. Van 't Hof, J. Timmer, & R. Van Est (Eds.), *Voorgeprogrammeerd: Hoe Internet ons leven leidt* (pp. 188–210). The Hague, The Netherlands: Boom/Lemma Uitgevers.

Shirky, C. (2008). *Here comes everybody: The power of organizing without organizations*. London, UK: Penguin Press.

Van den Boomen, M., Lehmann, A. S., Lammes, S., Raessens, J., & Schäfer, M. T. (2009). *Digital material: Tracing new media in everyday life and technology*. Amsterdam, The Netherlands: Amsterdam University Press.

Weng, J., Ling, E., Jiang, J., & He, Q. (2010). Twitterrank: Finding topic-sensitive influential Twitterers. *Proceedings of the Third ACM International Conference on Web Search and Data Mining* (pp. 261–270).

Wu, S., Hofman, J. M., Mason, W. A., & Watts, D. J. (2011). Who says what to whom on Twitter. *Proceedings 20th International Conference on the World Wide Web* (pp. 705–714), ACM, Hyderabad, India. Retrieved from http://research.yahoo.com/pub/3386

The Use of Twitter by Professional Journalists
Results of a Newsroom Survey in Germany

26
CHAPTER

Christoph Neuberger, Hanna Jo vom Hofe, and Christian Nuernbergk

newsroom survey shows: #journalists use Twitter for self-promotion, investigation, real-time coverage, and interaction with the public

In this chapter, the relationship between Twitter and professional journalism is discussed on the basis of a newsroom survey and related content analyses. Twitter-based communication has unique features which imply great relevance, but also some challenges for professional journalism:

- On a structural level, network analyses indicate that Twitter does not primarily function as a social network for establishing or maintaining contacts, but instead as a *network for disseminating information* and breaking news. This distinction is suggested by the dominance of one-sided relationships (Kwak, Lee, & Moon, 2010). A highly interlinked and nested network structure between Twitter accounts allows rapid forms of news diffusion, and affects the dynamics of information flow (Lerman & Ghosh, 2010).

- In terms of reach, Twitter still has a *limited direct reach* compared to other media and other social network sites. According to a study conducted by ARD and ZDF Media Research in 2012, only 4% of Internet users in Germany used Twitter at least seldom (Busemann & Gscheidle, 2012). In the US, 13% of all adults have ever used Twitter (Pew Research Center, 2012).

- Despite their small number, it is likely that Twitter users serve as important *multipliers for spreading information* communicated via the network to other channels. In Germany, Twitter users often work in media and communication professions (Busemann & Gscheidle, 2012). In the US, they often rely on the Internet for news reports (Pew Research Center, 2012; Purcell, Rainie, Mitchell, Rosenstiel, & Olmstead, 2010). More than half of American Twitter users (59%) also have already tweeted or retweeted news (Pew Research Center, 2012). Thus, it could be argued that Twitter users constitute an Internet avant-garde with a greater affinity for news.

ADOPTION OF TWITTER IN MAINSTREAM JOURNALISM

In recent years, the adoption of Twitter as a journalistic channel for the dissemination of information and the investigation of stories has gained public attention. Journalists experiment with Twitter in an attempt to find the best method for harnessing this new communications channel. In Germany, several events have also stimulated the debate about the moral dimension of Twitter usage. During the 2009 school shooting in Winnenden, students tweeted live from the scene of the killings. Furthermore, in 2010 the results of the German presidential election were "leaked" early on Twitter by a fake account using the name of actress Martina Gedeck, one of the electoral delegates. Twitter is part of the "new news ecosystem" in which the media and their audiences are mutually connected (Benkler, 2006). Journalism no longer has a centralised and powerful gatekeeping role as a mediator between news sources and the general public, as was the case during the era of traditional mass media. On the Internet, journalism functions more as a moderator and gatewatcher (Bruns, 2005; Neuberger & Nuernbergk, 2010). Twitter is clearly a multifunctional tool for public communication, but what specific uses does Twitter have in the world of journalism? Five dimensions for the use of Twitter can be identified:

1. Journalists *promote* their own websites. Tweets refer to website content and link to them.
2. Journalists conduct *real-time coverage* from the scene of current news events. They provide live reports via Twitter, directly from where the events are taking place.
3. Journalists *interact* with members of the public on Twitter.
4. They *monitor* audience reactions and follow-up communication to their reports.
5. Finally, they *investigate* stories and conduct research using Twitter.

Although Twitter offers a number of different possibilities of use for professional journalists, it is only one of many Internet-based and participatory platforms for communication. Yet, within the social media universe, it is still unclear on what kind of niche the use of Twitter might specialise from a journalistic viewpoint. Therefore, our research approach explores facets of the following overarching question: do newsrooms use Twitter in ways that fully tap into the potential of this particular public communications platform? Our attention was not limited to Twitter, as we also examined other forms of social media to evaluate the particular strengths of each medium.

NEWSROOM SURVEY

In order to explore Twitter use by news staff, we conducted a comprehensive newsroom survey. In May and June 2010, we surveyed editors-in-chief of Internet news departments headquartered in Germany. A number of national media listings were analysed to identify relevant departments offering regularly updated and relevant news content (for details, see Neuberger, vom Hofe, & Nuernbergk, 2011). In order for such departments to qualify, news content had to be updated at least once a day (actuality), and could not be limited to specific, single subjects or target groups (universality).

In the case of daily newspapers, we included all titles listed as independent media units on a regional and supra-regional level. In all other cases, only news departments with media products distributed nationwide or supra-regionally were selected. All news sites were visited to compile editorial contacts for our survey, and the actuality and universality of their Web content was briefly checked.

Table 26.1: Numbers and Response Rate (Evaluated Questionnaires) from the Survey of Internet Newsroom Staff, Compared across Different Types of Providers with Supra-Regional Distribution (Newsroom Survey, 2010)

	Total number	Response rate	Percentage
Daily newspapers (regional and supra-regional)	120	54	45.0
Weekly newspapers, public magazines	10	3	30.0
Broadcasting (TV/radio)	17	11	64.7
Internet-only providers	10	2	20.0
Total	157	70	44.6

Overall, 70 newsroom directors participated in the online survey; the response rate was approximately 45%. Table 26.1 provides an overview of the response rate and the number of identified departments which supervise relevant news sites on the Web.[1] The following discussion presents the key findings of this study.

WAYS OF USING TWITTER AND RULES FOR ITS USE

Journalists made some use of Twitter in almost all of the news departments. Only two of the surveyed news departments indicated that they avoided using Twitter. Compared to other social media, such as weblogs, this reflects a very high usage rate. Forty-one percent of the news departments began tweeting in the first half of 2009, the period during which Twitter made its first successful entry into German journalism. At this time, public debates about the relevance of tweeting increased significantly. Especially tweets about the sudden death of Michael Jackson in June 2009 sparked attention. In their estimation of the importance of Twitter for news publishing activities and interactions with users, news staff rated Twitter second among various social media—just after Facebook (see Figure 26.1). Yet, despite this finding, the actual emphasis placed on Twitter should not be overestimated:

- In over half of the news departments surveyed (57%), Twitter is used by less than a quarter of staff members.
- Moreover, almost two thirds (64%) of the surveyed news departments regard Twitter as "relatively unimportant" to their daily work.

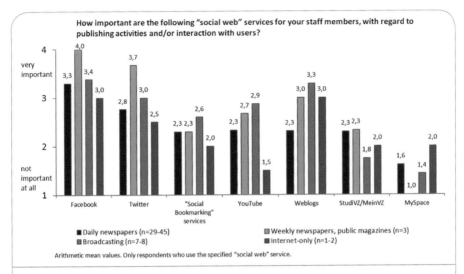

How important are the following "social web" services for your staff members, with regard to publishing activities and/or interaction with users?

■ Daily newspapers (n=29-45) ■ Weekly newspapers, public magazines (n=3)
■ Broadcasting (n=7-8) ■ Internet-only (n=1-2)

Arithmetic mean values. Only respondents who use the specified "social web" service.

Figure 26.1: Importance of Different "Social Web" Services for Publishing and/or Interaction with Users among News Staff, Compared across Different Types of Providers (Newsroom Survey, 2010)

How do news staff members make use of Twitter? Results show that almost all media organisations used Twitter to attract readership (97%), for investigative purposes (94%), and for monitoring audience responses (91%). About two thirds of participants said they used Twitter to interact with users (66%) and for live coverage and breaking news (63%). In the following, we comment on these patterns of use and describe some of our findings in more detail.

1. *Website promotion:* While almost all news departments use Twitter to advertise their own Internet content, they obtain only a small portion of their online readership this way. Of the newsrooms surveyed, 93% estimate the proportion of users that came to their website via Twitter at less than 10 percent. The promotional effectiveness of Twitter is thus not perceived to be particularly strong. It seems that only the cumulative effect of content posted to different social Web platforms might lead to a significant increase in visitation rates. Here, "breaking news" appears to be the most effective in fuelling public interest (see Figure 26.2). On account of its speed and mobile accessibility, Twitter is ideally suited for commenting immediately on breaking stories. Based on the particular experience of the editors of the surveyed daily newspa-

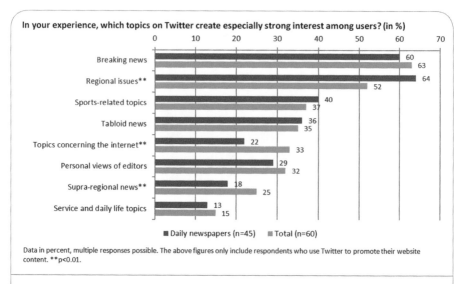

Figure 26.2: News Providers' Estimation of Which Topics Create Strong Interest among Users (Newsroom Survey, 2010)

pers, their audiences also seem to be especially interested in local and regional topics.

2. *Real-time, on-location coverage:* News departments most often provide "live" reporting of recurring events, such as sports events and awards presentations (see Table 26.2). An additional, open-ended question in our survey indicates that live reporting of elections via Twitter also appears to occur at times. Almost half of the news departments had previously provided multiple Twitter reports about unexpected, adverse events, such as accidents, disasters, and acts of violence. In reporting live events, the newsroom directors are in strong agreement that careful reporting should always take precedence over immediate timeliness (91%).

3. *Interaction with the audience:* Journalists are not only able to read what other people write on Twitter, they can also use it to interact with members of their audience. Nearly three quarters of the news departments (72%) that employ Twitter for exchanging comments with their users reported receiving an average of up to ten user inquiries and comments per day on Twitter. About a fifth of the news departments (21%) respond to incoming tweets. According to the survey participants, messages from users are useful, as they often alert news staff to breaking news,

Table 26.2: Percentage of Topics Which Were Posted in Real Time on Twitter by News Departments (Newsroom Survey, 2010)

How often has your newsroom used Twitter for real-time coverage of the following topics?	Never	Once	Several times
Recurring events (sporting events, award ceremonies, etc.)	16	13	71
Topics of high relevance to the public	32	18	50
Complex events without a distinct location (elections, demonstrations, etc.)	24	26	50
Unexpected news with negative consequences (accidents, disasters, acts of violence)	40	16	45
Press conferences, conventions, conferences	45	18	37
Events from which the public is excluded	71	8	21
Any story that can be reported on location	63	16	21

* Data in percent, n=38. Only respondents that use Twitter for real-time coverage.

such as fire department calls and traffic accidents, or provide ideas for articles. Through their interaction with the public, news departments hope to reach new readers, develop audience loyalty, learn about mistakes in their reporting, and obtain a sense of how readers react to their news coverage (see Figure 26.3).

4. *Monitoring of responses:* Journalists can use Twitter to observe comments about their articles, even if they do not contact the readers who post them. These comments may provide new story ideas or suggestions for improving their work. The observation of reader reactions occurs on a regular basis in half of the news departments (see Figure 26.4).

5. *Investigation:* Compared to other computer-based research tools, Twitter is relatively unimportant for investigative purposes. Only 12% of the news departments reported "frequently" using Twitter for research. Alongside search engines and Web catalogues, social media such as blogs, social networks, and "social news" services are more popular than Twitter for the purpose of research. Twitter is hardly ever used for gathering facts and background information or cross-checking information, but is used instead for "soft" research goals (see Figure 26.4), that is, to gain a picture of moods and trends and to uncover topical ideas, or to identify sources and eyewitnesses who can then be interviewed. Beyond these goals, there is also a wish on the part of journalists to observe the "Twitter phenomenon" itself and report about it.

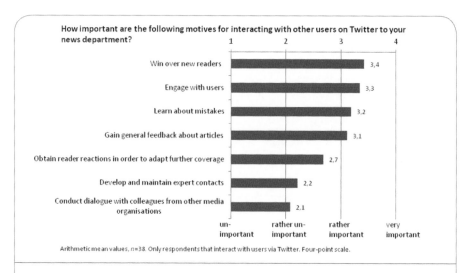

Figure 26.3: Relative Importance of Motives for Interacting with Other Users on Twitter (Newsroom Survey, 2010)

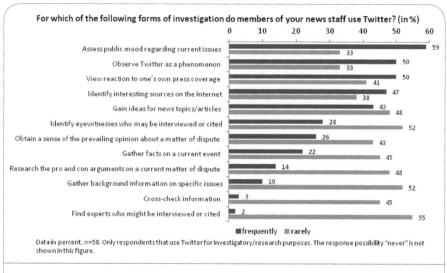

Figure 26.4: Specific Uses of Twitter for Investigative Purposes (Newsroom Survey, 2010)

Did newsroom staff formulate special rules for the use of Twitter? In the news departments that were surveyed, there is a widespread consensus that Twitter should be supplemented by additional types of research (97%), and that one should only rely on websites whose providers are well-known and considered credible (91%). However, only about three quarters of news staff (77%) consider it necessary to make contact with a Twitter author in terms of confirming the reliability of the source.

Another question is whether we are approaching a "Twitterisation" of journalism (analogous to a "Googleisation") as a result of the fact that this service is so quick and inexpensive and may thus supplant more appropriate research methods (see Table 26.3). Somewhat more than a third of the editors-in-chief surveyed at least partially support the notion that journalists may be failing to make use of better methods.

Table 26.3: Statements on Journalistic Investigation Using Twitter (Newsroom Survey, 2010)

In your opinion, to what extent are the following statements about Twitter true?	absolutely true	true	rather not true	not true at all
Twitter is better suited as a tool to quickly find information about unexpected events than for other forms of research. (n=58)	12	52	26	10
Twitter leads journalists to forego more comprehensive forms of investigation that would be more appropriate. (n=56)	5	32	38	25

* Data in percent. The possible response, "I do not know"/"I cannot tell" was not included in the analysis.

EXPERIENCES AND SKILLS

Most news departments in Germany now have experience with Twitter and have established rules for its use. Such rules pertain to the use of Twitter in preparing news reports, interacting with users, and research. With these rules, news departments hope to address those potentially negative aspects of Twitter that have been cited in public debates, including the acceleration of coverage and the lack of thorough research. On the one hand, nearly two thirds (62%) of those surveyed completely or somewhat agree with the opinion that the use of Twitter places increased time pressure on journalists.

On the other hand, 58% of the respondents reject the contention that it is impossible to maintain journalistic standards when using a platform that only permits 140 characters of text. In this regard, the prevailing opinion is that tweets offer sufficient space for maintaining good journalistic principles.

Further evidence for the fact that quality standards have been developed for working with Twitter is the existence of news departments whose social Web activities are universally considered to be of high quality. At least 29% of those surveyed indicated in response to an open question that the websites of news magazine *Der Spiegel* and of regional newspaper *Rhein-Zeitung* conducted work of exemplary quality in their use of Twitter. Although applicable rules and models may exist, 60% of the editorial directors consider that the competence of their staff in matters related to Web 2.0 technologies needs serious improvement. Only 7% suggested that there was no need for improvement. Daily newspapers in particular noted competence deficiencies. "Learning by doing" and informal dialogue with colleagues are the most common ways through which journalists acquire competence in this area.

CONTENT ANALYSIS OF POPULAR TWEETS AND USER RECOMMENDATIONS FOR NEWS SITES

Our survey of news departments was only one part of a larger project which aimed to study Twitter's relationship with journalism. In addition, the relative importance of different subjects, authors, and links to news sites on Twitter was explored by means of two quantitative content analyses. Compared to well-known journalistic selection criteria, these analyses of Twitter content may point to potential similarities as well as differences. In the following, we will present a short overview of the methodology used and key results. First, the further dissemination of tweets by means of retweets can be interpreted as an indicator of the relative popularity of particular subjects and authors (Cha, Haddadi, Benevenuto, & Gummadi, 2010; Romero, Galuba, Asur, & Huberman, 2010). In its "Top Tweets" lists, Twitter assembles those tweets that were forwarded most frequently in a certain time period. On this basis, our first quantitative analysis explored the 963 tweets listed by Twitter Germany's "Top Tweets" account (@toptweets_de) between 12 February 2010 and 17 May 2010. In short, empirical findings show that among the authors with real names whose tweets made it onto the list, more than one third (35%, $n=421$) had their own German Wikipedia entry, which allows us to conclude that these are well-known personalities. Also among the most frequently represented Twitter users in the "Top Tweets" list are a number of well-known Internet activists and journalists. Representatives from the media accounted for a third of the "Top Tweets" (34%), and were thus the second largest group, just after ordinary citizens (43%). Among the subsequent proportions, civil-society actors (5%) and economic actors (4%), as well as representatives from other social areas like sportsmen and celebrities (6%) follow. While other political parties are only represented in exceptional cases in the "Top Tweets", representatives of the German pirate party reach a group share of 5%. In a second content analysis, we used the external Twitter search engine backtweets.com to determine how frequently tweets incorporated links to journalistic websites, and what subjects were covered by these recommended articles. The search for these linking patterns was limited to those 157 news sites (main units) that constituted the scope of our news department survey (outlined above). All in all, backtweets.com identified 355,000 tweets linking to the selected group of news websites throughout April 2010. The subsequent content analysis of these tweets was based on a specific quota sample. The inclusion of news sites in this sample was in proportion to their share of in-links in the total data set (for details, see Neuberger et al., 2011).

Tweets linking to news sites can be divided into different forms of recommendations: first, Twitter *users* have the opportunity to recommend news sites by linking to them or their articles. Second, *editorial recommendations* for news content (i.e., editorial self-promotion) also occurs on Twitter. Some newsrooms have established forms of automated publishing, while others tweet specific recommendations in selected cases only. Twitter users may decide whether to include a link in their recommendation tweets or not. Posting a link does not necessarily mean that a user endorses the linked article, actor, or event: users can post critical assessments of linked content or actors as well.

In this part of the study, we aimed to investigate what type of news gathering and filtering Twitter users conduct in their tweets. In general, the findings show that the distribution of links corresponds to a power law distribution, with a small number of news providers jointly accounting for the majority of all links. The 20 websites with the largest number of links accounted for 74% (n=354,794) of all in-links; the top fifth of all analysed sites accounted for 82%. The centralisation of incoming links to a few top sites also reflects the sites' prominence in terms of reach. German IVW ranking, which measures Web traffic in terms of visits, and tweet in-links ranking show a robust correlation.

Nearly one fifth of our quota sample (19%, n=993) consisted of editorial tweets; the majority of recommendations were tweeted by users. User tweets with links to news sites are mostly devoid of value judgements: 90% (n=807) of all counted links were not embedded into an evaluative context. Only 10% of all recommendations are accompanied by comments directed to the news site, the article of interest, or the event or actors covered in the linked article.

The second purpose of this analysis was to determine what kinds of topics on the news sites were selected for recommendations. In comparison to other possible topics, political issues dominate—recommendation-based filtering and user-led news gathering is biased. Political subjects (35%) were more prevalent than business subjects (15%); media and Internet subjects accounted for only ten percent. Societal, cultural, and lifestyle topics were even less popular. It would appear that these link recommendations have an impact on how news departments assign priority to various stories.

These two content analyses reveal that both in the further transmission of tweets (through retweets) and in links to professional news websites there is a marked dominance of subjects and authors with a close relationship to 'hard' journalism. Thus, we cannot conclude from these findings that Twitter is generating a fundamentally new structure of relevance which diverges from the traditional news values established by mainstream journalism.

CONCLUSIONS

The survey results reveal that news departments in Germany consider Twitter—like other social media—to be a channel for mutual exchange with the public and with news sources. Moreover, the findings sketch out many complementary relations among Twitter and journalism. Nevertheless, this should not be overestimated: in Germany, Twitter use in journalism is mostly limited to special circumstances and niches. On Twitter, patterns of social navigation by users very often lead to already published news. In effect, this creates a dynamic space for follow-up communication and for recommendations of journalistic content. Only in exceptional cases do lay communicators on Twitter report exclusively on unexpected events and subjects. In journalism, Twitter is used for self-promotion, investigation, real-time coverage, and interaction with the public. In this way, journalism is contributing to the formation of a "networked public sphere". According to a model proposed by Benkler (2006), the Internet offers a chance to link two different levels of the public sphere: the broad public sphere of mass media and the small public sphere (or the "personal publics"; see Chapter 1 by Schmidt in this volume) of lay communicators. Thus, the increasing permeability between levels with different grades of public visibility expands citizens' opportunities to participate in opinion-shaping processes and to have a greater impact upon them.

NOTES

1   The study was commissioned by the Media Authority of North Rhine-Westphalia (LfM).

REFERENCES

Benkler, Y. (2006). *The wealth of networks*. New Haven, CT: Yale University Press.

Bruns, A. (2005). *Gatewatching: Collaborative online news production*. New York, NY: Peter Lang.

Busemann, K., & Gscheidle, C. (2012). Web 2.0: Habitualisierung der Social Communitys [Web 2.0: Development of habits in social communities]. *Media Perspektiven*, (7–8), 380–390.

Cha, M., Haddadi, H., Benevenuto, F., & Gummadi, K. P. (2010). Measuring user influence in Twitter: The million follower fallacy. In *Proceedings of the Fourth International AAAI Conference on Weblogs and Social Media (ICWSM)* (pp. 10–17). Washington, DC: Association for the Advancement of Artificial Intelligence. Retrieved from http://an.kaist.ac.kr/~mycha/docs/icwsm2010_cha.pdf

Kwak, H., Lee, C., Park, H., & Moon, S. (2010). What is Twitter, a social network or a news media? In *Proceedings of the 19th International World Wide Web (WWW) Conference* (pp. 591–600). Raleigh, NC. Retrieved from http://an.kaist.ac.kr/~haewoon/papers/2010-www-twitter.pdf

Lerman, K., & Ghosh, R. (2010). Information contagion: An empirical study of spread of news on Digg and Twitter social networks. In *Proceedings of the International AAAI Conference on Weblogs and Social Media (ICWSM)*. Washington, DC: Association for the Advancement of Artificial Intelligence. Retrieved from http://arxiv.org/pdf/1003.2664v1.pdf

Neuberger, C., & Nuernbergk, C. (2010). Competition, complementarity or integration? *Journalism Practice*, 4(3), 319–332. doi: 10.1080/17512781003642923

Neuberger, C., vom Hofe, H., & Nuernbergk, C. (2011). *Twitter und Journalismus: Der Einfluss des 'Social Web' auf die Nachrichten* [Twitter and journalism: The influence of the 'social Web' on news] (3rd ed.). Düsseldorf, Germany: Landesanstalt für Medien Nordrhein-Westfalen (LfM).

Pew Research Center. (2012). *In changing news landscape, even television is vulnerable: Trends in news consumption 1991–2012*. Washington, DC: Pew Research Center. Retrieved from http://www.people-press.org/files/legacy-pdf/2012%20News%20Consumption%20Report.pdf

Purcell, K., Rainie, L., Mitchell, A., Rosenstiel, T., & Olmstead, K. (2010). *Understanding the participatory news consumer: How Internet and cell phone users have turned news into social experience*. Pew Internet & American Life Project. Washington, DC: Pew Research Center. Retrieved from http://www.pewinternet.org/~/media//Files/Reports/2010/PIP_Understanding_the_Participatory_News_Consumer.pdf

Romero, D. M., Galuba, W., Asur, S., & Huberman, B. A. (2010). Influence and passivity in social media. In Hewlett Packard (Ed.), *Social computing lab*. Retrieved from http://www.hpl.hp.com/research/scl/papers/influence/influence.pdf

Twitter as an Ambient News Network

27
CHAPTER   Alfred Hermida

 #BREAKING: when news events happen, Twitter's
ambient news function becomes crucial

As a high school pupil, Morgan Jones had already shown a keen interest in news.
He had been editor for *RJ Voice*, the newspaper of the Regis Jesuit High School
(a private, Catholic college in Aurora, Colorado), as well as been involved with
the school radio station, RJ Radio (Jones, n.d.). The 18-year-old, self-described
techie was spending the 2012 summer break with his parents in Denver, before
going to Rice University to study engineering. In the early hours of 20 July 2012,
he was playing the fantasy role-playing video game Oblivion when he saw a
note on Facebook from a local TV station about shootings at a movie theatre
(Herman, 2012).

It later emerged that the reports referred to a shooting at a midnight screen-
ing of *The Dark Knight Rises* in Aurora, Colorado, in which 12 people were
killed and 58 wounded. The suspected gunman, 24-year-old James Holmes,
was arrested shortly after the rampage. On that night in July, Jones tuned in
to the Aurora police scanner and started posting updates under his username,

Integ3r, on Reddit, the largest Internet message board in the world with 35 million monthly users (Shaer, 2012).

The timeline provided a minute-by-minute account of the tragic events of the night (Integ3r, 2012). Jones pulled in fragments of information gleaned from police, from mainstream media outlets, and from messages and photos shared on Twitter by people at the cinema, adding new details and correcting old information as he went along. "I stayed up all night, and I am exhausted now, but it feels like I'm helping out people who need to know this stuff", he said the following day (quoted in Herman, 2012).

The night of the Aurora tragedy, Jones performed some of the communication functions that had previously been limited mainly to professional journalists working for media institutions with the structures and technologies to gather, process, and distribute the news. The cinema shootings were emblematic of how news flows in a networked media ecosystem, facilitated by social media such as Twitter. Social media services like Twitter have developed into platforms for news storytelling, becoming integral to any major news event, from the Sichuan earthquake in China in May 2008 to the Iranian election protests in June 2009 to the 2011 uprising in Egypt (Hermida, 2011, p. 672).

By 2012, 340 million tweets were being posted daily (Twitter, 2012), on topics ranging from the mundane to the comical to the momentous. Clearly not all of the content could be considered to be news in the public interest, but likewise, neither is everything published in magazines or other media formats. Twitter provides a distribution network for firsthand news accounts by eyewitnesses in near real time. Sometimes, this happens inadvertently—as in the case of Sohaib Athar, a Pakistani software engineer who unknowingly live-tweeted the U.S. raid on Osama Bin Laden's compound (Butcher, 2011).

Concurrently, Twitter serves as a channel for the distribution of material from journalists and the mainstream media. Exchanges around news events circulate as reports, rumours, and speculation are shared on the network and are challenged, contradicted, or corroborated. Surges in tweets are often linked to major news events, like the torrent of conjecture ahead of President Obama's announcement of the death of Bin Laden. There were more than 4,000 tweets per second on the topic by the time the president finally appeared on TV (Hermida, 2011, pp. 671–672).

Twitter facilitates the instant, online dissemination and reception of short fragments of information from sources outside the formal structures of journalism, creating social awareness streams that provide a constantly updated, live representation of the experiences, interests, and opinions of users. Such uses of Twitter can be considered "ambient journalism" (Hermida, 2010a, 2010b), a telemediated practice powered by networked, always-on communications technologies and media systems of immediacy and instantaneity.

It builds on the term "ambient news", which refers to the easy availability of news through a host of media platforms, such as electronic billboards in public spaces such as train stations or free commuter newspapers that convey news and information produced by professional media (Hargreaves & Thomas, 2002, p. 44). "Ambient news" works much like ambient music that plays in the background (Crawford, 2009, p. 528); a listener will tune in to the music when there is a change in tone or style that catches their attention. Applied to Twitter, "when important news breaks and spreads across the Twittersphere, shifts in tone and topical focus of incoming tweets may cause that user to pay attention to the story" (Bruns & Burgess, 2012, p. 2).

Ambient journalism concerns the collection, selection, and dissemination of news by both professional and non-professional para-journalists, where users undertake some of the institutional tasks commonly associated with the journalist. These tasks range from an individual sending a message about a breaking news event to alerting their online social network about a story in the mainstream media to curating the flow of information in real time. Users become part of the flow of news, reframing or reinterpreting a message through networked platforms that extend the dissemination of news through social interaction, introducing hybridity in news production and news values (Chadwick, 2011; Papacharissi & de Fatima Oliveira, 2012). As Hardey (2007) stressed, the key differentiator is that digital media technologies such as Twitter are "*inherently* social so that users are central to both the content and form of all material and resources" (p. 870).

Drawing on the research in new literacies (Lankshear & Knobel, 2007, 2011; Prinsloo, 2005), this chapter aims to contextualise and unpack the interplay between social media—specifically Twitter—and emergent paradigms in journalism (Hermida et al., forthcoming; Lotan et al., 2011; Papacharissi & de Fatima Oliveira, 2012). It frames Twitter and journalism in the light of what the renowned U.S. foreign policy analyst and Editor-at-Large of *The American Interest* magazine, Walter Russell Mead (2012), described as the disruptive and painful "transformation from late-stage industrial society to early-stage information society" (para 32). This approach diverges from a bias in journalism studies that sees it mostly drawing on the routines and practices of print media (Deuze, 2008).

TWITTER AND NEWSROOMS

Since its launch in 2006, Twitter has made significant inroads into newsrooms, with journalists rapidly embracing the service to report breaking news, to gather information, to connect with sources, and to drive traffic to websites (Ahmad, 2010; Farhi, 2009; Hermida, 2011). The adoption of Twitter has largely mirrored the path of earlier new media technologies such as blogging, with journalists extending established norms and routines as they incorporate novel tools and techniques into daily practice. In his study of the use of social media in mainstream media, Newman (2009) concluded that "the use of new tools has not led to any fundamental rewrite of the rule book—just a few tweaks round the edges" (p. 39).

Within newsrooms, a common practice has been to use Twitter as a channel to promote content and attract readers to a news website, as social media provides a ready-made free distribution network. Messner, Linke, and Eford (2012) found that leading U.S. newspapers and TV stations mostly used their official Twitter accounts to send out links to the latest news stories on their websites. In his study of TV stations in San Antonio, Texas, Blasingame (2011) found that many newsrooms automatically generated a tweet with a link when a new story was published online.

On an individual level, journalists are navigating some of the tensions that emerge as the affordances of Twitter interact with established journalistic conventions. Some studies indicate that journalists are extending traditional practices to social media, using new tools to do old things. Research into the use of tweets by Dutch and U.K. newspapers suggests that newsrooms are using tweets from ordinary people to represent the *vox populi*, especially among the British tabloid press (Broersma & Graham, 2012, p. 411). In some cases, journalists are approaching Twitter as another newsgathering tool, just as they would have used the telephone in the past. A 2011 survey of 500 journalists in 15 countries found that almost half used Twitter to source angles for a story (Oriella PR Network, 2011). Only a third said that they used social media to verify information, relying instead on traditional sources such as PR agencies and corporate spokespeople. But there is also some evidence of journalists pushing the boundaries of accepted practice: for example, Lasorsa, Lewis, and Holton (2012) found that the journalists they studied offered personal opinions on Twitter, straying from traditional norms of objectivity.

As with every new communication technology, there is a process of negotiation, as established ways of working rub up against new affordances. Social media platforms such as Twitter have a communicative structure that can chafe with the notion of the journalist as the professional who works "full-time to access, select and filter, produce and edit news, which is then distributed via the media to network members" (Domingo et al., 2008, p. 329). Social media bring together interpersonal communication, content production, immediacy, and large-scale distribution in a way that blurs the line between the public and private, and does not have the spatial, social, or temporal boundaries of print or broadcast media (boyd, 2010; Bruns & Burgess, 2012).

JOURNALISM THROUGH THE LENS OF NEW LITERACIES

Technological innovations from the telegraph to the telephone to Twitter have informed the norms and practices of journalism. As new technologies for information and communication appear, so does the demand for new skills, knowledge, and understanding. Media theorist Neil Postman (1992) noted how new technologies "alter the character of our symbols: the things we think with. And they alter the nature of community: the arena in which thoughts develop" (p. 20). In other words, new literacies emerge as communication tools evolve.

Literacies refer to the ability of individuals to interpret and communicate information in a meaningful way in order to participate in society. The nature of these literacies changes as a result of shifting social contexts, and the development and use of new media. New literacies researchers separate the technical and cultural aspects of literacy (Kalantzis & Cope, 2011; Lankshear & Knobel, 2007, 2011). The technical aspect refers to the properties of digital media, combining written, oral, and audiovisual modalities of communication using screen-based, networked services and devices. The cultural aspect refers to the mind-set that informs these literacies. The mind-set is the general way of thinking about the world, based on a set of assumptions, beliefs, and values that shape actions and reactions.

Lankshear and Knobel (2011) suggested that there are two mind-sets. They locate the two mind-sets in the historical development of society, labelling one as the modern/industrial paradigm and the other as the postmodern/post-industrial/knowledge society paradigm (p. 53). In the first mind-set, the world is uniform, monolithic, enclosed, individualised, stable, and linear. In the second mind-set, the world is distributed, open, collaborative, dynamic, and non-linear. Lankshear and Knobel applied this framework to explore how the shift to a post-industrial, knowledge society mind-set reconfigures everyday social practices around communication.

Part 2: Perspectives and Practices #practices: journalism

The industrial paradigm can be applied to print journalism, which developed at a time when access to the machinery of news production, publication, and provision was expensive. The newspaper is a stable and fixed product with a specific purpose—to provide news, information, and entertainment on a daily basis. Once it is printed, the hierarchy of information on the page cannot be changed. The story on the page is linear in nature, with a beginning, middle, and end. The story is usually written by an individual reporter, identified by the byline, though the material will have gone through various editorial layers. The journalist is identified by their attachment to a professional news organisation, where the processes of identifying, gathering, filtering, processing, and publishing are in the hands of a select few. The newsroom itself is an enclosed physical space that claims authority and expertise in the production of news. Outsiders are not welcomed, except through narrow, controlled channels such as letters to the editor. The news is shaped into a product that is pushed out to audiences, through home delivery or via newsstands.

Seen through the lens of new literacies, journalists have a mind-set rooted in the modern industrial period. As a wide body of research indicates, newsrooms have predominantly adapted digital media to existing norms and practices, rather than taking on the affordances of new forms of communication that challenge established ways of working (Lasorsa, 2012; Lasorsa et al., 2012; Robinson, 2007; Singer, 2005; Singer et al., 2011). By and large, journalism practices have become more technologised, with reporters doing old things in new ways, rather than negotiating the transition to a post-industrial knowledge society.

In order to function effectively, journalists have always required knowledge of both the technical and cultural aspects of literacy. What is new about Twitter are the differences in the technical and cultural aspects of the platform, compared to print or broadcast. The technical dimensions include the 140-character limit; the follower-followee structure system; the use of URL-shortening services; and the platform's markers for mutual exchange and conversation, such as the hashtags and mentions. But it can be a bewildering space without an understanding of the norms and conventions of Twitter. This bewilderment might account for some early derision from journalists, such as when *The New York Times* columnist Maureen Dowd (2009) described Twitter as "a toy for bored celebrities and high-school girls". Viewed through an industrial mind-set, Twitter is a shambolic, messy, and noisy torrent of seemingly everyday details of life. These negative attributes are transformed into positive attributes when viewed through the new literacies of a post-industrial mind-set, revealing instead a complex, networked communications environment.

Twitter exhibits core values of new literacies—"interactivity, participation, collaboration, and the distribution and dispersal of expertise and intelligence" (Lankshear & Knobel, 2011, p. 76). Participation is prioritised over publication, sharing over owning, change over stability, abundance over scarcity, relationships over information delivery. As an ambient news network, Twitter offers a mix of information and comment usually associated with current reality, but without an established order. In contrast with the print newspaper or the TV broadcast, anyone can publish and distribute at anytime, outside of the formal constraints of traditional journalism, with no established editorial structures or processes. The content flows continuously in near real time. Twitter breaks with the classic, narrative structure of journalism, and instead creates multi-faceted, fragmented, and fluid news experiences. Journalism shifts from being a product to a process with no end state.

AMBIENT NEWS BEYOND JOURNALISM

As an open platform, Twitter itself offers a space for the co-construction of news. Writing before the widespread adoption of Twitter, Manuel Castells (2007) described the rise of mass self-communication that is *"self-generated in content, self-directed in emission, and self-selected in reception by the many that communicate with many"* (p. 248). Studies into the use of Twitter during the Arab Spring indicate how the production and dissemination of news is becoming ambient, as citizens become part of the flow of news, reframing or reinterpreting a message. (See also Chapter 28 by Bruns & Burgess on the uses of Twitter for crisis communication in natural disasters, in this volume.)

The hybrid nature of news production on Twitter was highlighted in a study by Lotan et al. (2011) into information flows on Twitter during the Tunisian and Egyptian uprisings. The researchers sought to identify who were the influential voices on the network in amplifying and spreading news of the protests across the world. Of particular interest was the symbiotic relationship between mainstream media and actors outside of the formal structures of journalism. They found that journalists and activists served as primary sources of information. Activists were more likely to retweet content, as were bloggers, serving as clearing houses (2011, p. 1390). Lotan et al. concluded that bloggers, activists, and journalists co-constructed the news on Twitter, fashioning "a particular kind of online press" (2011, p. 1400).

Twitter provides a newsroom for this "online press" that is open, distributed, and collective, in contrast to traditional models of the newsrooms as enclosed, concentrated, and exclusive spaces. The networked nature of the newsroom affects what becomes newsworthy. Papacharissi and de Fatima Oliveira (2012) examined the news values and forms of news displayed in tweets sent during the Egyptian uprising and subsequent fall of President Hosni Mubarak over January and February 2011, finding that the types of events reported and the tone of coverage on Twitter mirrored the news values of traditional media. But they also argued that there was evidence of values specific to an ambient news network that collide with journalistic conventions.

On Twitter, the news was characterised by values of instantaneity, solidarity, and ambience (Papacharissi & de Fatima Oliveira, 2012, p. 273). Events *became* news, because details of what was happening were disseminated instantly and were repeatedly shared across the network. Papacharissi and de Fatima Oliveira (2012) suggested that "as individuals constantly tweeted and retweeted observations, events instantly turned into stories" (p. 274). The constant flow of messages broke with journalistic conventions by blending fact, opinion, and emotion to the extent that the researchers found it hard to separate one from the other. The authors concluded that the pace, frequency, and tone of messages created an ambient, always-on system where users gain an emotive, immediate sense of the drama unfolding, but without the fact-checking and arms-length reporting associated with traditional news. They posited the concept of "affective news streams" to explain "how news is collaboratively constructed out of subjective experience, opinion, and emotion within an ambient news environment" (p. 274).

Moreover, collaborative co-construction of the news affects the frames applied to events. By selecting specific facts or sources, journalists frame the news "in such a way as to promote a particular problem definition, causal interpretation, moral evaluation, and/or treatment recommendation for the item described" (Entman, 1993, p. 52). However, Meraz and Papacharissi's (2013) research on the Arab Spring suggested that framing on Twitter is negotiated on the network through interactions between journalists, activists, and citizens in an "organic, ad hoc manner" (p. 159).

JOURNALISM IN AN AMBIENT NEWS NETWORK

Emerging research points to new forms of journalism that move away from journalism as a framework to provide reports and analyses of events through linear narratives composed after an event. An increasingly popular format is the "live blog" or "live update" page. Live blogs weave together reports from professional journalists and information gleaned from social networks such as Twitter, together with commentary and analysis in near real time, generating a multidimensional, temporal, and fast-moving news experience. They have been rapidly adopted by leading news organisations, including the BBC, *The Guardian, The New York Times,* and CNN. Thurman and Walters (2012) suggested that the format is "increasingly the default form for covering major breaking news stories, sports events, and scheduled entertainment news" (p. 1).

The live blog differs from conventional reporting in both its technical and cultural aspects. A live blog is made up of timestamped entries in reverse chronological order, with multiple forms of media, numerous links to material elsewhere on the Internet, and signposted content from third parties (Thurman & Walters, 2012). Through the lens of new literacies, live blogs are more collaborative, open, fluid, and less author-centric than other forms of journalism. The conventional newspaper story strives to convey a definitive and authoritative account of an event. Live blogs present an iterative and incremental account. The editor of the BBC News website, Steve Herrmann, has talked about live blogs as a way to reflect "the unfolding truth in all its guises" (as quoted in Newman, 2009, p. 9), while for *Guardian* reporter Matthew Weaver the format offers "a more fluid sense of what's happening" (as quoted in Bruno, 2011, p. 44). The imperative to provide constantly updated information is not without its consequences. Thurman and Walters (2012) suggested that it may be hard for journalists to adapt existing fact-checking practices to live blogging.

The live-blog format creates a locus for journalists to bring in material curated from the real-time flow of information on Twitter and other social media services, and integrate it within the confines of a news organisation's website. The imperative to harvest Twitter is at its peak in the hours after breaking news events where there is a news vacuum due to the lack of professional journalists on the scene. In the immediate aftermath of the earthquake in Haiti in January 2010, major news organisations relied heavily on messages, photos, and video streaming on social media until their own reporters arrived on the scene, hours or even days later (Bruno, 2011). The "Twitter effect", as Bruno called it, allows newsrooms "to provide live coverage without any reporters on the ground, by simply newsgathering user-generated content available online" (Bruno, 2011, p. 8).

One of the more noteworthy aspects of Twitter, though, is how journalists can operate on the platform itself, bypassing the need for a home on a news website. At the Online News Association Conference in San Francisco in September 2012, company CEO Dick Costolo was asked, "so, how does it feel to be the voice of the press in the 21st century?" (Silverman, 2012). The question was somewhat in jest, but it underscored how Twitter has matured as a networked newsroom where news is filtered, discussed, contested, and verified. At the time of writing, the best-known example of a media professional operating in this ambient news network was Andy Carvin.

CONCLUSION: TWITTER AS THE NEWSROOM

A social media strategist with National Public Radio (NPR) in Washington, DC, Carvin rose to prominence during the uprisings in the Middle East at the end of 2010 and start of 2011—the "Arab Spring". Over the course of the Arab Spring, he emerged as a key hub on Twitter for news from the region, amassing tens of thousands of followers, including other journalists and news outlets. Carvin shared images and video, exchanged messages, mediated discussions, and turned to his followers to help him translate, verify, and put into context the endless amount of data flowing across the network. He tweeted up to 16 hours a day, seven days a week (Farhi, 2011, para 3). Over one weekend in August 2011, when the rebels in Libya pushed into the capital Tripoli, he sent out 1,200 messages (Sonderman, 2011). Through Twitter, Carvin fashioned a rich and dynamic tapestry of the dramatic upheavals in the region.

In their analysis of Carvin's sourcing practices on Twitter, Hermida, Lewis, and Zamith (forthcoming) advance that his coverage of the Arab Spring hints at a new journalistic paradigm at play. They suggest that the Carvin case study points to a shift away from the traditional, journalistic gatekeeping function (Shoemaker, 1991) towards gatewatching (Bruns, 2005), where the journalist evaluates, highlights, and publicises relevant information plucked from social awareness streams on Twitter.

For Carvin, Twitter was effectively his workplace. He has described his network of followers as "my editors, researchers & fact-checkers. You're my news room" (Carvin, 2012). Carvin described his work as "another flavour of journalism" (as quoted in Farhi, 2011, para 15). His approach was in line with a new literacies framework, demonstrating not just technical competency, but an understanding of the emerging norms and practices of Twitter. Carvin was not simply broadcasting, but was immersed in the culture of a media environment that privileges relationship over information delivery, interacting and conversing with others to co-construct the news. This approach can have positive professional results: emerging research suggests that journalists who adopt more discursive strategies, conversing with the public, tend to receive the most retweets and mentions (Meraz & Papacharissi, 2013).

Twitter as a newsroom exposes the tentative process through which the news is constructed, as information, rumour, and speculation are authenticated or denied in a recurrent cycle. Journalism shifts from being a finite story with the fixed endpoint of publication to being an iterative process through which information is dissected, discarded, or disseminated in near real time. In an ambient news network, then, the journalist serves as a pivotal node which is "trusted to authenticate, interpret, and contextualize information flows on social awareness streams, drawing on a distributed and networked newsroom where knowledge and expertise are fluid, dynamic, and hybrid" (Hermida, Lewis, & Zamith, forthcoming, n.p.). This and the other examples discussed in this chapter show how the new literacies of networked media are shaping, and are being shaped by, journalism in this current phase of its co-evolution with technological and societal change.

REFERENCES

Ahmad, A. N. (2010). Is Twitter a useful tool for journalists? *Journal of Media Practice*, *11*(2), 145–155.

Blasingame, D. (2011). Gatejumping: Twitter, TV news and the delivery of breaking news. *#ISOJ Journal: The Official Journal of the International Symposium on Online Journalism*, *1*(1), n.p. Retrieved from http://online.journalism.utexas.edu/ebook.php

boyd, d. (2010). Social network sites as networked publics: Affordances, dynamics, and implications. In Z. Papacharissi (Ed.), *Networked self: Identity, community, and culture on social network sites* (pp. 39–58). New York, NY: Routledge.

Broersma, M., & Graham, T. (2012). Social media as beat: Tweets as a news source during the 2010 British and Dutch elections. *Journalism Practice*, *6*(3), 403–419. doi: 10.1080/17512786.2012.663626

Bruno, N. (2011). Tweet first, verify later? How real-time information is changing the coverage of worldwide crisis events. Oxford, UK: Reuters Institute for the Study of Journalism, University of Oxford. Retrieved from http://reutersinstitute.politics.ox.ac.uk/fileadmin/documents/Publications/fellows__papers/2010-2011/TWEET_FIRST_VERIFY_LATER.pdf

Bruns, A. (2005). *Gatewatching: Collaborative online news production*. New York, NY: Peter Lang.

Bruns, A., & Burgess, J. (2012). Researching news discussion on Twitter. *Journalism Studies*, *13*(5–6), 801–814. doi:10.1080/1461670X.2012.664428

Butcher, M. (2011, 1 May). Here's the guy who unwittingly live-tweeted the raid on Bin Laden. *TechCrunch*. Retrieved from http://techcrunch.com/2011/05/01/heres-the-guy-who-unwittingly-live-tweeted-the-raid-on-bin-laden/

Carvin, A. (2012, 26 Mar.). 'I don't just have Twitter followers. You're my editors, researchers & fact-checkers. You're my news room. And I dedicate this award to you' [Twitter post]. Retrieved from http://twitter.com/acarvin/status/184424440757624832

Castells, M. (2007). Communication, power and counter-power in the network society. *International Journal of Communication, 1*(1), 238–266.

Chadwick, A. (2011). The political information cycle in a hybrid news system: The British Prime Minister and the 'Bullygate' affair. *International Journal of Press/Politics, 16*(1), 3–29.

Crawford, K. (2009). Following you: Disciplines of listening in social media. *Continuum: Journal of Media & Cultural Studies, 23*(40), 525–535.

Deuze, M. (2008). Toward a sociology of online news? In C. Paterson & D. Domingo (Eds), *Making online news: The ethnography of new media production* (pp. 45–60). New York, NY: Peter Lang.

Domingo, D., Quandt, T, Ari Heinonen, A., Paulussen, S., Singer, J. B., & Vujnovic, M. (2008). Participatory journalism practices in the media and beyond. *Journalism Practice, 2*(3), 326–342. doi: 10.1080/17512780802281065

Dowd, M. (2009). To tweet or not to tweet. *The New York Times.* Retrieved from http://www.nytimes.com/2009/04/22/opinion/22dowd.html

Entman, R. M. (1993). Framing: Toward clarification of a fractured paradigm. *Journal of Communication, 43*(4), 51–58. doi: 10.1111/j.1460-2466.1993.tb01304.x

Farhi, P. (2009, Apr./May). The Twitter explosion. *American Journalism Review.* Retrieved from http://ajr.org/article_printable.asp?id=4756

Farhi, P. (2011, 12 Apr.). NPR's Andy Carvin, tweeting the Middle East. *The Washington Post.* Retrieved from http://www.washingtonpost.com/lifestyle/style/npr-andy-carvin-tweeting-the-middle-east/2011/04/06/AFcSdhSD_story.html

Hardey, M. (2007). The city in the age of Web 2.0: A new synergistic relationship between place and people. *Information, Communication and Society, 10*(6), 867–884.

Hargreaves, I., & Thomas, J. (2002, Oct.). New news, old news. ITC/BSC. Retrieved from http://legacy.caerdydd.ac.uk/jomec/resources/news.pdf

Herman, J. (2012). How 18-year-old Morgan Jones told the world about the Aurora shooting. *Buzzfeed.com.* Retrieved from http://www.buzzfeed.com/jwherrman/how-18-year-old-morgan-jones-told-the-world-about

Hermida, A. (2010a). From TV to Twitter: How ambient news became ambient journalism. *M/C Journal, 13*(2). Retrieved from http://journal.media-culture.org.au/index.php/mcjournal/article/viewArticle/220

Hermida, A. (2010b). Twittering the news. *Journalism Practice, 4*(3), 297–308.

Hermida, A. (2011). Tweet the news: Social media streams and the practice of journalism. In S. Allan (Ed.), *The Routledge companion to news and journalism* (2nd ed., pp. 671–682). Abingdon, UK: Routledge.

Hermida, A., Lewis, S., & Zamith, R. (forthcoming). Sourcing the Arab Spring: A case study of Andy Carvin's sources on Twitter during the Tunisian and Egyptian revolutions. *Journal of Computer-Mediated Communications.*

Integ3r. (2012, 20 July). Comprehensive timeline: Aurora massacre. Retrieved from http://www. reddit.com/r/news/comments/wv8t1/comprehensive_timeline_aurora_massacre/

Jones, M. (n.d.). About. *Facebook.com*. Retrieved from https://www.facebook.com/integ3rpublic/ info

Kalantzis, M., & Cope, B. (2011). The work of writing in the age of its digital reproducibility. In S. S. Abrams & J. Rowsell (Eds.), *Rethinking identity and literacy education in the 21st century* (pp. 40–87). New York, NY: Teachers College Press.

Lankshear, C., & Knobel, M. (2007). Sampling 'the new' in new literacies. In M. Knobel & C. Lankshear (Eds.), *A new literacies sampler* (pp. 1–24). New York, NY: Peter Lang.

Lankshear, C., & Knobel, M. (2011). *New literacies: Everyday practices and social learning* (3rd ed.). Maidenhead, UK: Open University Press.

Lasorsa, D. (2012). Transparency and other journalistic norms on Twitter. *Journalism Studies, 13*(3), 402–417. doi:10.1080/1461670X.2012.657909

Lasorsa, D., Lewis, S., & Holton, A. (2012). Normalizing Twitter: Journalism practice in an emerging communication space. *Journalism Studies, 13*(1), 19–36. doi: 10.1080/1461670X.2011.571825

Lotan, G., Graeff, E., Ananny, M., Gaffney, D., Pearce, I., & boyd, d. (2011). The revolutions were tweeted: Information flows during the 2011 Tunisian and Egyptian revolutions. *International Journal of Communication, 5*, 1375–1405.

Mead, W. R. (2012, 23 Sep.). Americans turn on MSM: What does it mean? *Via Meadia. The American Interest.* Retrieved from http://blogs.the-american-interest.com/wrm/2012/09/23/americans-turn-on-msm-what-does-it-mean/

Meraz, S., & Papacharissi, Z. (2013). Networked gatekeeping and networked framing on #egypt. *International Journal of the Press and Politics, 18*(2), 138–166. doi:10.1177/1940161212247447

Messner, M., Linke, M., & Eford, A. (2012). Shoveling tweets: An analysis of the microblogging engagement of traditional news organizations. *#ISOJ: The Official Research Journal of the International Symposium on Online Journalism, 2*(1), 76–90. Retrieved from http://online.journalism.utexas.edu/ebook.php

Newman, N. (2009). The rise of social media and its impact on mainstream journalism. Reuters Institute for the Study of Journalism Working Paper. Oxford, UK: University of Oxford. Retrieved from http://thomsonreuters.com/content/news_ideas/white_papers/media/487784

Oriella PR Network. (2011). The state of journalism in 2011. Retrieved from http://www.orielladigitaljournalism.com

Papacharissi, Z., & de Fatima Oliveira, M. (2012). Affective news and networked publics: The rhythms of news storytelling on #Egypt. *Journal of Communication, 62*(2), 266–282. doi:10.1111/j.1460-2466.2012.01630.x

Postman, N. (1992). *Technopoly: The surrender of culture to technology.* New York, NY: Vintage Books.

Prinsloo, M. (2005). The new literacies as placed resources. *Perspectives in Education, 23*(4), 87–98. Retrieved from http://web.uct.ac.za/depts/educate/download/05thenewliteracies.pdf

Robinson, S. (2007). 'Someone's gotta be in control here': The institutionalization of online news and the creation of a shared journalistic authority. *Journalism Practice, 1*(3), 305–321. doi: 10.1080/17512780701504856

Shaer, M. (2012, 8 July). Reddit in the flesh. *New York Magazine*. Retrieved from http://nymag.com/news/features/reddit-2012-7/

Shoemaker, P. J. (1991). *Gatekeeping*. Newbury Park, CA: Sage.

Silverman, C. (2012, 21 Sep.). Twitter CEO says curation tools for newsrooms are coming. *Poynter*. Retrieved from http://www.poynter.org/latest-news/mediawire/189297/twitter-ceo-says-curation-tools-for-newsrooms-are-coming/

Singer, J. B. (2005). The political j-blogger: 'Normalizing' a new media form to fit old norms and practices. *Journalism*, 6(2), 173–198. doi: 10.1177/1464884905051009

Sonderman, J. (2011, 22 Aug.). NPR's Andy Carvin tweets 1,200 times over weekend as rebel forces overtake Tripoli. *Poynter*. Retrieved from http://www.poynter.org/latest-news/mediawire/143580/nprs-andy-carvin-tweets-1200-times-over-weekend-as-rebel-forces-overtake-tripoli/

Thurman, N., & Walters, A. (2012). Live blogging—Digital journalism's pivotal platform? A case study of the production, consumption, and form of live blogs at *Guardian.co.uk* . *Digital Journalism*, 1(1), 82–101. doi: 10.1080/21670811.2012.714935

Twitter. (2012, 21 Mar.). Twitter turns six. Retrieved from http://blog.twitter.com/2012/03/twitter-turns-six.html

Crisis Communication in Natural Disasters

The Queensland Floods and Christchurch Earthquakes

28
CHAPTER   Axel Bruns & Jean Burgess

from #qldfloods to #eqnz, Twitter plays an important role in #crisiscomms and emergency management #SMEM

Over the past decade, social media have gone through a process of legitimation and official adoption, and they are now becoming embedded as part of the official communications apparatus of many commercial and public-sector organisations—in turn, providing platforms like Twitter with their own sources of legitimacy. Arguably, the demonstrated utility of social media platforms and tools in times of crisis—from civil unrest and violent crime through to natural disasters like bushfires, earthquakes, and floods—has been a crucial driver of this newfound legitimacy. In the mid-2000s, user-created content and 'Web 2.0' platforms were known to play a role in crisis communication; back then, the involvement of extra-institutional actors in providing and sharing information around such events involved distributed, *ad hoc*, or niche platforms (like Flickr), and was more likely to be framed as 'citizen journalism' or 'crowdsourcing' (see, for example, Liu, Palen, Sutton, Hughes, & Vieweg, 2008, on the then-emerging role of photo-sharing in disasters). Since then, the dramatically increased take-up of mainstream social media platforms like Facebook and Twitter means that the pool of potential participants in online crisis communication has broadened to include a much larger proportion of the general population, as well as traditional media and official emergency response organisations.

The growing, multidisciplinary field of crisis informatics engages with a range of perspectives on social media in emergency management and crisis communication: from specific media practices like photo-sharing (Liu et al., 2008); to the role of mobile media in social support and emotional resilience (Hjorth & Kim, 2011); and, most relevantly for this chapter, in-depth analyses of the dynamics and characteristics of microblogging in crisis situations, particularly in natural disasters (Kongthon, Haruechaiyasak, Pailai, & Kongyoung, 2012; Murthy & Longwell, 2012; Qu, Huang, Zhang, & Zhang, 2011; Sakaki, Toriumi, & Matsuo, 2011; Sinnappan, Farrell, & Stewart, 2010), as well as research on issues of trust and veracity in such cases (Spiro et al., 2012; Starbird, Muzny, & Palen, 2012). In this chapter, we draw on our research into the uses of Twitter, and particularly of Twitter hashtags, during the 2010–2011 Queensland floods and Christchurch earthquakes (see also Bruns & Burgess, 2012; Bruns, Burgess, Crawford, & Shaw, 2012) to demonstrate patterns of Twitter-based communication during natural disasters, and to highlight further challenges for crisis communication research in social media, and for the practical application of Twitter during future crisis events.

The year 2011 was something of an *annus horribilis* for crises—from the January floods in south east Queensland, Australia, through the destructive February earthquake in Christchurch, New Zealand, to the March earthquake and tsunami on the east coast of Japan, and extending further through the unrest of the Arab Spring and the riots in London and the wider United Kingdom (see also Chapter 29 by Vis et al., in this volume); and in late 2012, the U.S. East Coast was rocked by Hurricane Sandy. These events have shown, each in their own way, how social media are used to disseminate breaking news, coordinate responses, monitor new developments, and express sympathy; news and emergency organisations around the world are now regularly incorporating social media into their crisis activities.

In all these cases, Twitter—as part of a broader media ecology including word-of-mouth, broadcast radio and television, the websites of mainstream news organisations, and official emergency organisations—has filled a significant mediating and coordinating function. Due to the specific communicative affordances of the Twitter platform (see also Chapter 2 by Bruns & Moe, in this volume), it lends itself especially well to the dissemination of breaking news from a range of sources, essentially in real time, to a wide network of users who can rapidly form an *ad hoc* public around the event or issue (Bruns & Burgess, 2011): when news with a high degree of perceived global interest breaks on Twitter, it travels around the world with unprecedented speed. Here, it should be noted that as crises emerge, one or several Twitter hashtags (like #sandy or #qldfloods) usually emerge with them: as Twitter users realise that it is important to share the latest crisis information quickly and effectively, they seek to establish a unified hashtag as a reliable marker of tweets which relate to the crisis.

While such processes do not always proceed smoothly, and may result (especially for crises which attract a large Twitter user base) in one or more competing options, network effects—that is, the preferential use of those hashtags which users already encounter in large volumes in their incoming Twitter feeds—nonetheless do tend to produce a very small number of key hashtags over time. Indeed, earlier adopters will often encourage other Twitter users who have important information to use 'their' hashtags, as in the following example from the February 2011 Christchurch earthquake, directed at a key emergency organisation:

> @NZcivildefence please use #eqnz hashtag. Thanks.

The combination of widespread global interest with the intensity and rapid sequence of events that characterise an emergency situation can produce uncertainty around the trustability of information, leading to concerns about rumours and misinformation—perhaps especially so during the early stages of an emerging crisis situation, at a time when the full facts have yet to be established. For news and emergency organisations, as much as for everyday users seeking to draw on Twitter to inform themselves about the current situation, therefore, the problem of effectively monitoring the flow and evaluating the veracity of crisis-related information on Twitter becomes paramount; in the first place, this begins by examining the patterns of information dissemination and user interaction in relevant communicative spaces on Twitter (see, for example, Mendoza, Poblete, & Castillo, 2010; Spiro et al., 2012).

MAKING SENSE OF #QLDFLOODS AND #EQNZ

Although hashtags such as #qldfloods (for the Queensland floods) and #eqnz (for the earthquakes in Christchurch, New Zealand) emerge as mechanisms for the gathering of *ad hoc* publics, made up of directly affected locals, emergency, media, and other government, NGO, and commercial organisations, and a wider audience of domestic and international Twitter users who are interested in tracking the event, the patterns of communication which take place amongst these rapidly accumulated publics—whose participants are largely unlikely to have been aware of one another beforehand—are far from random or disorganised. Here, as in most other forms of interpersonal and mass communication, network effects apply, and a small number of key accounts will quickly emerge as leading drivers of the communicative exchange.

This is true in the first place for how active participation is distributed across the user base of hashtag participants: usually, a comparatively small number of users will be regularly and committedly engaged in sharing crisis information, while a much larger number of Twitter users will participate only in retweeting or commenting on crisis tweets from time to time, almost randomly. This 'long tail' (Anderson, 2006) of users remains important as—especially through its retweeting of information—it enables crisis information to reach a much wider audience than the hashtag or its key contributors would be able to do by themselves, but such users largely act as amplifying 'repeater stations' for crisis information only, not as original sources of the information. At the same time, it is also common for some of the most active Twitter accounts in crisis contexts to serve deliberately as pure retweet accounts: the most active Twitter

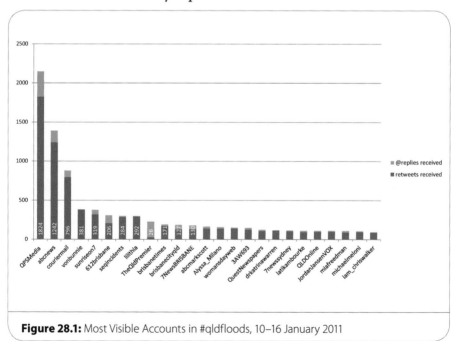

Figure 28.1: Most Visible Accounts in #qldfloods, 10–16 January 2011

account during the Queensland floods, for example, was @thebigwetfeed, virtually all of whose tweets were retweets. Whether operated manually or using a retweet bot, such accounts simply pass along all hashtagged tweets which they encounter, thereby fulfilling a function similar to the hashtag itself: Twitter users who wish to receive all hashtagged tweets pertaining to a specific issue or crisis now have the choice between subscribing to the hashtag itself or following the retweet account.

While an identification of the most *active* contributors to a hashtag provides a measure of interest and commitment, therefore, it does not necessarily point to the most important sources of information. Rather, it is useful instead to establish which accounts are the most *visible* in a hashtag, as measured by the number of @replies and retweets they receive: this approach draws on the communicative actions of the hashtag participant base itself, and examines which Twitter accounts are seen by that community itself as most worthy of engaging with (for an overview of the methods used in such analyses, see Chapter 6 by Bruns & Stieglitz, in this volume). What emerges from such analyses is a clear indication of the systematicity of Twitter communication even during acute crisis events: in most cases, a small group of key accounts receive the lion's share of attention from their peers. Figure 28.1 shows this distribution of attention

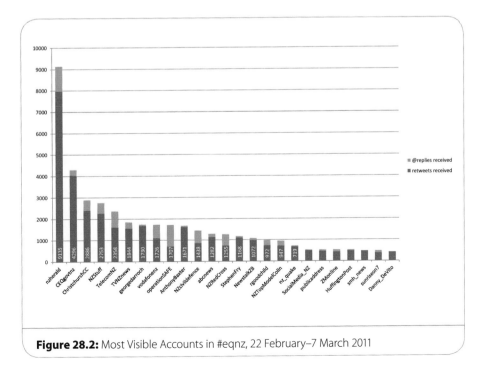

Figure 28.2: Most Visible Accounts in #eqnz, 22 February–7 March 2011

for the main week of the January 2011 Queensland floods, for example, while Figure 28.2 presents a similar picture for the fortnight following the February 2011 Christchurch earthquake.

What is immediately observable from these analyses is that traditional sources of authority remain central to crisis communication on Twitter, too. The Queensland floods field is led by the accounts of Queensland Police (@QPSMedia); the national public-service broadcaster ABC (@abcnews); and local newspaper *Courier-Mail* (@couriermail). The corresponding lead accounts for Christchurch are the *NZ Herald* newspaper (@nzherald); the Canterbury Earthquake Authority (@CEQgovtnz); and the Christchurch City Council (@ChristchurchCC). In both cases, many of the accounts which follow are operated by media, emergency, and other key organisations, too. This includes local radio stations (@612brisbane), or the personal accounts of political leaders (@TheQldPremier), and those of journalists and media executives (like ABC Managing Director @abcmarkscott or political journalist @latikambourke). There are also relevant corporate accounts (landline and mobile-phone providers @TelecomNZ and @vodafonenz); NGOs (@operationSAFE, @NZRedCross), and accounts of individuals in such organisations (like *Google.org*'s @anthonybaxter).

At the same time, several accounts of non-affiliated, individual users also appear in this list of the most visible participants. It is in this context that the value of Twitter and other social media as platforms for the dissemination of situational information from the scene of the disaster is most clearly revealed: a number of such accounts appear in prominent positions because they used Twitter to share their photos or video footage from the crisis event, or provided other important details about the current situation on the ground. Where such accounts providing important firsthand information can be identified, in close to real time, by drawing on the collective curation of crisis information by participants in topical hashtags, this may also be of immediate operational value to the emergency authorities themselves: in such cases, the Twitter community helps to highlight and thereby bring to the attention of authorities the latest important updates from the crisis area. Importantly, this distribution of attention may change over time, depending on the current crisis context (see Bruns & Burgess, 2012). During the early stages of an unexpected crisis event, on the one hand, news organisations may be highly featured, as local as well as remote Twitter users seek to understand the situation; later on, on the other hand, attention may shift towards emergency and support organisations, as the attention of more remote audiences to the crisis fades and local users turn to the question of how to deal with the aftermath and long-term effects of the disaster.

In several cases, however, highly visible accounts are visible not because they are able to share new updates from the disaster area, but simply because their comments on the situation were retweeted in sympathy by their already established, large networks of followers. This accounts for the presence of international celebrities such as actress @Alyssa_Milano or actor and comedian @StephenFry in the two cases above—each used their Twitter account to express their sympathies for affected locals, and were joined in such sentiments by their followers as they retweeted these messages. While such phenomena demonstrate the rapidly established global reach of hashtags which address localised natural disasters and other crises, these tweets necessarily have no immediate operational value; emergency organisations seeking to draw on Twitter as an additional source of crisis intelligence must find approaches to filter out this kind of material.

But even peripheral accounts like these can help—along with more closely involved Twitter users—by lending their follower base to the amplification of news and information about the crisis event, and to raising awareness of the associated hashtag(s). Retweeting of messages—which, as Figures 28.1 and 28.2 show, is the most important driver of visibility for the leading accounts—disseminates them well beyond the hashtag itself, making the tweets, the originating accounts, and the hashtag as such visible not only to followers of the original senders or the hashtag, but also to the many more Twitter users who follow any one of the retweeting accounts. On average, for example, every one of @QPSMedia's 72 hashtagged tweets during the main week of the 2011 Queensland floods crisis received some 25 (manual) retweets; this means that these tweets did not only reach the thousands of users following the @QPSMedia account or the #qldfloods hashtag at the time, but also the potentially many tens of thousands of additional users who followed the various retweeters, but were as yet unaware of @QPSMedia or #qldfloods itself. While the hashtag itself already enables Twitter users to make their tweets potentially visible to an audience which is much larger than their established retinue of followers, the retweeting of such hashtagged tweets by other users to *their* followers amplifies that reach even further, possibly by a substantial factor. (This demonstrates the interweaving of the various layers of Twitter communication which Bruns & Moe discuss in Chapter 2, in this volume.)

Finally, additional amplification is also achieved by the cross-media flow of information between Twitter and other media channels. Although this remains difficult to track empirically, anecdotal evidence from the Queensland floods event documents, for example, how updates tweeted by the @QPSMedia account directly from situation briefings with Queensland Premier Anna Bligh and the heads of emergency services organisations were copied verbatim (misspellings included) into the on-screen news tickers inserted into live TV coverage of the floods, and "retweeted" again as viewers passed on the information seen on the news tickers to their own Twitter followers. Similarly, of course, crisis news and information which originated outside of Twitter was also passed on in this manner—Twitter and other social media platforms, therefore, are not separate from, but increasingly deeply embedded into the overall crisis communication infrastructure, where they complement rather than replace existing channels.

CONCLUSION: USING TWITTER IN CRISIS MANAGEMENT

The widespread use of Twitter to share information about unfolding crisis situations, and the initial successes of various emergency organisations in experimenting with the use of social media as additional channels for crisis communication, have led to a growing interest in developing more comprehensive strategies for such approaches. Two major areas of activity must be distinguished here: on the one hand, emergency services are interested in using Twitter as an additional means to *communicate* their messages; on the other hand, they are also exploring the potential to *monitor* valuable situational information from directly affected locals in crisis-affected areas.

In the first place, crisis communication efforts by emergency services crucially depend on a thorough understanding of Twitter as a communicative environment, also established through regular use of the platform outside of crisis contexts. Emergency organisations must build a visible Twitter presence during non-crisis periods so that at times of crisis, Twitter users will be able to find and follow them quickly and effectively; users may be considerably less prepared to trust an account (and retweet its messages) if it is new to Twitter and has as yet failed to build up a strong track record. In developing a Twitter presence, it is also important for an emergency service organisation to follow the established standards of the platform—that is, to use appropriate hashtags as required to maximise the visibility of its messages, and to engage with other Twitter users as they correspond with it through @replies. Such activities build trust and increase the likelihood that the account's messages are disseminated further through retweets.

This also increases the authority of emergency service accounts, and improves their ability to address and correct possible misinformation. During the Queensland floods, for example, the Queensland Police Service Twitter account instituted a series of tweets which directly engaged in the debunking of potentially dangerous rumours about the current situation in the disaster area, responding to them one by one with the dedicated hashtag #mythbuster (Bruns et al., 2012). Such tweets were amongst the most widely retweeted @QPSMedia messages during the Queensland floods crisis, considerably helping to limit the spread of misinformation. Initiatives of this kind can only be successful if they come from an account which already has strong traction in the overall community of Twitter users following an unfolding crisis event, however.

The monitoring of crisis-relevant information on Twitter by emergency services must similarly build on a sophisticated understanding of how social media are used during crisis situations. Automatic monitoring of relevant hashtags and keywords on Twitter may be able to pinpoint a range of potential relevant tweets sent by users, which may then be evaluated further by assigning differing levels of trust to users and tweets depending on a range of parameters—for example, the past track record of users (the extent of their past Twitter activity, their number of followers, tweeting styles, or location information), or the correlation of information between different user accounts. Such evaluations are far from straightforward. A very recently established Twitter account may be considered to be less reliable because of its lack of an established track record, but it may also have been set up specifically to provide disaster-related information. Tweets from several users reporting a bushfire may be seen as reliable if the senders are separate from one another, but not if the users are retweeting one another. In most contexts, automated crowdsourcing of situational information will be unable to replace the manual evaluation of such information altogether; using Twitter as an *additional* information source can, however, provide useful further detail on local circumstances in the disaster area, especially if emergency services staff have yet to reach the area or are insufficient in numbers to cover the entire space.

Potentially, such automated approaches to identifying and highlighting crisis information may also play an important role in the early detection of crisis situations. Twitter has rapidly become an important medium, especially for the dissemination of breaking news, including news about natural disasters and other crises. Although relatively simplistic in its implementation, Twitter's own "trending topics" feature often provides a useful pointer to emerging issues on the global, national, and local scale. Taking a similar approach to the identification of trending themes, but using a more sophisticated set of measures which are able to describe the ebb and flow in Twitter communication in more comprehensive detail, it may be possible to detect the weak signals of an impending crisis even before they appear in the "trending topics", and certainly before the Twitter community settles on a unified hashtag to coordinate the further discussion of the event (for examples of early conceptual and experimental work on such "real-world event" detection by "social sensors", see Becker, Naaman, & Gravano, 2011; Sakaki et al., 2010).

This, however, would also depend on comprehensive access to the full "firehose" of Twitter data, from which such trends would need to be extracted. By contrast, much current research into crisis communication on Twitter (including our own studies of #qldfloods and #eqnz, as outlined above) proceeds by evaluating user activities within a range of clearly established hashtags. While such approaches are useful for tracking the further development of crises once they have been clearly recognised by the wider Twitter user base, and for tracking the further aggregation, structuration, and eventual dissipation of a community of users following the disaster, they are unable to shed sufficient light on the early, formative stages of such crisis communication efforts on the platform. They provide very little information on where the very first—possibly unhashtagged—tweets reporting the crisis events originated; how they were shared; and how, through these processes, an overall awareness of the crisis began to grow.

To track such early developments, it would be necessary to monitor the Twitter firehose on an ongoing basis for any small-scale signs of increased user activity that might be indicative of a potential emergency situation. Such an approach would need to distinguish—for example on the basis of keywords—natural disasters and other emergencies from other forms of trending topics (such as breaking news of a non-crisis nature), and to identify the key users, keywords, and eventually hashtags which come to be associated with the event. If such weak signals of emerging crisis events can be detected and highlighted to emergency services authorities with any degree of accuracy, this would substantially boost their ability to respond in close to real time to crisis situations as they are reported on Twitter.

REFERENCES

Anderson, C. (2006). *The Long Tail: Why the future of business is selling less of more*. New York, NY: Hyperion.

Becker, H., Naaman, M., & Gravano, L. (2011). Beyond trending topics: Real-world event identification on Twitter. In *Proceedings of the Fifth International AAAI Conference on Weblogs and Social Media (ICWSM'11)*. Retrieved from http://sm.rutgers.edu/pubs/becker35-icwsm2011.pdf

Bruns, A., & Burgess, J. (2011). The use of Twitter hashtags in the formation of ad hoc publics. In *6th European Consortium for Political Research General Conference*, August 25–27, 2011, University of Iceland, Reykjavik. Retrieved from http://eprints.qut.edu.au/46515/

Bruns, A., & Burgess, J. (2012). Local and global responses to disaster: #eqnz and the Christchurch earthquake. In *Proceedings of Earth: Fire and Rain—Australia and New Zealand Disaster and Emergency Management Conference, Conference* (pp. 86–103). April 16–18, 2012, Brisbane, Australia. Retrieved from http://snurb.info/files/2012/Local%20and%20Global%20Responses%20to%20Disaster.pdf

Bruns, A., Burgess, J., Crawford, K., & Shaw, F. (2012). *#qldfloods and @QPSMedia: Crisis communication on Twitter in the 2011 south east Queensland floods*. Brisbane, Australia: ARC Centre of Excellence for Creative Industries and Innovation, Queensland University of Technology. Retrieved from http://eprints.qut.edu.au/48241/

Hjorth, L., & Kim, K. H. Y. (2011). Good grief: The role of social mobile media in the 3.11 earthquake disaster in Japan. *Digital Creativity, 22*(3), 187–199. doi: 10.1080/14626268.2011.604640

Kongthon, A., Haruechaiyasak, C., Pailai, J., & Kongyoung, S. (2012). The role of Twitter during a natural disaster: Case study of 2011 Thai flood. In *Proceedings of PICMET'12, Technology Management for Emerging Technologies* (pp. 2227–2232).

Liu, S. B., Palen, L., Sutton, J., Hughes, A. L., & Vieweg, S. (2008). In search of the bigger picture: The emergent role of on-line photo sharing in times of disaster. In F. Friedrich & B. Van de Walle (Eds.), *Proceedings of the 5th International ISCRAM Conference*, Washington, DC. Retrieved from https://www.cs.colorado.edu/~palen/Papers/iscram08/OnlinePhotoSharingISCRAM08.pdf

Mendoza, M., Poblete, B., & Castillo, C. (2010). Twitter under crisis: Can we trust what we RT? In *Proceedings of the First Workshop on Social Media Analytics (SOMA '10)* (pp. 71–79). Retrieved from http://snap.stanford.edu/soma2010/papers/soma2010_11.pdf

Murthy, D., & Longwell, S. A. (2012). Twitter and disasters: The uses of Twitter during the 2010 Pakistan floods. *Information, Communication & Society*. Retrieved from http://dx.doi.org/10.1080/1369118X.2012.696123

Qu, Y., Huang, C., Zhang, P., & Zhang, J. (2011, Mar.). Microblogging after a major disaster in China: A case study of the 2010 Yushu earthquake. In *Proceedings of the ACM 2011 Conference on Computer Supported Cooperative Work (CSCW 2011)* (pp. 25–34). Retrieved from http://personal.stevens.edu/~rchen/readings/yushu.pdf

Sakaki, T., Okazaki, M., & Matsuo, Y. (2010). Earthquake shakes Twitter users: Real-time event detection by social sensors. In *Proceedings of the 19th International Conference on the World Wide Web* (pp. 851–860). ACM. doi:10.1145/1772690.1772777

Sakaki, T., Toriumi, F., & Matsuo, Y. (2011). Tweet trend analysis in an emergency situation. In *Proceedings of the Special Workshop on Internet and Disasters (SWID 2011)*. Retrieved from http://conferences.sigcomm.org/co-next/2011/workshops/SpecialWorkshop/papers/1569500807.pdf

Sinnappan, S., Farrell, C., & Stewart, E. (2010). Priceless tweets! A study on Twitter messages posted during crisis: Black Saturday. *Proceedings of ACIS 2010*. Paper 39. http://aisel.aisnet.org/acis2010/39

Spiro, E. S., Fitzhugh, S., Sutton, J., Pierski, N., Greczek, M., & Butts, C. T. (2012). Rumoring during extreme events: A case study of Deepwater Horizon 2010. In *Proceedings of the 3rd Annual ACM Web Science Conference* (pp. 275–283). doi:10.1145/2380718.2380754

Starbird, K., Muzny, G., & Palen, L. (2012). Learning from the crowd: Collaborative filtering techniques for identifying on-the-ground twitterers during mass disruptions. In L. Rothkrantz, J. Ristvej, & Z. Franco (Eds.), *Proceedings of the 9th International ISCRAM Conference*, Vancouver, Canada, April 2012. Retrieved from http://www.iscramlive.org/ISCRAM2012/proceedings/148.pdf

Twitpic-ing the Riots
Analysing Images Shared on Twitter during the 2011 U.K. Riots

29 CHAPTER	Farida Vis, Simon Faulkner, Katy Parry, Yana Manyukhina, and Lisa Evans

 during the 2011 #ukriots, image sharing helped get the word out about the situation on the streets

Crisis events like natural disasters and civil disobedience can be intensely visual, and it is often through images that we come to know and remember them. Photography has been a long-standing medium for the recording of such events, especially after the establishment of photojournalism during the early and mid-twentieth century. With the popularisation of digital cameras in combination with the development of social media, large amounts of user-generated imagery is typically produced in response to crisis events and circulated within wider media ecologies. Of particular importance is the widespread use of camera phones with a networked capacity, making these devices not merely cameras, but 'communication-connection device[s]' (Cruz & Meyer, 2012, p. 214). This technology enables a wide range of bystanders to create images on the spot to either send to mainstream media organisations, upload to image-sharing platforms, or share elsewhere online. In the London bombings (July 2005), camera phones were used by witnesses and people involved in the events to produce images that were posted online and sent to the BBC and *The Guardian* newspaper. It was also around this time that image-sharing site, Flickr (founded in 2004), began to come to prominence as a platform for sharing images of crises, produced using camera phones, amongst other devices (Liu, Palen, Sutton, Hughes, & Vieweg, 2009). More recently, Twitter has displaced Flickr, in terms of its real-time, image-uploading role (Burgess, 2011).

Research that includes discussions of images of crisis events on Twitter is already emerging, and in part extends crisis communication research into online photo-sharing. For example, Murthy (2011) noted the uploading to Twitter of a picture of the US Airways jet that crashed on the Hudson River in January 2009, while Sarcevic et al. (2012) discussed image content in their study of medical Twitter users during the 2010 Haiti earthquake, noting that 85 percent of these users uploaded or tweeted links to images. In their analysis of the 2011 Pukkelpop festival incident in Belgium, Terpstra, de Vries, Stronkman, and Paradies (2012) emphasised the immediate and after-the-fact evidential value of tweeted photographs. In contrast, Reuter, Marx, and Pipek (2012, p. 8) noted that due to the collapse of the mobile phone network, images could only be uploaded *after* a fatal incident killing 21 at the 2010 Love Parade in Germany. Bruns, Burgess, Crawford, and Shaw (2012, p. 7; see also Burgess, 2011, and Chapter 28 by Bruns & Burgess in this volume) noted the importance of images in their research on the role of Twitter during the 2011 Queensland Floods, observing that one in every five shared links was to an image. In response to this literature, we encourage further research that focusses on Twitter images as a subject in their own right. Such work requires the development of new approaches to Twitter images, combining the discussion of their basic content with the study of other aspects of their function and meaning. This chapter is exploratory in nature, and addresses these issues through examining image production and sharing practices on Twitter during the 2011 U.K. riots.

The riots began with a protest on 6 August 2011 over the killing by police of a young Black man (Mark Duggan) in Tottenham on 4 August. This protest took a destructive turn when police vehicles and a Double Decker bus were set on fire, and was followed by four further nights of confrontations with police, looting, and destruction of property across London and the UK. In the absence of an official enquiry into their causes, a ground-breaking project, 'Reading the Riots', was established between *The Guardian* and a number of U.K. academics, including the current first and last two authors, to better understand the riots. As part of the project, Twitter's role in the riots was investigated, and 2.6 million tweets (donated by Twitter) were analysed. Because we worked with the same data set based on a series of riot hashtags, images included in this chapter were subject to the same limitation. To partially overcome this, search engine Topsy was also used to compare image-share frequencies beyond the confines of this data set.

The next section describes the methods we used. We then move on to discuss the findings. We discuss a number of general findings, followed by more detailed readings of a set of images in the 'bus' subcategory. We end the chapter with a brief conclusion drawing on a range of concepts we see as relevant for the further development of research in this area.

METHODS

In order to distinguish images, we first identified all links in the data set. A total of 10,001 unique, mostly shortened links were extracted, which had been shared 19,315 times during the four days. Links were resolved and coded as follows:

1. Image-sharing platform;
2. Video-sharing platform;
3. Social-media platform;
4. Mainstream media—riot coverage;
5. Mainstream media—other;
6. Alternative media;
7. Blogs (included in Technorati top 100);
8. Blogs—other;
9. Websites—news-focussed;
10. Websites—other;
11. Spam;
12. Broken link.

We then organised the images into categories according to their basic, denotative content following a conventional content analysis approach. All links pointing to image-sharing platforms were thus coded as follows:

1. Police car (burning, attack, and aftermath);
2. Bus (burning, aftermath, and altered image);
3. Other vehicle (burning, attack, and aftermath);
4. Building (burning, aftermath, before and after shots);
5. Looting (in the act, aftermath, trophy shots);
6. Screenshots (TV screens);
7. Street scenes;

8. Police;
9. Arrests;
10. Image of text (screen grab other than TV screen, sign, newspaper front page);
11. Riot cleanup;
12. Unclear;
13. Other;
14. Excluded (not about riots, not single still image, broken link, image removed, etc.).

Mustafaraj, Finn, Whitlock, and Metaxas (2011) reminded us of a 'vocal' minority, compared to a relatively 'silent' majority in a 'long tail' data distribution. With this in mind, two sets of files, containing all image links shared more than once (n= 433, 4620 shares) and those shared only once (n=374, 374 shares) were kept separate. Both sets were double-coded, and inter-coder reliability was calculated using ReCal (Freelon, 2010). For the multiple shares Scott's pi was 0.824 and 0.838 for the single shares, which are both satisfactory.

We then focussed on images of the burning bus. Looking at these led us to make further subdivisions within the 'bus' category and to think about how the images were produced and by whom, and how they functioned in relation to general ideas about photographic eyewitnessing. Particularly useful for our speculations on this subject was Cruz and Meyer's (2012, p. 204) redefinition of photography as a "socio-technical network", which, for the burning bus images, involves relationships between an event, technologies that enable the production and distribution of images of this event, and a set of meanings that relate not only to the event itself, but also to the very practices involved with its visual representation. In relation to the latter, it seemed important to think about long-standing and widespread cultural commitments to photography as a key medium through which the reality of significant events can be recorded and witnessed. We would suggest that such beliefs in the evidential and witnessing capacity of photography relate in general to what people believe they are doing when they use cameras to document their visible surroundings. With the widespread use of ready-to-hand camera phones, anyone can now potentially engage in photographic acts of eyewitnessing through the sharing of images of what they have seen (Mortensen, 2011a, p. 63; Mortensen, 2011b, p. 8). This eyewitnessing is not necessarily informed by any sense of responsibility or moral position (Mortensen, 2011a, p. 72). Rather it is more likely defined by a desire to give a pictorial form to seeing as part of personal experience, and to contribute to the online circulation of acts of seeing. Mortensen also suggested that in an age of hyper-mediation, there is a slippage from eyewitnessing as something that is direct to eyewitnessing as something experienced through mediation (2011a, p. 70). This can be related to the complex relationships between actual and virtual space involved with the production and distribution of Twitter images. Images of crisis events are made at one point in actual space, and then uploaded to Twitter from where the images can be seen by multiple spectators located anywhere in the world. These spectators can then mark their own acts of eyewitnessing by retweeting the images.

Table 29.1: Top Ten Image Links

	Image URL	Description	Our Corpus	Topsy
1	http://twitpic.com/623gp0	'The moment the bus went up in flames'	357	452
2	http://twitpic.com/62m6nx (removed)	'Tottenham looter may regret posing with his loot'	303	3,761
3	http://yfrog.com/kjg6vp	Carpet shop before and after fire (via Google maps)	218	236
4	http://www.flickr.com/photos/56312368@N04/sets/72157627372500124/	Tottenham Riots August 2011—a set on Flickr	180	199
5	http://yfrog.com/gysv8fpj	Arrest of very young looter caught outside Subway	96	281
6	http://twitpic.com/6289b1	Carpet shop on fire (same as 3)	87	347
7	http://yfrog.com/kf4rlauj	Police car, its window smashed in with bricks	81	483
8	http://yfrog.com/z/kffpgozj	Carpet shop on fire (same as 3)	76	80
9	http://yfrog.com/h8mt9hlaj	HMV (record store) being looted	69	660
10	http://hashalbum.com/tottenham	A collection of Tottenham photos from Twitter	66	229

FINDINGS

In our corpus, one in four shared links (26 percent) was a still image on an image-sharing platform, compared to one in ten that were videos (10 percent). Table 29.1 highlights the top ten most shared image links, and shows how often

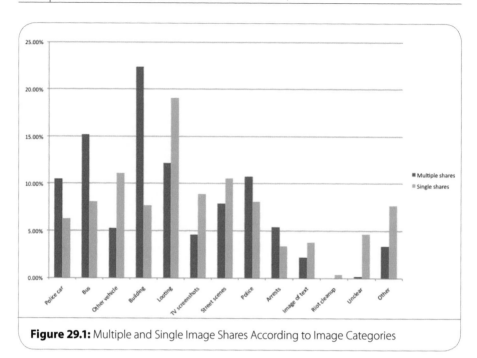

Figure 29.1: Multiple and Single Image Shares According to Image Categories

they were shared in our database compared to Topsy. Significant discrepancies can be seen, for example: the second most shared image link in our database is a looter posing with his loot. Although the image was quickly removed, it was widely shared: 303 times in our database, but more than ten times that according to Topsy (3,761 shares).

A total of 3,466 multiply shared riot images and 235 single shares were kept and included in the final analysis. Figure 29.1 highlights differences between these two data sets for the remaining 13 categories. Most notably, images of burning police cars were shared more in the multiples, as were images of the bus and especially of buildings. However, acts of looting, its aftermath or 'trophy shots' were more often single shares (19 percent compared to 12.2 percent). Images of TV screens were on the whole shared proportionately more frequently as single shares. Combining results (Figure 29.2), images of burning buildings, specifically of the Carpet Right in Tottenham, was the most shared category (794). They frequently showed 'before' and 'after' shots, often included in the same image, along with pictures of buildings on fire. One picture in particular is worth highlighting. The third most shared link (Table 29.1) is a 'before' image of the Carpet Right, using footage from Google Street View. This mapping technology gives users access to 360-degree, street-level imagery. The image includes, in the bottom right corner, the familiar graphic depiction of a person, allowing you to navigate the view you wish to explore. Although the Twitter user chose the viewing position and shared the image through Yfrog, the original image data was created by one of Google's "numerous data collection vehicles" using their R5 "panoramic camera system" (Anguelov et al., 2010, pp. 32–33). This thus raises an interesting question about the production of the image and how we might consider the use of Google Street View photography and image appropriation more widely within crisis communication. The rest of this chapter focusses on the images in the second most shared category (545), the setting on fire of the Double Decker bus, which was also the subject of the single most shared individual image link in our data set.

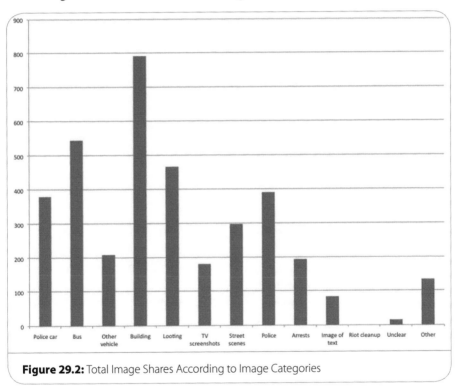

Figure 29.2: Total Image Shares According to Image Categories

THE BURNING BUS

To identify all bus images, links from the 'bus' category and those depicting the bus in TV screenshots were identified. The resulting 57 images were printed out and arranged on a wall (Figure 29.3) to inductively identify subcategories.

During this process, a number of things became clear: the TV screenshots were all from Sky News (mostly from live broadcasts), and included a time-

Figure 29.3: Image Wall

stamp so that we could establish a timeline. The earliest Sky News image was recorded at 22:56 p.m., the last at 23:23 p.m., showing the various stages of the burning of the bus within this 27-minute window. A number of images ('smoking bus', 'bus consumed by fire', 'Call of Duty', and 'Other') were subsequently identified as cropped shots of these news images, even though news organisation interfaces and logos were not identified and (TV) screens were not readily visible at first. These results highlight the value of seeing the images side by side in this way. In the Sky News screen shots, we distinguish between those where the image clearly shows the TV itself, placing the viewing of the event within a domestic setting, and those that are more likely a screen grab from a (tablet) computer screen where news was consumed online, making the location of viewing and screenshot production more difficult to identify. In these instances, the image production and sharing could potentially have taken place on the same device. Table 29.2 shows the bus image categories, along with the total number of unique URLs identified per category, details of shares in our database, and data obtained from Topsy. In some cases, image categories concern identical content: for example, for those labelled, 'bus on fire', we found

13 instances of the same image. The Sky News images were not identical in content, but rather, they all depicted the origin of the content, the live TV coverage. They were spread across 14 URLs, shared 64 times in the corpus, but 394 times according to Topsy. What is more, combined with the cropped TV screens, 123 of the bus image shares in our corpus are derived from live TV, which accounts for 20 percent of the bus category.

Table 29.2: Bus Image Categories

	Bus Image Category by Name	Example URL	# of Unique Image URLs	Our Corpus	Topsy
1	Moment bus went up in flames	http://twitpic.com/623gp0	4	363	475
2	Smoking bus (air full of smoke)	http://twitpic.com/622qvx	11	40	140
3	Bus on fire (engulfed in flames)	http://twitter.yfrog.com/klkd3bzj	13	72	184
4	Sky News—TV visible in shot	http://pics.lockerz.com/s/127296239	4	6	11
5	Sky News—TV not visible in shot	http://twitpic.com/622k0g	10	58	383
6	Burning bus, police, and crowd	http://twitpic.com/6240hq	2	6	18
7	Bus consumed by fire	http://twitpic.com/623a1c	2	7	8
8	'Call of Duty' ('smoking bus')	http://pics.lockerz.com/s/127295300	6	10	20
9	Aftermath (carcass of burnt bus)	http://twitpic.com/62ax23	3	41	154
10	Other	http://twitter.yfrog.com/h0d9uknj	2	2	9

Returning to the most shared image (Table 29.1)—'the moment the bus went up in flames', uploaded on Twitpic—the producer is a user called @Heardinlondon, whose website highlights a keen interest in amateur street photography. This user also has a Flickr 'pro' account, and links to the same bus image on Flickr are also shared widely. On Flickr it is part of a set, 'Tottenham Riots August 2011', which contains 29 images for the nights of 6 and 7 August.

The Exchange Image File data presented by Flickr, recording the device that produced the image, shows that nine images were shot on an Apple iPhone 3GS and 20 on a Canon Powershot 3X10 IS. Three images from this set also appear on Twitpic, two uploaded via e-mail and one, 'the moment the bus went up in flames', directly onto the site, viewed 17,555 times at the time of writing. Two additional images appear on Twitpic, which predate the Flickr set, most probably the first two images of the riots this user took. The first Twitpic, of a line of police officers with police vans in the distance behind them, seems to confirm this in the text: 'Kids on the street in Tottenham have just told me this is revenge for the police shooting the guy in the minicab'. Cross-referencing these images between platforms enables us to begin to address the complexity of their production, uploading, and sharing. On the one hand, the Canon SLR camera used for most of the Flickr set controls the production of these images. They can be viewed on the device, but the camera cannot directly distribute them online. The images taken on the iPhone, on the other hand, are produced on the device, can be viewed there, and can be digitally distributed through it, highlighting it as an all-in-one communication-connection device. We are thus dealing with different photographic practices that reflect the status of this person as not simply a bystander, but as somebody adopting a self-conscious role as a reporter. A second set of Flickr images—'Tottenham Riots, the morning after'—contains 86 images, of which many are also shared through Twitpic, seeming to further confirm this. Moreover, if we look at how these interlinked modes of sharing were received and discussed in tweets, what stands out is that the Twitpic link to the bus image is mainly retweeted and thus not much discussed, but that the Flickr one includes positive assessments of the photographs as 'reportage', 'excellent photojournalism', and 'excellent photography and initial reporting'. Such comments affirm the relevance of recent discussions of 'citizen journalism' and the now fuzzy dividing line between those who are and are not seen as photojournalists when they take pictures of such events. It also affirms the existence of commitments to the value of photographic reportage as the documentation and eyewitnessing of reality.

A different type of eyewitnessing activity can be observed in the images of Sky News (and those derived from it). As discussed in the introduction, eyewitnessing in this instance involves a mediated and spatially removed relationship to the unfolding crisis event. One technology, live TV, allows for still-image creation so that Twitter users can say and show: 'This is happening right now!' In a number of cases, those uploading these screenshots are journalists, such as Jonathan Haynes from *The Guardian*, who uploads three different screenshots, the first at 22:56 p.m., stating: 'Here is the bus on fire in #Tottenham yfrog.com/kiqi4bxj' (shared 12 times in our corpus; 139 times according to Topsy). His images were thought to be computer/iPad screen grabs, different from four separate images by other users where TVs are clearly visible. This image production suggests the value of this simultaneous, removed eyewitnessing and the significance of the turning of a live TV stream into a still, shareable image. This reminds us of the way in which John Berger (1972) noted the significance of the act of photography in general in terms of the statement: '*I have decided that seeing this is worth recording*' (p. 179). As Berger observed, photography 'is the process of rendering observation self-conscious' (1972, p. 180). This observation is even more relevant when what is documented in the photograph is partly the status of the image-maker as a media spectator. Here, the eyewitness is both a spectator of mainstream media news and an image-maker who utilises the camera phone as a communication-connection device to produce images and distribute them through Twitter. The emerging literature on Twitter in relation to TV viewing and audience engagement through 'second screen devices' has so far predominantly focussed on entertainment (Lochie & Coulton, 2012). Our study may have identified another significant activity specifically linked to breaking news and crisis communication that mobilises audiences around hashtags, but crucially also includes active image production of TV screens and the subsequent sharing of these on Twitter.

Finally, the 'Call of Duty' image is a reworking ('meme') of the 'smoking bus' image, which we suggest is a cropped TV screenshot, again implying a form of removed eyewitnessing. It is reworked through the addition of a yellow banner representing police tape ('police line do not cross') starting from the bottom left, going diagonally across the image and ending three quarters up towards the top right corner. An additional text at the top states 'Call of Duty', with the words 'Tottenham Warfare' at the bottom, both in large, white capitals. This reference to the popular video game series *Call of Duty: Modern Warfare* relies on the audience being familiar with the popular cultural reference, suggesting possibilities for first and third shooter action in the Tottenham 'war zone'. Such an image indicates that when dealing with Twitter images there is a need for approaches that are sensitive to intertextual relationships.

These different 'bus' subcategories have started to highlight the multiple ways in which images of the burning Double Decker bus have been made, uploaded, shared, circulated, and discussed on Twitter. We briefly draw some conclusions below.

CONCLUSION

This chapter reports on an exploratory study of images that were tweeted during the 2011 U.K. riots, focussing primarily on the burning of the Double Decker bus on the first night. Our study considers the tweeting of different types of images. Many were produced through digital cameras at the scene of the event, then shared via Twitter. Other images were appropriated in one way or another, such as the Carpet Right image and the 'Call of Duty' adaptation. The TV screenshots are in themselves key images of the event, and we have pointed to practices of remote witnessing, as well as to how this might be related to the growing work on mobile 'second screen devices' in relation to Twitter. These different kinds of images attest to the multifarious ways Twitter users create and mobilise images as means of communicating their experiences and thoughts. This points to the need for the development of adaptable responses on the part of researchers interested in Twitter images. Future research will also need to deal with the changing online environment in which Twitter functions as a site for viewing images directly uploaded to it through image-sharing services and as a hub for interaction with images located elsewhere. Such research will also need to deal with the different temporalities involved with these functions: sometimes Twitter images function as real-time mediations of crisis events, and sometimes Twitter is a conduit through which images uploaded elsewhere can be viewed after the fact.

Our discussion dealing with photographic eyewitnessing also suggests continuities between Twitter images and long-standing discourses about the realism of photography. Relationships between digital-imaging technologies, Twitter, and other online image-sharing platforms are complex and dynamic, but the enduring idea of the veracity of the photographic image is a crucial part of these dynamics. As Bernd Stiegler (2008) has pointed out, photographs continue to be taken as 'visual reflections of reality', and 'neither the alterations photography is currently experiencing within various media nor digitalisation has changed any of this' (p. 194). This suggests the need to address how the medium of photography as a set of historically generated practices and discourses informs the making and use of Twitter images, and at the same time, how Twitter contributes to the reconfiguration of photography as a socio-technical network. It is our position that approaches that deal with such considerations can only enrich examinations of Twitter images and open up the interdisciplinary context of Twitter research in new ways.

NOTE

1   See http://www.guardian.co.uk/uk/series/reading-the-riots

REFERENCES

Anguelov, D., Dulong, C., Filip, D., Frueh, C., Lafon, S., Lyon, R., . . . Weaver, J. (2010). Google Street View: Capturing the world at street level. *Computer, 43*(6), 32–38.

Berger, J. (1972). *Selected essays and articles: The look of things.* Middlesex, UK: Pelican Books.

Bruns, A., Burgess, J., Crawford, K., & Shaw, F. (2012). *#qldfloods and @QPSMedia: Crisis communication on Twitter in the 2011 South East Queensland Floods.* Brisbane, Australia: ARC Centre of Excellence for Creative Industries and Innovation. Retrieved from http://cci.edu.au/floodsreport.pdf

Burgess, J. (2011, 6 Mar.). Image sharing in the #qldfloods. *Mapping Online Publics Blog.* Retrieved from http://mappingonlinepublics.net/2011/03/06/image-sharing-in-the-qldfloods/

Cruz, E. G., & Meyer, E. T. (2012). Creation and control in the photographic process: iPhones and the emerging fifth moment of photography. *Photographies, 5*(2), 203–221.

Freelon, D. (2010). ReCal: Intercoder reliability calculation as a Web service. *International Journal of Internet Science, 5*(1), 20–33.

Liu, S. B., Palen, L., Sutton, J., Hughes, A. L., & Vieweg, S. (2009). Citizen photojournalism during crisis events. In S. Allan & E. Thorsen (Eds.), *Citizen journalism: Global perspectives* (pp. 43–63). New York, NY: Peter Lang.

Lochie, M., & Coulton, P. (2012). Sharing the viewing experience through second screens. In *Proceedings of EuroITV 2012, 10th European Conference on Interactive TV and Video* (pp. 199–202), Berlin, Germany.

Mortensen, M. (2011a). The eyewitness in the age of digital transformation. In K. Andén-Papadopoulos & M. Pantti (Eds.), *Amateur images and global news* (pp. 63–75). Bristol, UK: Intellect.

Mortensen, M. (2011b). When citizen photojournalism sets the news agenda: Neda Agha Soltan as a Web 2.0 icon of post-election unrest in Iran. *Global Media & Communication, 7*(1), 4–16.

Murthy, D. (2011). Twitter: Microphone for the masses? *Media, Culture & Society, 33*(5), 779–789.

Mustafaraj, E., Finn, S., Whitlock, C., & Metaxas, P. (2011, October). Vocal minority versus silent majority: Discovering the opinions of the long tail. Paper presented at IEEE SocialCom Conference, Boston, MA.

Reuter, C., Marx, A., & Pipek, V. (2012). Crisis management 2.0: Towards a systematization of social software use in crisis situations. *International Journal of Information Systems for Crisis Response and Management, 4*(1), 1–16.

Sarcevic, A., Palen, L., White, J., Starbird, K., Bagdouri, M., & Anderson, K. (2012). 'Beacons of hope' in decentralized coordination: Learning from on-the-ground medical twitterers during the 2010 Haiti earthquake. Retrieved from http://www.cs.colorado.edu/~palen/Home/Crisis_Informatics_files/Sarcevic-et-al-HaitiMedicalTwitterers.pdf

Stiegler, B. (2008). Photography as the medium of reflection. In R. Kelsey & B. Stimson (Eds.), *The meaning of photography* (pp. 194–197). New Haven, CT: Yale University Press.

Terpstra, T., de Vries, A., Stronkman, R., & Paradies, G. L. (2012). Towards a realtime Twitter analysis during crises for operational crisis management. Retrieved from http://www.iscramlive.org/ISCRAM2012/proceedings/172.pdf

Twitter in Scholarly Communication

30
CHAPTER   Merja Mahrt, Katrin Weller, and Isabella Peters

twitter might act as an #altmetrics indicator for scholarly communication, but Twitter use remains rare in most scientific disciplines

TWITTER IN THE ECOLOGY OF SCHOLARLY COMMUNICATION

Since the emergence of the personal website, and later scholarly blogging, academics have used the Internet for both strictly scientific and self-promotional purposes. Microblogging via Twitter is one such example, with guidelines recently emerging for its effective use in scholarly communication. Such guidelines see microblogging as useful for diverse academic purposes and contexts (Herwig, Kittenberger, Nentwich, & Schmirmund, 2009; Mollett, Moran, & Dunleavy, 2011). Building and maintaining professional networks is one of the core uses, especially around conferences. Individual researchers, as well as group research projects and institutions, may use Twitter for advertising their own research, events, publications, or other updates, much in the same way as other commercial, political, or societal actors do in their marketing and PR efforts via Twitter (Kortelainen & Katvala, 2012; Sammer & Back, 2011). A well-connected Twitter account and the use of pertinent hashtags help increase visibility, both for one's own research and a given field in general. Twitter can also be used for internal communication, to let people know what others in a project or department are doing. In addition, Twitter can serve as a personal archive of information that one once found worth sharing and would like to access later on, for instance, through the use of URLs in tweets.

This chapter discusses the prevalence of Twitter usage among scholars in different countries and disciplines, before presenting selected cases from research on academics who tweet, Twitter usage around conferences, and the use of URLs in tweets. While beyond the scope of this chapter, Bruns and Burgess (2012) and several chapters in the present volume deal with challenges of using Twitter *data* in scholarly research (e.g., Bruns & Stieglitz, Chapter 6 in this volume; Gaffney & Puschmann, Chapter 5 in this volume; Puschmann & Burgess, Chapter 4 in this volume). The complementary role of Twitter in teaching is reviewed by van Treeck and Ebner in Chapter 31 in this volume.

THE UPTAKE OF TWITTER AMONG SCHOLARS

No fully comprehensive studies exist on how, why, or in what ways scholars use Twitter. A number of surveys among scholars from select American and European countries reported that only around 1 or 2% of the respective respondents use Twitter at all (Bader, Fritz, & Gloning, 2012; Gerber, 2012; Harley, Acord, Earl-Novell, Lawrence, & King, 2010). Scholars at a Finnish university used "mini blogs" more frequently (14%; Gu & Widén-Wulff, 2011), but compared to other social media, only a few respondents perceived them as being useful. Ponte and Simon (2011) found the highest proportion of Twitter users (18%) in a survey of 349 European scholars, making it still the least popular Web 2.0 application. Surveys of academic Twitter users usually employ Web-administered questionnaires, and rely on self-selected samples of scholars. The representativeness of the results is therefore difficult to determine. Some samples have clear biases by age as well as academic disciplines. Moreover, Web-based surveys, where recruitment and promotion of the survey take place almost entirely via electronic means, usually underrepresent those individuals who are more reluctant to use online media.

In contrast to such studies, Priem, Costello, and Dzuba (2011) attempted to draw a random sample of academic Twitter users. They selected five British and American universities, and searched Twitter for accounts of the universities' entire academic staff. Of the roughly 5,800 scholars thus identified (and who had sufficiently unique names), 230 could be matched to a Twitter profile, of which 145 were active. This results in a ratio of one in 40 scholars (or 2.5%) using Twitter—although it probably underestimates true activity, as only scholars who revealed their identity on their Twitter profile could be included. This alternative method still establishes Twitter participation as rare among scholars, but no differences across disciplines or levels of seniority could be observed. Interestingly, however, scholars who used Twitter did not necessarily do so in a professional context; 60% of the scholars' tweets did not pertain to job-related activities.

Although only used by a minority of scholars, Gerber (2012) found Twitter to be among the most well-known Web 2.0 applications; however, four out of five scholars had a decisively negative opinion of the microblogging service. Scholars seem to clearly distinguish between different kinds of digital media use. They embrace services that to them have clear advantages and/or can be easily integrated into the workflow (e.g., e-mail, academic search engines, and databases), but are more reluctant towards newer or more 'experimental' forms. Some scholars fear that using Twitter or other social media would prove to be a waste of their time, for which they would receive no professional recognition, and that this activity might even harm their professional reputations (Bader et al., 2012).

In spite of the vast majority of scholars' notable reluctance to integrate Twitter into their ecologies of academic communication, the service has become vital to the communication in some fields of research. It appears that Twitter is more popular in scholarly disciplines that are themselves related to the Web and computer-mediated communication (for example, the semantic Web research community; see Letierce, Passant, Breslin, & Decker, 2010). Here, but also in other domains, scholars engage in microblogging enthusiastically—and with considerable success. For instance, a poll on which academic Twitter users to follow that was conducted by the London School of Economics in August 2011[1] revealed over 500 popular tweeting scholars, some of whom have several thousands of followers.

USE CASE: CONFERENCES

A key use case of Twitter in scholarly environments is tweeting during conferences. Scholars are more likely to use Twitter at a conference than in everyday use (Ross, Terras, Warwick, & Welsh, 2011) and to take up tweeting when attending conferences (Reinhardt, Ebner, Beham, & Costa, 2009). Academic conferences are an ideal setting for using Twitter as a backchannel, i.e., as a separate communication channel from the main event's formal communication activities (McCarthy & boyd, 2005; Ross et al., 2011). Tweets are used to take notes or record thoughts; share information (also with non-attendees); engage in discussions before, during, and after attending an event (Reinhardt et al., 2009; Ross et al., 2011); and, more generally, pick up conference chatter (Letierce et al., 2010). But backchannels during live events may also have negative effects, including distraction and partial attention, disrespectful comments, and the formation of cliques (McCarthy & boyd, 2005). The unavailability of WiFi or power sockets, confusion about the correct hashtag, or multiple hashtags may further constrain Twitter use during conferences.

A number of large-scale analyses of conference tweets (usually collected based on conference hashtags) provide insights into communication behaviour, user networks, key users and key topics, activities over the course of time, and shared resources. They have revealed different communication patterns for different conferences: levels of user *participation*, for example, can be determined by comparing registered participants (if available) with the number of unique Twitter users in the data set. In the cases observed so far, the participation rate varied from only 1.4% (Desai et al., 2012) to a maximum of about twice as many Twitter users than registered participants (Ross et al., 2011). On such occurrences, Twitter apparently serves as a platform to widen the reach of a conference far beyond the actual participants on site.

Activity is a second criterion for comparing Twitter use at conferences. Table 30.1 shows activity measured as tweets per user for a list of different conferences, ranging from an average of 3 to almost 17 tweets per user. Looking at the distribution of tweets per user, one frequently finds that only a small number of users write the majority of tweets around a given conference, while others only tweet once (e.g., Ross et al., 2011). Following up on this, different studies have analysed the behaviour of such *key users* as well as *connectedness of users* based on retweets or @messages, for instance by computing hubs and authority scores (Letierce et al., 2010). Puschmann, Weller, and Dröge (2011) illustrated how connectedness can change over the course of an event. If one considers @messages mainly as a means for interaction, and retweets as a means for information distribution, some conference communities seem to be more focussed on talking to each other while others share information with a broader audience.

The *distribution of tweets over time* during a conference can lead to interesting insights about the most significant or resonant events within it. Letierce et al. (2010) identified spikes in activities and mapped them to single events: as might be expected, Twitter users were most active during the conference key-

Table 30.1: Comparison of Key Twitter Metrics for Different Conferences

Conference (Source)	No. of Tweets	No. of Users	Tweets per User	% of @ messages	% of Retweets	% of Tweets with URLs
#dcmi2009 (Dröge, Maghferat, Puschmann, Verbina, & Weller, 2011)	146	27	5.4	5.5%	25.3%	19.9%
#dh09 (Ross et al., 2011)	1,732	169	10.2	n.a.*	n.a.*	n.a.*
#drh09 (Ross et al., 2011)	274	23	11.9	n.a.*	n.a.*	n.a.*
#edmedia (Ebner & Reinhardt, 2009)	1,595	177	9.0	n.a.	n.a.	n.a.
#estc2009 (Letierce et al., 2010)	322	75	4.3	14.3%	15.2%	11.8%
#geoinst (Dröge et al., 2011)	1,673	99	16.9	24.3%	8.3%	14.8%
#iswc2009 (Letierce et al., 2010)	1,444	273	5.3	27.1%	20.2%	35.8%
#kidneywk11 (Desai et al., 2012)	993	172	5.8	n.a.	24.8%	42.9%
#mla09 (Dröge et al., 2011)	1,929	369	5.2	13.3%	21.4%	27.2%
#online09 (Letierce et al., 2010)	2,245	507	4.4	25.2%	18.8%	22.3%
#thatcamp 2009 (Ross et al., 2011)	2,568	187	13.7	n.a.*	n.a.*	n.a.*
#www2010 (Dröge et al., 2011)	3,358	903	3.7	7.5%	33.4%	39.9%

* Ross et al. (2011) only provided aggregated values for the three conferences in the data set: 66% @messages, 10% retweets, and 24% tweets with URLs.

notes, the awards, and closing sessions. However, related external events that happen independently from the actual conference can also produce peaks in activity if they affect the scholarly interests of that community. During the 2009 International Semantic Web Conference (#iswc2009), for instance, *The New York Times* released a data set for sharing which many of the attendees, mainly computer scientists, commented on via Twitter (Letierce et al., 2010). Analyses of timelines can also be used to study the *contents* of conference tweets. Stankovic, Rowe, and Laublet (2010) tried to automatically map tweet contents to particular sub-events of a conference (e.g., specific talks or sessions). They identified topics that were tweeted a lot without being present in talks in order to indicate trends. Inversely, the supposedly central topics of a conference may go unnoticed on Twitter (Desai et al., 2012).

The manual coding of tweets allows for deeper analyses of their content—for instance, with regard to topic or communicative purpose. Ross et al. (2011) manually categorised 43% of a sample of conference tweets as "jotting down notes". Other studies have distinguished between the use of "informative" (referring to conference topics) and "non-informative" messages (e.g., opinions or advertisements). The proportion of informative versus uninformative tweets can vary considerably from one event to another (Desai et al., 2012; Dröge et al., 2011), but given that this research field is still in development, it is unclear whether such differences are due to, for example, the nature of the event, traditions of a discipline, or communicative routines of the participants. As tweets often include little commentary, Ross et al. (2011) concluded that Twitter is used for establishing an online presence rather than for encouraging a participatory conference culture. The fact that many conference tweets contain URLs (Table 30.1) suggests that attendees like to use Twitter for the dissemination of additional information.

So far, studies on microblogging during conferences hint towards differences in the Twitter practices of different disciplines. Conferences related to (digital) humanities appear to have lower percentages of URLs and retweets compared to those from computer science, while the latter tend to have less @messages. However, based on the relatively small and hardly representative set of conferences examined so far, one can only speculate about the reasons for the apparent differences. In addition to diverging communicative traditions, conference size and format may equally influence tweeting styles; small and rather informal events like #geoinst and #thatcamp (both digital humanities; THATCamp had about 100, and the conference of the Institute for Enabling Geospatial Scholarship had less than 500 attendees) see higher numbers of tweets per user, for example. Lastly, microblogging practices may also develop over time, while Twitter adoption rates or familiarity with the medium, its potentials, and limitations change as well. Instead of studying an arbitrary set of conferences, it would be important for the advancement of this research field to examine subsequent events in a conference series for which a baseline has been established.

However, there are currently no explicit attempts to investigate the landscape of Twitter usage during scholarly conferences more comprehensively (and considering the slow uptake of Twitter in many research communities, possibly for good reasons). Instead, the focus of research on conference tweets has recently shifted to ways of automatically extracting additional information from tweets and detecting conference highlights for (non-)attendees.

USE CASE: URLS AS CITATIONS

Like other social media, Twitter has been discussed with regard to alternative measures for scholarly impact ("altmetrics"; Priem, Taraborelli, Groth, & Neylon, 2010). Retweeting or referencing content via URLs may, in fact, be seen as an act of citation, and social media-based citing behaviour positively affects traditional indicators of scholarly influence, i.e., download rates or citations in scholarly publications (Eysenbach, 2011; Priem, Piwowar, & Hemminger, 2012; Shuai, Pepe, & Bollen, 2012). As shown for conference tweets, tweeting URLs is frequent among scholars, and it seems more common than among other groups of users. Of a random sample of 720,000 tweets, 22% contained URLs (boyd, Golder, & Lotan, 2010), while this proportion was 55% in an eight-month sample of tweets collected from roughly 600 academic users (Weller & Puschmann, 2011). Typical scholarly practices performed via tweets are information, resource, and media sharing (Veletsianos, 2012), as well as recommending literature (Ebner & Schiefner, 2008), although neither study explicitly stated whether these practices necessitate the use of URLs.

Following a similar approach to citing as in traditional publications, Priem and Costello (2010) analysed 2,322 tweets from scholars and searched for Twitter citations, i.e., links to peer-reviewed articles, to determine the impact of traditional publications in social media. It was shown that 6% of tweets in the sample were Twitter citations, half of them linking directly to the referenced articles. More than half of the directly linked articles (56%) were open-access articles. Moreover, Twitter citations happened fast: 39% occurred within one week after publication of an article, 15% on the same day.

Peters, Beutelspacher, Maghferat, and Terliesner (2012) studied Twitter practices of scholarly bloggers (affiliated with universities or other research institutions). About one in three of a total of 50,019 tweets contained a URL. An analysis of the top-level domains of link destinations showed that tweeting science bloggers most often linked to their own blog posts or those of colleagues on the same blogging platform. More popular link destinations were online news outlets (1%), Twitter-centred services (e.g., twitpic.com, 2%), or other media channels (e.g., youtube.com, 2%; friendfeed.com, 2%). In an analysis of 3,631 conference tweets, Weller, Dröge, and Puschmann (2011) manually categorised URLs based on the link destinations (e.g., blog, media, slides, publication, etc.). The rank-frequency distribution of URLs was highly skewed, with more than half of URLs appearing only once in the data set. URLs tweeted at the 2009 conference of the Modern Language Association (#mla09; literary studies) mainly pointed to blog posts or press articles, while Twitter users at the World Wide Web conference (#www2010; computer science) frequently linked to scholarly publications and presentation slides.

The results of the presented studies indicate that scholars use Twitter to quickly distribute information on relevant, often open-access publications, and to facilitate their retrieval. They also promote their own work, not necessarily from traditional scholarly outlets, but also from social media. Furthermore, although URLs are frequently added to tweets, Twitter citations of scholarly publications in the stricter sense are rare, and are only performed by few people. Thus, using such indicators in altmetrics to measure the impact of publications in social media may lead to false impressions of a paper's or author's popularity. This may even be aggravated by the common practice of linking to own publications (i.e., self-citation; Weller & Peters, 2012), while the short life-span of shortened URLs puts the stability of such metrics into question (Weller et al., 2011).

While posting URLs is popular during conferences as well as in everyday scholarly communication, the respective studies again seem to indicate differences between disciplines with regard to preferred formats of the material linked to. Tweets sent around the selected literary studies conference referred to a diverse set of resources, while computer scientists at WWW 2010 focussed on original publications or slides. It is possible that scholars from the humanities refer to blogs more often because they see them as a space for explaining and discussing content and would like to invite others to join in the debate. However, the exploratory state of the literature only allows for tentative interpretations of the apparent differences between disciplines.

CONCLUDING REMARKS

Twitter is not uniformly used across all academic disciplines and fields. Although the service is widely known and actively researched, considerable parts of the academic community are reluctant to use Twitter at all or in relation to their work. As long as Twitter use remains rare in their academic environment, scholars are not likely to take up tweeting. Yet, for some fields and/or specific occasions like academic conferences, Twitter has become a part of the communicative ecologies of scholars. It facilitates exchange among existing networks of scholars, e.g., via sharing URLs, but also allows adding contacts in an informal and low-threshold manner, or simply helps people in getting into conference mood. In the end, scholars have to decide for themselves whether Twitter offers sufficient benefits to devote time to tweeting. However, the range of studies on Twitter use in scholarly communication documents that the platform can be integrated into scholarly practices in a multitude of ways.

Subsequently, using Twitter for the evaluation of scientific output is equally explored, and the number of alternative impact indicators based on social media has exploded in recent years. Twitter-specific altmetrics actually reveal patterns comparable to traditional evaluation metrics (Eysenbach, 2011), but tweeting behaviour is still too understudied to determine the validity of Twitter metrics. In future, indicators should be carefully scrutinised instead of using social media data on scholarly communication and referencing practices simply because they are available.

The studies reviewed in this chapter stem from a variety of scholarly fields, including humanities, computer science, and health science. Thus, the respective data may be based on different underlying methods for data collection and analysis. In addition, data sets are often compiled on a study-specific basis. Since data exchange between researchers is rare, Twitter data studies are difficult to reproduce and to compare, but replications of existing studies would be useful in confirming patterns of practices beyond single events. Given the often quantitative and 'big data' oriented rationales of research on academic Twitter use, qualitative and more interpretative approaches into the how and why of scholarly Twitter behaviour may be another fruitful direction for future research (Veletsianos, 2012).

NOTE

1   http://blogs.lse.ac.uk/impactofsocialsciences/2011/09/02/academic-tweeters-your-suggestions-in-full/

REFERENCES

Bader, A., Fritz, G., & Gloning, T. (2012). Digitale Wissenschaftskommunikation 2010–2011: Eine Online-Befragung [Digital scholarly communication 2010–2011: An online survey]. Retrieved from http://geb.uni-giessen.de/geb/volltexte/2012/8539/

boyd, d., Golder, S., & Lotan, G. (2010). Tweet, tweet, retweet: Conversational aspects of retweeting on Twitter. In *Proceedings of the 43rd Annual Hawaii International Conference on System Sciences,* Kauai, Hawaii (pp. 1–10). Washington, DC: IEEE Computer Society.

Bruns, A., & Burgess, J. (2012). Notes towards the scientific study of public communication on Twitter. In A. Tokar, M. Beurskens, S. Keuneke, M. Mahrt, I. Peters, C. Puschmann, . . .Weller, K. (Eds.), *Science and the Internet* (pp. 159–169). Düsseldorf, Germany: Düsseldorf University Press.

Desai, T., Shariff, A., Shariff, A., Kats, M., Fang, X., Christiano, C., & Ferris, M. (2012). Tweeting the meeting: An in-depth analysis of Twitter activity at Kidney Week 2011. *PLoS ONE* 7(7). Retrieved from http://www.plosone.org/article/info:doi/10.1371/journal.pone.0040253

Dröge, E., Maghferat, P., Puschmann, C., Verbina, J., & Weller, K. (2011). Konferenz-Tweets: Ein Ansatz zur Analyse der Twitter-Kommunikation bei wissenschaftlichen Konferenzen [Conference tweets: An approach for the analysis of Twitter communication at academic conferences]. In J. Griesbaum, T. Mandl, & C. Womser-Hacker (Eds.), *Information und Wissen: Global, sozial und frei? Proceedings des 12. Internationalen Symposiums für Informationswissenschaft* (pp. 98–110). Boizenburg, Germany: VWH.

Ebner, M., & Reinhardt, W. (2009). Social networking in scientific conferences: Twitter as tool for strengthen a scientific community. In *Proceedings of Workshop Science 2.0 for TEL, ECTEL 2009. 4th European Conference on Technology Enhanced Learning, EC-TEL 2009,* Nice, France. Berlin, Germany: Springer.

Ebner, M., & Schiefner, M. (2008). Microblogging—More than fun? In I. Arnedillo-Sánchez & P. Isaías (Eds.), *Proceedings of IADIS Mobile Learning Conference,* Algarve, Portugal (pp. 155–159).

Eysenbach, G. (2011). Can tweets predict citations? Metrics of social impact based on Twitter and correlation with traditional metrics of scientific impact. *Journal of Medical Internet Research,* 13(4). doi: 10.2196/jmir.2012

Gerber, A. (2012). Online trends from the first German trend study on science communication. In A. Tokar, M. Beurskens, S. Keuneke, M. Mahrt, I. Peters, C. Puschmann, . . .Weller, K. (Eds.), *Science and the Internet* (pp. 13–18). Düsseldorf, Germany: Düsseldorf University Press.

Gu, F., & Widén-Wulff, G. (2011). Scholarly communication and possible changes in the context of social media: A Finnish case study. *Electronic Library,* 29(6), 762–776.

Harley, D., Acord, S. K., Earl-Novell, S., Lawrence, S., & King, C. J. (2010). *Assessing the future landscape of scholarly communication: An exploration of faculty values and needs in seven disciplines.* Berkeley, CA: Center for Studies in Higher Education. Retrieved from http://escholarship.org/uc/item/15x7385g

Herwig, J., Kittenberger, A., Nentwich, M., & Schmirmund, J. (2009). *Microblogging und die Wissenschaft: Das Beispiel Twitter* [Microblogging and academia: The example of Twitter]. Retrieved from http://epub.oeaw.ac.at/ita/ita-projektberichte/d2-2a52-4.pdf

Kortelainen, T., & Katvala, M. (2012). 'Everything is plentiful—Except attention': Attention data of scientific journals on social web tools. *Journal of Informetrics,* 6(4), 661–668.

Letierce, J., Passant, A., Breslin, J., & Decker, S. (2010). Understanding how Twitter is used to spread scientific messages. In *Proceedings of the WebSci10: Extending the Frontiers of Society On-Line.* Raleigh, NC. Retrieved from http://journal.webscience.org/314/2/websci10_submission_79.pdf

McCarthy, J., & boyd, d. (2005). Digital backchannels in shared physical spaces: Experiences at an academic conference. In W. Kellogg, S. Zhai, G. van der Veer, & C. Gale (Eds.), *Extended abstracts of the Conference on Human Factors and Computing Systems (CHI 2005)*, Portland, OR (pp. 1641–1644). New York, NY: ACM.

Mollett, A., Moran, D., & Dunleavy, P. (2011). *Using Twitter in university research, teaching and impact activities: A guide for academics and researchers*. London, UK: London School of Economics. Retrieved from http://blogs.lse.ac.uk/impactofsocialsciences/files/2011/11/Published-Twitter_Guide_Sept_2011.pdf

Peters, I., Beutelspacher, L., Maghferat, P., & Terliesner, J. (2012). Scientific bloggers under the altmetric microscope. In *Proceedings of the 75th Annual Meeting of the American Society for Information Science and Technology*, Baltimore, MD. Retrieved from https://www.asis.org/asist2012/proceedings/Submissions/305.pdf

Ponte, D., & Simon, J. (2011). Scholarly communication 2.0: Exploring researchers' opinions on Web 2.0 for scientific knowledge creation, evaluation and dissemination. *Serials Review, 37*(3), 149–156.

Priem, J., & Costello, K. L. (2010). How and why scholars cite on Twitter. In C. Marshall, E. Toms, & A. Grove (Eds.), *Proceedings of the 73rd ASIS&T Annual Meeting on Navigating Streams in an Information Ecosystem*, Pittsburgh, PA (Article No. 75). New York, NY: ACM.

Priem, J., Costello, K. L., & Dzuba, T. (2011). *Prevalence and use of Twitter among scholars*. Poster presented at Metrics 201: Symposium on Informetric and Scientometric Research, New Orleans, LA. Retrieved from http://jasonpriem.org/self-archived/5uni-poster.png

Priem, J., Piwowar, H. A., & Hemminger, B. M. (2012). *Altmetrics in the wild: Using social media to explore scholarly impact*. Retrieved from http://arxiv.org/abs/1203.4745

Priem, J., Taraborelli, D., Groth, P., & Neylon, C. (2010). *Altmetrics: A manifesto*. Retrieved from http://altmetrics.org/manifesto

Puschmann, C., Weller, K., & Dröge, E. (2011). *Studying Twitter conversations as (dynamic) graphs: Visualization and structural comparison*. Poster presented at General Online Research '11, Düsseldorf, Germany. Retrieved from http://files.ynada.com/posters/gor11.pdf

Reinhardt, W., Ebner, M., Beham, G., & Costa, C. (2009). How people are using Twitter during conferences. In V. Hornung-Prähauser & M. Luckmann (Eds.), *Creativity and innovation competencies on the Web. Proceeding of 5. EduMedia Conference*, Salzburg, Austria (pp. 145–156).

Ross, C., Terras, M., Warwick, C., & Welsh, A. (2011). Enabled backchannel: Conference Twitter use by digital humanists. *Journal of Documentation, 67*(2), 214–237.

Sammer, T., & Back, A. (2011). Towards microblogging success factors: An empirical survey on Twitter usage of Austrian universities. *Proceedings of the 6th Mediterranean Conference on Information Systems*, Limmassol, Cyprus. Retrieved from http://aisel.aisnet.org/mcis2011/40/

Shuai, X., Pepe, A., & Bollen, J. (2012). *How the scientific community reacts to newly submitted preprints: Article downloads, Twitter mentions, and citations*. Retrieved from http://arxiv.org/abs/1202.2461

Stankovic, M., Rowe, M., & Laublet, P. (2010). Mapping tweets to conference talks: A goldmine for semantics. In *Proceedings of the Third Social Data on the Web Workshop SDoW2010*, Shanghai, China. Retrieved from http://sdow.semanticweb.org/2010/

Veletsianos, G. (2012). Higher education scholars' participation and practices on Twitter. *Journal of Computer Assisted Learning, 28*(4), 336–349.

Weller, K., Dröge, E., & Puschmann, C. (2011). Citation analysis in Twitter: Approaches for defining and measuring information flows within tweets during scientific conferences. In *Proceedings of Making Sense of Microposts Workshop* (#MSM2011), Crete, Greece. Retrieved from http://ceur-ws.org/Vol-718/paper_04.pdf

Weller, K., & Peters, I. (2012). Citations in Web 2.0. In A. Tokar, M. Beurskens, S. Keuneke, M. Mahrt, I. Peters, C. Puschmann, . . . Weller, K. (Eds.), *Science and the Internet* (pp. 211–224). Düsseldorf, Germany: Düsseldorf University Press.

Weller, K., & Puschmann, C. (2011). Twitter for scientific communication: How can citations/references be identified and measured? In *Proceedings of the Poster Session at the Web Science Conference 2011*, Coblence, Germany. Retrieved from http://journal.webscience.org/500/

How Useful Is Twitter for Learning in Massive Communities?

An Analysis of Two MOOCs

31

CHAPTER   Timo van Treeck and Martin Ebner

 microblogging can help make learning more engaging and interactive, especially in #moocs

The use of Web technology in education has constantly increased over the last few years. After the initial introduction of so-called learning management systems (Helic, Maurer, & Scerbakov, 2004), a considerable shift to more interactive technologies gradually occurred. Web 2.0—coined for the first time by O'Reilly (2010)—services such as weblogs, wikis, and podcasts have become more and more common in today's lectures in higher education (Augar, Raitman, & Zhou, 2004; Evans, 2007; Luca & McLoughlin, 2005). In the last three years, social media platforms such as Facebook, Twitter, and Google+ have attracted millions of users, including many students (Ebner, Nagler, & Schön, 2011). As a consequence of the world becoming more and more connected, the idea of opening online courses to anyone who is interested in them (referred to as Massive Open Online Courses, or MOOCs), had emerged. According to McAuley, Stewart, Siemens, and Cormier (2010), a MOOC "integrates the connectivity of social networking, the facilitation of an acknowledged expert in a field of study, and a collection of freely accessible online resources" (p. 4). More importantly, in this context, MOOCs are usually spread across the world through existing social networks, mostly Facebook and Twitter. Consequently, these platforms are not only used before, but also during and after a lecture session. Afterwards, educational data-mining methods, or, more precisely, learning analytics methods (Duval, 2010; Long & Siemens, 2011) can be used to analyse course data and attempt to make predictions regarding learning outcomes.

In this chapter, we concentrate on an analysis of Twitter usage surrounding a German-language MOOC that could indicate future trends in technology-enhanced learning. Our research focusses on the Twitter stream accompanying the course, and asks how Twitter is used and for what purposes by the heavy Twitter users; by the educators, organisers, guest speakers in the course; and if tweets from "outside" get into the stream.

USE OF MICROBLOGGING IN EDUCATION

Microblogging platforms are part of an increasing number of social software tools that feature opportunities for information management; interaction; and communication, identity, and network management (Ebner & Lorenz, 2012; Koch & Richter, 2008). Twitter is the most frequently used and well-known microblogging platform worldwide.

Due to Twitter's large number of users and its interactive nature (Ebner & Schiefner, 2008; McFedries, 2007), different ideas, concepts, and educational approaches for Twitter use in the classroom have appeared. Ebner (2013) pointed out six different uses of Twitter in education:

- Enhancing interaction in mass education through the use of Twitter walls.
- Discussion beyond face-to-face lectures by using a specific Twitter hashtag.
- Exchanging lecture content by collecting Internet resources using a defined Twitter hashtag.
- Documentation and information retrieval, with the help of specific Web applications that collect tweets automatically.
- Enhancing academic conferences by using Twitter as an online backchannel.
- Connecting with researchers, teachers, and learners with similar interests based on Twitter's recommendations (see also Chapter 30 by Mahrt, Weller, & Peters, in this volume).

These uses can be combined with different methods of designing teaching and learning. For example:

- Using Twitter as a communication channel to support different phases of think–pair–share (Barkley, Kross, & Howell Major, 2004): Students work on a question alone, in pairs, and then in the plenum of the lecture. By doing so, each pair-group can share their discussion results via Twitter; some do so by articulating their results during the lesson.
- Use a moderator instead of a Twitter wall: At each session, a student chooses aspects or questions from the Twitter stream that he or she thinks to be of interest to the audience of the lecture, and therefore supports the offline discussion.
- Use a lead learner in sessions: Learners can be assigned one-time roles as experts on the topic of the session; they will try to find additional resources for the lecture on the Web and send them via Twitter.
- Use Twitter in the way a fishbowl session is constructed (Barkley et al., 2004): Experts can get into the inner circle of the discussion by bringing forth arguments on Twitter; the inner circle of the discussion is held within the course itself.

In the case of MOOCs, the use of Twitter follows the principles of hashtag usage. A hashtag defined in advance by the organiser of a class has to be used within each tweet relating to the course. A simple search can possibly help organisers, learners, and other participants to follow up on the communication and information stream, get in touch with others, exchange information, or simply discuss topics concerning the course. A very important detail is the use of retweets. boyd, Golder, and Lotan (2010) mentioned the social relevance of just resending or copying a tweet to followers beyond spreading information. On the other hand, Ebner et al. (2010) pointed out that a massive use of retweeting and copying can make it hard to follow the stream on a certain topic, e.g. when dealing with a conference or a topic in a MOOC which might lead to decreased attention from readers.

As the use of Twitter is in the centre of the activities in the special teaching concepts of a MOOC, the following study can help to understand questions regarding teaching and learning by analysing the tweets of the MOOC.

DESCRIPTION OF THE STUDY

For the current study, two German-language MOOCs were selected that had a special focus on e-learning. One of the online classes, which was conducted by studiumdigitale in cooperation with Jochen Robes (weiterbildungsblog), the Gesellschaft für Medien in der Wissenschaft (GMW) and the Zentrum für

Lehrerbildung und Schul- und Unterrichtsforschung (ZLF) in 2011, covered the future of learning and ran over the course of 11 weeks.[1] Every week, an expert gave an input talk, and participants discussed its topic via Twitter or by writing individual blog posts about their experiences and opinions. In 2012, the same team (together with the association eteaching.org, Institut für Wissensmedien and MMKH) organised a second course. This time, the course followed the outcomes of the Horizon report (Johnson, Adams, & Cummins, 2012) and its predicted future trends in technology-enhanced learning. The course topic was trends in e-teaching, and lasted 8 weeks.[2] Every one or two weeks in this course, an expert gave a short introduction intended to help participants in further discussions. Both courses were named OpenCourse (OPCO), followed by the year when they were conducted (#OPCO11, #OPCO12).

In our analysis, we collected all tweets over the observation time (from one week before the respective courses started until one week after the courses ended) with twitterSTAT. This tool, programmed at the Graz University of Technology, Austria, is able to archive tweets containing a predefined hashtag in a database. Afterwards, an automated structural analysis can be performed. For example, an analysis of the number of different users of the stream, as well as the number of tweets per user, can be carried out. Furthermore, each word of each tweet is extracted, collected, and summarised. A visualisation of the outcomes can be generated to provide a quick overview of the archive, and the results can be further analysed with the help of other semantic profiling tools (De Vocht, Selver, Ebner, & Mühlburger, 2011; Softic, 2012; Thonhauser, Softic, & Ebner, 2012).

In addition to the quantitative analysis, the tweets containing #OPCO12 were analysed in detail with a qualitative approach using the SOLO taxonomy (Biggs & Collis, 1982), which has been used before to study learning-related discussions in forums. The SOLO taxonomy has five dimensions that structure the relationships shown in communication acts: pre-structural, unistructural, multistructural, relational, and extended abstract. This gives an insight regarding the complexity of the communication: Do the tweets contain information which refers to no other information (pre-structural)? Are the users building simple connections to another concept or another tweet (unistructural)? Are they taking more than one reference into account (multistructural)? Do they evaluate the relation, or do they try to think further (extended abstract)? In addition, the tweets are categorised according to the actions that were introduced by the organisers to the participants in the MOOC. These actions were explained on the MOOC home page: aggregate, remix, repurpose, and feed forward. Further categories were inductively extracted from the material: questions, answers, retweets, and retweets with comments. Additionally, we interpreted whether the content of the tweets had some affective or evaluating aspects. Furthermore, the tweets were classified with regard to whether they related to general aspects regarding the format of MOOCs; the organisation of the course; the different topics of the OPCO; hints for interesting tools or techniques (for the role of new tools in building a learning environment in a MOOC, see van Treeck, 2012); or some kind of self-marketing for products, papers, etc.

The first 1,000 tweets of #OPCO12 were categorised in this manner, in order to determine the communication strategies used and the documented learning through microblogging in this course format.

RESULTS

The results of the analysis can be differentiated into two categories: general statistical analysis for both MOOCs, and a language-based analysis of the tweets for OPCO12.

GENERAL STATISTICAL ANALYSIS

Table 31.1 gives an overview of both courses. In the year 2012, the number of total tweets was nearly halved, despite having almost the same number of users. Therefore, the quantity of average tweets per user decreased from 10 to 7. What is remarkable is the stable percentage of retweets (about 30%), as well as the number of tweets from the top 10 (about 33%) and top 20 users (about 50%). In other words, half of all tweets were sent by about 6% of the participants. Finally, it can be pointed out that about 30% of the users participated in both courses.

LANGUAGE-BASED ANALYSIS

The first 1,000 tweets with the hashtag #OPCO12 then were analysed on the language basis with regard to questions such as: What topics can be found? Did the tweets deal with information only, or also have an affective aspect? Can indicators for interactions be found, like formulating questions and answers?

Within the first 1,000 tweets of OPCO12, a structure of topics can be analysed, giving insight into the question of for which aspects there had been the most tweets. The top ten topics get high scores on tweet numbers, because they were mostly repeated (not only retweeted), without any significant change in the

Table 31.1: General Statistics of OPCO 2011 and OPCO 2012

	OPCO 2011	OPCO 2012
Total number of tweets within the observation time	4,085	2,431
Total number of Twitter users	393	367
Tweets per user (average; rounded)	10	7
Total number of retweets	1,181 (29%)	734 (30%)
Retweets per user (average)	3	2
Top 10 most active users' share of all tweets	1,428 (35%)	810 (33%)
Top 20 most active users' share of all tweets	2,132 (52%)	1,139 (47%)
Three most frequently mentioned keywords	live, lernen, learning	learning, mobile, online
Three most widely used additional hashtags	#schulmeister, #edublogs, #surfingkant	#mooc, #elearning, #av
Twitter users in both courses	111	

amount of content dealing with the organisation of the course. This includes videos, which the organisers developed for the course; general information about the start of #OPCO12; the starting time of an online event on the topic of tablet computing; links to recordings of online events; and information that a user has viewed them (and posted the link to the recording). Another topic of the mostly unaltered, repeated tweets was live events from other contexts which started shortly after the online events of #OPCO12 and were related to the main topics of the courses.

Probably it is very easy to find the motivations for this behaviour. The participants might have felt the need to help other users have a good start in the course and, therefore, shared the most relevant tweets in the beginning: When does one have to be where online? What are the basic rules of the course? What is special in the course (e.g., is there a possibility to repeat missed inputs because they were recorded)?

Only two categories of the most repeated tweets directly related to the topics of #OPCO12, but these were unusual, in that they referred to assignments that had to be performed by the participants, such as reading certain texts before a session, or in that they promoted apps developed by a university. The tweets that were aimed at promoting learner activities, then, often were done with different approaches to the activities: informing about the need to read (texts before live events), reminding of the activity and explaining the importance of the activity to the learning experience in a humorous way. This kind of support for learning seemed to be honoured by retweeting a lot.

After taking a closer look at the 10 most repeated tweets of the first 1,000 tweets of #OPCO12, we found that 83 (68%) of the retweets (including the original tweet) were related to course organisation, while 15% involved topics from the course. These tweets were mostly initiated by the organisers or by guest speakers in the course (7 out of 10).

When looking at the timeline of #OPCO12, it seems that the number of tweets on organisational aspects decreased and became less important over time, compared to the tweets about the topics of the course. This was the case for the first 1,000 tweets that were analysed. In the starting week there are—as was expected—very many tweets about the organisation of the MOOC. They decrease very much already in the second week. To have a glimpse at the long tail of the tweets, we analysed tweets without the top 50 Twitter users. The relation between tweets with organisational aspects and tweets dealing with the topics of the course is more or less constant, when the top 50 Twitter users are not included. This changes only in the starting week, where the top 50 Twitter users, for the only time, sent more tweets about organisational aspects (183) than about topics (60) of the course. In the same period, the other users very much reduce the organisational tweets in relation to the overall tweets of this week. They made only 2% of tweets, but 27% in the week before and 25% in the week thereafter (Table 31.2).

Nevertheless, the tweets that dealt with the topics of the course only made up 39% of the analysed 1,000 tweets of the course. Some of the tweets seemed to have no connection with the course topics, and just focussed on mutually interesting aspects for people enrolled in a course about the future of learning or trends in learning. These tweets constituted about 18% of the first 1,000 tweets of the course.

On the other hand, about 6% of the tweets were coded as questions and answers, which represents direct interaction. These interactions were made by 51 users, of whom 23 are among the 50 most active Twitter users of the course. Questions and answers pertained to topics of the course (47%), course organisation (31%), suggestions for tools to use (12%), and the course format of MOOCs (2%). Looking at the interactions that cover the topic of the course (which are about 6% of the tweets), around 40% of these questions and answers were made by individuals who are not among the 50 most active Twitter users of #OPCO12.

The tweets were also analysed for affective aspects, such as expressing pleasure on finding or doing something, or expressing a judgement on materials

Table 31.2: Topics of Tweets from OPCO12 Overall and without Top 50 Twitter Users, Percentage of Tweets of Each Week by the Respective Users

	All Twitter Users		Without Top 50	
	Organisation of the Course	Topics of the Course	Organisation of the Course	Topics of the Course
Pre-course week	8/41 (20%)	17/41 (41%)	3/11 (27%)	7/11 (64%)
Starting week	186/378 (49%)	91/378 (24%)	3/160 (2%)	31/160 (19%)
Second week	48/210 (23%)	104/210 (50%)	19/75 (25%)	41/75 (55%)

or positions. At least 156 out of 1,000 tweets (16%) were put into this category, but they were not retweeted or answered more often than other tweets. Some of these affective aspects where mentioned with regard to enjoying taking part in the course or the live event. Users called for other users to join in or asked who else took part. Probably, in this way, users tried to support the feeling of being a course group, and to enhance social awareness and/or social integration as a motivating aspect of class participation.

One major aspect of MOOCs can be that they foster (social) serendipity (Buchem, 2011; van Treeck, 2012), which means that they are open to unexpected irritations, information, and discussions (e.g., from outside), because the whole conversation can be accessed by anybody interested or following the Twitter stream of one of the participants. To find out whether Twitter users from outside the class considered the course activity to be interesting, we counted users that sent only one tweet—a retweet that included #OPCO11 or #OPCO12 (Table 31.3).

Table 31.3: Social Serendipity: Tweets Likely to Come from Users Not Participating in the Course

	OPCO 2011	OPCO 2012
Retweets by users with only a single tweet under the hashtag	3%	4%

The experiment to analyse the tweets using the SOLO taxonomy did not generate any results, as it was not possible to clearly match the tweets and the interactions of the Twitter users to the taxonomy. A conclusion might be drawn, therefore, that Twitter communication, even in course format, is very different from forum communication where SOLO taxonomy could be used.

To summarise, the stream of the OPCO12 course was structured into broader categories (Table 31.4). It shows that a major part of the tweets (70%) is directly related to the course, nearly half of these tweets relating to topics of the course (39%) and to course organisation (31%). But there is also a large number of tweets that could not be interpreted as connected to the course, as they simply addressed interests that users participating in this kind of course might have.

Table 31.4: Parts of #OPCO12 Stream

Related to the topic of the course	39%
Related to course organisation	31%
No visible connection to course topic	17%
Related to MOOCs in general	4%
Related to software tools/platforms	4%
Self-marketing	4%
Total	99%

DISCUSSION

Having analysed the two MOOC Twitter streams (#OPCO11 and #OPCO12), the following findings seem noteworthy:

- The number of retweets was surprisingly high and also stable in both courses. Similar to Weller, Dröge, and Puschmann (2011), this micro-blogging characteristic (because there is no *like* or *share* button as in other social media platforms, e.g., Facebook) constituted a considerable amount of the entire Twitter stream. About one third of the stream was just repetition for any participant who followed the stream permanently. On the other hand, this might be helpful to attract more users, or to allow casual participation. Nevertheless, expressing agreement and sharing information are essential parts of social media.
- According to Table 31.1, only a small number of users are responsible for a large part of the tweets. Six per cent of users posted half of all the messages. There is a long-tail effect, also described by Brown and Adler (2008), in which only a few participants engage in online activities

in an interactive way. On the other hand, it must be noted that many others seem to be reached by these tweets, regardless of whether they are heavy Twitter users or not. Brown and Adler (2008) also pointed out that, similar to a big business platform like Amazon, the long-tail effect does not only mean that few users are sending out most of the tweets; it also means that learning 2.0 is attracting many people with only one or two tweets in the long tail.

- As Table 31.1 illustrates, about one third of the participants were active in both courses. Because both courses were on different topics, there seems to be interest for a lot of people. Therefore, it can be stated that the concept of the MOOC is a promising one, at least for a special target group that did take advantage of the first MOOC and therefore joined the second.

- At the beginning of the course, many tweets concerned organisational aspects of the MOOC (starting time, links to a live event, announcements of participation, etc.); later, these tweets decreased in number, and more tweets about topics from the MOOC emerged (Table 32.2). As in other course settings, the participants needed some time to organise themselves, and took responsibility for this by themselves. This is indicated by the fact that the same relation between organisational tweets and topic tweets can be found in both the top 50 Twitter users and the less active users—after the first course week.

- Tweets from organisers or experts that asked for activities from the participants prompted many retweets and comments from other users. It seems that the users appreciated being active, or at least wanted to support calls for activities by retweeting them.

- Questions and answers in the tweets were mainly sent by the top Twitter users, but 40% of them also came from less active Twitter users. So it is not only interesting to have a look at the Twitter users who sent a lot of tweets, but also at the long tail of users only sending a handful or less.

- The use of Twitter allows contact with unknown people—even people who were not enrolled in the course. A small percentage (3% and 4% in OPCO 2011 and OPCO 2012, respectively; see Table 31.3) of all tweets were retweets of users who did not send other tweets. Social serendipity, through meeting unexpected people, is thus possible.

CONCLUSION

In this study, the use of microblogging in education was explored, and a closer look was taken at the Twitter stream of two massive open online courses (MOOCs). Our cursory analysis shows that further research is needed to better understand how social media can be integrated into learning and teaching. It is the interpretation of the concrete messages that leads further; statistical analysis can only help to explore relations. A categorisation of the tweets revealed that although there were many similar or even identical tweets in the Twitter streams of OPCO 2011 and OPCO 2012, there was also a large number of tweets in which the topics of the course were addressed, and even with questions and answers discussing them. Activity calls prompted by some tweets resulted in heavy traffic, which also shows that interaction was appreciated in this course format. And, although only 6% of the users posted half of the messages, the content of the organisational tweets and tweets on course topics were the same as those from the other users. Therefore, it can be concluded that microblogging can play a relevant role in educational contexts, especially in open online courses, by making learning more interactive and engaging, but the potentials of microblogging in other teaching formats (such as regular lecture classes or field projects, etc.) have yet to be explored.

NOTES

1 http://blog.studiumdigitale.uni-frankfurt.de/opco11/

2 http://opco12.de/willkommen-zum-opencourse-trends-im-e-teaching/

REFERENCES

Augar, N., Raitman, R., & Zhou, W. (2004). Teaching and learning online with wikis. In R. Atkinson, C. McBeath, D. Jonas-Dwyer, & R. Phillips (Eds.), *Beyond the comfort zone: Proceedings of the 21st ASCILITE Conference* (pp. 95–104), Perth, Australia.

Barkley, E. F., Kross, K. P., & Howell Major, C. (2004). *Collaborative learning techniques: A handbook for college faculty*. San Francisco, CA: Jossey-Bass.

Biggs, J. B., & Collis, K. F. (1982). *Evaluating the quality of learning: The SOLO-Taxonomy (Structure of the observed learning outcome)*. New York, NY: Academic Press.

boyd, d., Golder, S., & Lotan, G. (2010). Tweet, tweet, retweet: Conversational aspects of retweeting on Twitter. In *Proceedings of the 43rd Annual Hawaii International Conference on System Sciences, Kauai, Hawaii* (pp. 1–10). Washington, DC: IEEE Computer Society.

Brown, J. S., & Adler, R. P. (2008). Open education, the long tail and learning 2.0. *EduCause Review, 43*(1), 16–33.

Buchem, I. (2011). Serendipitous learning: Recognizing and fostering the potential of micro-blogging. *Form@re, 74* . Retrieved from http://formare.erickson.it/wordpress/it/2011/serendipitous-learning-recognizing-and-fostering-the-potential-of-microblogging/

De Vocht, L., Selver, S., Ebner, M., & Mühlburger, H. (2011). Semantically driven social data aggregation interfaces for research 2.0. In *Proceedings of the 11th International Conference on Knowledge Management and Knowledge Technologies* (Article 43; pp. 43:1–43:10), Graz, Austria.

Duval, E. (2010). Attention please! Learning analytics for visualization and recommendation. In *Proceedings of LAK11: 1st International Conference on Learning Analytics and Knowledge 2011.* Retrieved from https://lirias.kuleuven.be/bitstream/123456789/315113/1/la2.pdf

Ebner, M. (2013). The influence of Twitter on the academic environment. In B. Patrut, M. Patrut, & C. Cmeciu (Eds.), *Social media and the new academic environment: Pedagogical challenges* (pp. 293–307), Hershey, PA: IGI-Global.

Ebner, M., & Lorenz, A. (2012). Web 2.0 als Basistechnologien für CSCL-Umgebungen [Web 2.0 as basic technologies for CSCL environments]. In J. Haake, G. Schwabe, & M. Wessner (Eds.), *CSCL-Lernumgebungen* (pp. 97–111). Munich, Germany: Oldenbourg.

Ebner, M., Mühlburger, H., Schaffert, S., Schiefner, M., Reinhardt, W., & Wheeler, S. (2010). Get granular on Twitter—Tweets from a conference and their limited usefulness for non-participants. In N. Reynolds & M. Turcsányi-Szabó (Eds.), *Key competencies in the knowledge society* (pp. 102–113). Berlin, Germany: Springer.

Ebner, M., Nagler, W., & Schön, M. (2011). The Facebook generation—Boon or bane for e-learning at universities? In *Proceedings of the World Conference on Educational Multimedia, Hypermedia and Telecommunications* (pp. 3549–3557), Lisbon, Portugal.

Ebner, M, & Schiefner, M. (2008). Microblogging—More than fun? In I. Arnedillo Sánchez & P. Isaías (Eds.), *Proceedings of IADIS Mobile Learning Conference 2008* (pp. 155–159), Algarve, Portugal

Evans, C. (2007). The effectiveness of m-learning in the form of podcast revision lectures in higher education. *Computers & Education, 50*(2), 491–498.

Helic, D., Maurer, H., & Scerbakov, N. (2004). Knowledge Transfer processes in a modern WBT system. *Journal of Network and Computer Applications, 27*(3), 163–190.

Johnson, L., Adams, S., & Cummins, M. (2012). *The NMC Horizon report: 2012 higher education edition.* Austin, TX: New Media Consortium.

Koch, M., & Richter, A. (2008). *Enterprise 2.0: Planung, Einführung und erfolgreicher Einsatz von Social Software in Unternehmen* [Enterprise 2.0: Planning, introducing and successfully applying social software in enterprises]. Munich, Germany: Oldenbourg.

Long, P., & Siemens, G. (2011). Penetrating the fog: Analytics in learning and education. *EDUCAUSE Review Magazine, 46*(5), 31–40.

Luca, J., & McLoughlin, C. (2005). Can blogs promote fair and equitable teamwork? In *Proceedings of ASCILITE 2005: Balance, Fidelity, Mobility: Maintaining the momentum?* (pp. 379–385).

McAuley, A., Stewart, B., Siemens, G., & Cormier, D. (2010). *The MOOC model for digital practice* (Research report). University of Prince Edward Island. Retrieved from www.edukwest. com/wp-content/uploads/2011/07/MOOC_Final.pdf

McFedries, P. (2007). All a-Twitter. *IEEE Spectrum.* Retrieved from http://spectrum.ieee.org/ computing/software/all-atwitter

O'Reilly, T. (2010). What is Web 2.0? —Design patterns and business models for the next generation software. Retrieved from http://oreilly.com/Web2/archive/what-is-Web-20.html

Softic, S. (2012). *Towards identifying collaborative learning groups using social media.* Paper presented at the 15th International Conference on Interactive Collaborative Learning (ICL 2012), Villach, Austria.

Thonhauser, P., Softic, S., & Ebner, M. (2012). Thought bubbles: A conceptual prototype for a Twitter based recommender system for research 2.0. In *Proceedings of the 12th International Conference on Knowledge Management and Knowledge Technologies* (i-KNOW '12) (Article 32). New York, NY.

van Treeck, T. (2012). Do it yourself—Lernende gestalten ihre Online-Lernumgebung. [Do it yourself—Learners design their online learning environment]. In M. Ockenfeld, I. Peters, & K. Weller (Eds.), *Proceedings of the 2. DGI-conference* (pp. 449–452). Frankfurt am Main, Germany: DGI.

Weller, K., Dröge, E., & Puschmann, C. (2011). Citation analysis in Twitter: Approaches for defining and measuring information flows within tweets during scientific conferences. In M. Rowe, M. Stankovic, A. S. Dadzie, & M. Hardey (Eds.), *Making Sense of Microposts (#MSM2011), Workshop at the Extended Semantic Web Conference (ESWC 2011)* (pp. 1–12), Heraklion, Greece. CEUR Workshop Proceedings, Vol. 718.

Epilogue
Why Study Twitter?

EPILOGUE   Cornelius Puschmann, Axel Bruns,
Merja Mahrt, Katrin Weller, and Jean Burgess

 society in 140 characters: why Twitter research is
necessary and important #ftw

Each of the thirty-one contributions in this volume implicitly spells out its own answer to this question. Surprisingly perhaps even for such a highly interdisciplinary volume as this one, these answers vary considerably in their approaches, their objectives, and their underlying assumptions about the object of study. This diversity of scholarly perspectives on Twitter, barely half a decade since it first emerged as a popular platform, highlights its versatility. Beginning as a side project to a now-forgotten podcasting platform, rising to popularity as a social network service focussed around mundane communication and therefore widely lambasted as a cesspool of vanity and triviality by incredulous journalists (including technology journalists), it was later embraced by those same journalists, governments, and businesses as a crucial source of real-time information on everything from natural disasters to celebrity gossip, and from debates over sexual violence to Vatican politics.

Studies of Twitter not only use many approaches (from computational modelling to critical inquiry), they also analyse a very wide range of phenomena (from fandom to disaster preparedness), and follow many different, implicit assumptions about Twitter's core purpose. Is Twitter a site of public debate? A tool for journalism, activism, education, and public relations? A data source for scientists, pollsters, and marketers? The mass of data generated each day by a user base exceeding 500 million accounts around the globe alone makes it both fascinating and impossible to describe holistically. What Twitter is, to celebrities, activists, pundits, marketers, and private individuals, is futile to answer without overlooking, as danah boyd (2006) has put it with regard to blogging, "the efficacy of the practice". The practice of using Twitter signifies something to those who engage in it that is difficult to describe only in terms of the data that is produced. Twitter is a platform, a piece of infrastructure comparable to the Internet itself, and it does a wide range of things for a diverse network of user communities.

Why should we study Twitter? An obvious answer would be that it is a global phenomenon, growing in users and posts every day. Another is that it is increasingly entrenched in our media ecology, an instrument that few politicians, journalists, or marketers would want to miss. And yet another is that through Twitter, researchers gain access to huge volumes of data, a treasure trove of digital traces, waiting to be mined for precious insights into people's behaviours, their moods, their consumption patterns, their language, and their voting behaviours. All of these are excellent reasons, yet there may be even more important grounds on which social scientists should study Twitter, reasons that point to how social media platforms increasingly influence certain aspects of our lives, as we can increasingly access them whenever and wherever we want, and millions of individuals around the globe use them.

Twitter's embeddedness in everyday social and communicative interactions across so many nations of the developed world, and its role as a very public, global, real-time communications channel highlight the fact that it—alongside other major social media, like Facebook or YouTube—provides a window on contemporary society as such, at national and global levels. We named this collection *Twitter and Society* for that reason: because the interrelations between Twitter and society which the chapters in this volume explore and explain make this book not just a collection of articles in an emerging field of 'Twitter Studies', but one which is able to develop our understanding of social and societal trends at the present moment by bringing together work that happens to draw on Twitter as its primary locus of observation.

In doing so, we seek to connect with the broader stream of Internet research, which is concerned not simply with an exhaustive investigation of the next shiny new tool or technology, as valuable as such efforts may be in their own right (sometimes they are described somewhat derisively as 'toaster studies' by the Internet research community), but which seeks to discover the deeper patterns of user activity that tell us much more about users' interests, motivations, and attitudes, and that generate insights which exist independent of the specific communications platforms that may be popular at any one point. From mailing lists and newsgroups to Facebook and Twitter, for example, the use of computer-mediated communication platforms for social networking and community interaction now looks back on a 50-year history, and while its particular historical formations also shine a light on the politics of the specific platforms in each period, perhaps the more fascinating observation to be made from this history is that of humanity's relentless drive to communicate, to gather, exchange, and organise knowledge, and to develop the community structures that enable and sustain such processes. At its best, Internet research is able to reflect back to us, by studying these online processes, just who we are and how we work as a society or a range of societies, and how we operate differently in contexts ranging from everyday social life to high-stakes politics, from acute crisis events to televised mass entertainment, from activism to marketing.

In developing this collection, therefore, we have come back time and again to Richard Rogers's (2009) dictum that we ought to redefine our ambitions as scholars, from studying the Internet to "studying culture and society *with the Internet*" (p. 29). We hope that *Twitter and Society* has succeeded not just at presenting a collection of work on Twitter as such, but also in tracing Twitter's emerging role *in* society, documenting its growing impact *on* society, and exploring to what extent it is possible to use the study of Twitter as a lens through which we may observe contemporary society. By their nature, lenses amplify, skew, and distort what they depict, and we must not make the mistake of taking such observations simply at face value; Twitter is no more perfect a representation of contemporary societal structures and trends than newspapers, television, or any other popular medium is able to be. But studying society *with* Twitter can highlight different aspects of contemporary life from doing so through the lens of other media and communications tools, and it is the aggregate and productively contradictory picture which emerges from a combination of all of these observations which is ultimately of the greatest value. *Twitter and Society* seeks to make a contribution to that bigger picture.

Lenses can have blind spots, and undoubtedly this collection pays greater attention to some areas of Twitter research than it does to others. It leans slightly towards the humanities side of the humanities and social sciences continuum; and while international, it is far from complete in its inclusiveness—indeed, with its focus on the Twitter platform, it inherently excludes other microblogging services like Sina Weibo, which is hugely popular in China. Aside from those issues, the very flexibility of Twitter as a platform for public, interpersonal, and private communication means that its potential uses are vast and diverse, and that researchers are presented with an endless array of possible case studies and areas of investigation. Future editions of this and other books on Twitter research will face an even more difficult challenge of choosing and organising their chapters. This is a challenge to look forward to.

Significantly, the further development of Twitter studies also depends on the further development of Twitter itself, of course—and while several of the contributions to the present volume touch on the platform politics of Twitter (and Twitter, Inc.'s role in them), much more could and should be said about them still. As with any proprietary platform, Twitter usage practices by individual and corporate users, as well as Twitter research approaches, exist in a precarious state that is bound up with the technological choices, commercial fortunes, and internal politics of the company which operates the platform; we are all no more than guests here, with a limited ability to bend or ignore the rules which govern this space.

SOCIETY IN 140 CHARACTERS

At the very heart of Twitter's success there has always been a simple technical limitation that may initially look more like a bug than a feature: the restriction of messages to 140 characters. Originally, there were very simple reasons for this restriction, as Richard Rogers explains in the foreword to this collection: Twitter developers sought to ensure backwards compatibility with the 160-character limit of short messages from mobile phones. But its effects highlight the far-reaching consequences of design choices for a sociotechnical system that connects people through nothing but a set of very basic, software-encoded, communicative rules—what could be called Twitter's underlying sociotechnical grammar. Twitter opened up a world of impossible discourses through the restriction to 140 characters: discourses that could never have come to pass had the creators of the service not chosen to constrain users' ability to compose messages in this way.

The result was a medium whose style is closer to oral than written communication, and closer to synchronous messaging than asynchronous discussion threads. Much of the criticism of Twitter—for example, the widely cited Pear Analytics (2009) study categorising 40% of what is posted there as "pointless babble"—highlights how much of a break with then-current design trends for social media this represented. Such clichéd accusations of irrelevance show how uncomfortable those accustomed to platforms that privilege long, complex texts were and may still be with this concept, yet this focus on short, quick messaging is also precisely what makes Twitter so useful in particular contexts. This approach has scale built into it: it represents an ingenious solution to the problem of having to divide a limited amount of attention span across a very large number of communicators.

Such limitations to our ability to use the full range of available communicative tools often turn out to be highly productive of innovative new solutions: much as early e-mail users responded to the lack of visual cues, introduced by the text-only format, by inventing a canon of now ubiquitous emoticons and other paratextual markers, so do user-generated Twitter features such as @messaging and retweeting point to the gradual development of new communicative conventions, as users negotiated this new space for communal expression and interaction. These conventions, these user-initiated innovations, were necessary to order and organise discourse in specific, intelligible, and predictable ways, and only they have made Twitter what it is today.

It is far from surprising, therefore, that attempts to tinker with this winning, if restrictive, formula have met with little success in the past. Twitter client TweetDeck's 'long tweet' functionality—which enabled users to post longer messages whose first words would be tweeted alongside a *deck.ly* link to the rest of the message—was widely criticised as breaking up the fast and easy interactivity of Twitter conversations, and has quietly disappeared from view. Twitter, Inc.'s own attempt to streamline retweeting, by offering a one-click 'retweet button' that posted an existing tweet verbatim to one's own feed, failed to account for the fact that many users wanted to engage, by adding their own comments, with the messages they sought to retweet; continued use of such 'manual' retweeting by a large section of the Twitter user base has meant that even most of Twitter's own user interfaces now once again offer a choice between 'button' (verbatim) and 'manual' (editable) retweeting functionality.

This is not to claim that none of the interventions by Twitter, Inc. or third-party client providers stand a chance of being widely adopted by the Twitter user base, however. The sprawling ecosystem of URL shorteners—from market leaders such as *bit.ly* to bespoke services which offer a personalised short URL that incorporates the user's nickname or initials—demonstrates that tools and functionality which are widely seen as enhancing the Twitter experience are able to find a large audience, even in spite of the fact that Twitter, Inc.'s introduction of its own mandatory URL shortener, *t.co*, now makes the use of such additional shortening services unnecessary in principle. Similarly, there are many tools for sharing images, audio, and video materials which enjoy substantial popularity on Twitter—to such an extent that Twitter, Inc. has increasingly seen the need to offer its own, built-in functionality in order to retain user traffic on its own site, rather than lose it to third-party providers.

As this book goes to print, in fact, Twitter has just introduced its Vine service for sharing short videos. Vine introduces artificial constraints similar to Twitter's original 140-character limit by capping the length of its videos at a maximum of six seconds, thereby also carving out a different market niche from mainstream video-sharing sites such as YouTube; it remains to be seen whether this limitation will turn out to be similarly productive of innovative uses as the 140-character limit has proved for Twitter's text messages, or whether the considerably greater effort which must go into shooting, editing, and uploading videos means that sharing Vine videos remains a niche pursuit for a small section of the Twitter population only.

Yet other functionality developments, usually initiated by third-party providers, utilise the affordances of the Twitter platform for a range of increasingly more esoteric applications. From the early Twitter bots which reported on the state of the office coffee machine or the coming and going of household pets through cat flaps to attempts to use Twitter as a sensor network for earthquake detection, potential applications appear limited only by their developers' imaginations. Recent initiatives have sought to institute micropayment systems where tweets directed at a designated account result in funds transfers; or have encouraged residents in northeastern Japan to use Twitter to report the readings of household Geiger counters in order to create a more comprehensive picture of radioactive pollution following the 2011 nuclear meltdown in Fukushima.

Beyond the success or failure of individual initiatives, this proliferation of projects demonstrates that Twitter—like other network infrastructures before it—has become a backbone for a much wider range of manual and automated communicative exchanges than its inventors may have envisaged initially. While it would be no more difficult—and possibly more effective—to use the underlying Internet infrastructure itself to report current coffee machine status or local radiation levels, to do so in a public tweet affords this information wider visibility, and thereby also enables users other than original sender and intended recipient to access and systematically collect such information, if they choose to do so.

In addition to its role as a public, instant communications medium, therefore, Twitter has now also become a key source of open data on a wide range of personal and societal practices around the world, and the importance of this role must not be underestimated. This constitutes a somewhat problematic role, as not all Twitter users will be aware that their apparently 'private' exchanges with a handful of Twitter friends and followers are also visible to virtually anyone else online, unless they are conducted through direct messages or from 'protected' accounts; at the same time, where such 'big data' on large-scale user activity patterns on Twitter are being used while giving due consideration to ethical and privacy concerns, they enable entirely new approaches to studying society with the Internet.

It is especially in this way, ultimately, that the myriad of 140-character messages which are posted to the global Twitter network every day combine to offer a view of communicative trends in contemporary society which is unprecedented in its level of detail. Never before have researchers in the humanities and social sciences had access to such a rich tapestry of everyday, real-time communication—and in spite of the impressive steps already undertaken by the contributors to the present collection and by their many peers in related fields whose work we were unable to accommodate here, much more remains to be done to fully develop our suite of methodologies, tools, and conceptual frameworks for the study of Twitter and society—and beyond, of other platforms and other media futures.

Twitter, Inc.'s increasingly restrictive policies governing data access and data use pose a significant challenge to the future of work on Twitter, and potentially foreshadow a social data ecosystem ever more tightly controlled by corporate interests. However, as we have seen with the 140-character limit which is imposed on tweets, such constraints can sometimes lead to creative workarounds—hacks which in turn result in useful innovations that would not have emerged otherwise. From little things, big things do indeed grow. In this spirit is our hope that the current collection may serve as a stepping stone for fruitful future research.

REFERENCES

boyd, d. (2006). A blogger's blog: Exploring the definition of a medium. *Reconstruction, 6*(4). Retrieved from http://reconstruction.eserver.org/064/boyd.shtml

Pear Analytics. (2009). Twitter study—August 2009. Retrieved from http://www.pearanalytics. com/wp-content/uploads/2012/12/Twitter-Study-August-2009.pdf

Rogers, R. (2009). *The end of the virtual: Digital methods.* Amsterdam, The Netherlands: Vossiuspers UvA. Retrieved from http://www.govcom.org/publications/full_list/oratie_Rogers_2009_preprint.pdf

Notes on Contributors

FABIAN ABEL (@fabianabel) is working as Postdoctoral Researcher in the Web Information Systems group at Delft University of Technology, The Netherlands, and is performing research activities in the line of user modelling and personalisation on the Social Web. Fabian completed his PhD at L3S Research Center and was concerned with (distributed) user modelling on the Social Semantic Web, covering topics such as linking, integrating and enriching user models as well as personalised retrieval of social media content. In his young career, he has published more than 50 scientific articles in the field of Semantic Web, User Modelling and Personalisation.

JULIAN AUSSERHOFER (@boomblitz) is a Digital Media Researcher at the Institute for Journalism and PR at Graz University of Applied Sciences (FH JOANNEUM), Austria. He is also a PhD candidate at the University of Vienna, Department of Communication. Ausserhofer co-founded the Web Literacy Lab in Graz and acts as a member of the Board of the Open Knowledge Foundation in Austria. His research interests include the political use of social media, open (government) data, online publishing practices and data driven journalism. He blogs at http://ausserhofer.net/

NANCY BAYM (@nancybaym) is a Principal Researcher at Microsoft Research in Cambridge, MA, USA. She is the author of *Personal Connections in the Digital Age* (Polity), *Internet Inquiry* (co-edited with Annette Markham, Sage) and *Tune In, Log On: Soaps, Fandom and Online Community* (Sage). Her website is http://www.nancybaym.com/

MICHAEL BEURSKENS (@mbeurskens) is a Lecturer and Researcher in the field of Internet and Intellectual Property Law. His research covers general issues pertaining to the ownership of information, privacy law, and network neutrality. He teaches copyright law at the specialised LL.M.-program on information law. Beurskens holds both a Master's degree in intellectual property law received from Heinrich-Heine University and a general LL.M. from the University of Chicago. He passed both German state exams with honours, and is admitted to the New York State bar. He received his doctoral degree (Dr.iur., s.j.d.-equivalent) based on a thesis on capital market information. He is an active member of the German law blogging community and also supervises the faculty-run blogs at HHU. Beurskens is currently involved in the development of Web 2.0-based eLearning-technologies, including collaborative game design.

THOMAS BOESCHOTEN (@boeschoten) is a Master's student at the New Media and Digital Culture programme at Utrecht University, the Netherlands. He specialises in Twitter and other new media that use big data. After focussing on the use of Twitter by politicians and the Occupy movements, he is now a member of the 'Project X' Research Committee that is investigating the 2012 'Facebook' riots in the small northern village of Haren in the Netherlands. His website is at http://www.boeschoten.eu/

AXEL BRUNS (@snurb_dot_info) is an Associate Professor in the Creative Industries Faculty at Queensland University of Technology in Brisbane, Australia, and a Chief Investigator in the ARC Centre of Excellence for Creative Industries and Innovation (http://cci.edu.au/). He is the author of *Blogs, Wikipedia, Second Life and Beyond: From Production to Produsage* (2008), and *Gatewatching: Collaborative Online News Production* (2005), and a co-editor of *A Companion to New Media Dynamics* (2013, with John Hartley and Jean Burgess), and *Uses of Blogs* (2006, with Joanne Jacobs). His research blog is at http://snurb.info/; see http://mappingonlinepublics. net/ for more details on his current social media research.

JEAN BURGESS (@jeanburgess) is Deputy Director of the ARC Centre of Excellence for Creative Industries & Innovation (CCI) and an Associate Professor in the Creative Industries Faculty at Queensland University of Technology, Australia. She has published widely on social media, user-created content, and community-based co-creative media such as digital storytelling. Her books include *YouTube: Online Video and Participatory Culture* (Polity Press, 2009), which has been translated into Polish, Portuguese, and Italian; *Studying Mobile Media* (Routledge, 2012); and *A Companion to New Media Dynamics* (Wiley-Blackwell, 2013). Her current research focusses on methodological innovation in the context of the changing media ecology, and in particular on the development of computational methods for media and communication studies.

MARK DANG-ANH (@mdanganh) received his Magister Artium degree from the RWTH Aachen. He is a Research Assistant in the Department of Media Studies at the Institute of Linguistics, Media and Sound Studies, University of Bonn, Germany. He currently works on the project "Political Deliberation on the Internet: Forms and Functions of Digital Discourse Based on the Microblogging System Twitter", which is part of the Priority Program 1505 "Mediatized Worlds" funded by the German Research Foundation (DFG). His research interests lie at the intersection of linguistics and media studies, primarily focussing on political communication in online media.

MARTIN EBNER (@mebner) is Head of the Department for Social Learning at Graz University of Technology, Austria, and therefore responsible for all university-wide e-learning activities. He is an Associate Professor of Media Informatics and also works at the Institute for Information System Computer Media as Senior Researcher. His research focusses strongly on e-learning, mobile learning, learning analytics, social media, and the usage of Web 2.0 technologies for teaching and learning. Martin gives a number of lectures in this area as well as workshops and talks at international conferences. For publications as well as further research activities, please visit http://martinebner.at/

JESSICA EINSPÄNNER (@jeinspaenner) studied media & communications at Bonn University (Germany) and the National University of Singapore. She is currently working as a Research Fellow at Bonn University within the project "Political Deliberation on the Internet: Forms and Functions of Digital Discourse Based on the Microblogging Platform Twitter", which is part of the DFG priority program "Mediatized Worlds". She is also writing her doctoral thesis on "User Generated Privacy: Mechanisms of Constructing Privacy Online". Her main teaching and research areas are social media communication, political communication, and online journalism.

LISA EVANS (@objectgroup) was a Writer, Data Researcher, and Programmer for *The Guardian*, with interests including statistics, public spending, data journalism, and accounting. Much of this work was published on *The Guardian*'s Datablog. She is now investigating where money flows around the world with the Open Knowledge Foundation's (England) Open Spending project, and completing her degree in mathematics from the Open University. In the near future, she will be producing training materials and tools for the open spending community to help journalists and non-government organisations use open financial data in more meaningful ways.

SIMON FAULKNER (@simonfaulkner2) is a Senior Lecturer in Art History and Visual Culture at Manchester Metropolitan University, England. He has published on the subject of British art in the mid-twentieth century, and is the editor (with Anandi Ramamurthy) of *Visual Culture and Decolonisation in Britain* (2006). His current research is on relationships between visual culture and the Israeli–Palestinian conflict. This work includes the development of an artist/writer's book, *Between States*, with the Israeli artist David Reeb.

DEVIN GAFFNEY (@dgaff) is a Research Assistant at the Oxford Internet Institute, England, and recent graduate of its Master's Program in the Social Science of the Internet. Since 2009, he has concentrated on research surrounding Twitter's various legal, ethical, methodological, and practical aspects, and has most recently focussed on the measurable role of geographic distance in interactions between users. Beyond this work, he has also worked on assessing the impact of Twitter on the 2009 Iran Election and 2011's Arab Spring, and is most recently working on assessing the merits of social media influence metrics, and the various issues surrounding projections of quantifiable methods on online users and activity.

ALEXANDER HALAVAIS (@halavais) is Associate Professor of Social and Behavioral Sciences at Arizona State University, where he researches the role of social media in social learning. He is also the president of the Association of Internet Researchers, and technical director of the Digital Media and Learning Hub at the University of California. His work investigates the use of social media by activists and others hoping to create social change. His most recent book is *Search Engine Society*, and his upcoming book examines new forms of participatory surveillance.

STEPHEN HARRINGTON (@_StephenH) is a Senior Lecturer in Journalism, Media and Communication at the Queensland University of Technology (QUT), Australia. His research has focussed mainly on the changing relationships between television, journalism, politics, and popular culture, and, in particular, understanding the qualitative impact of these changes in terms of public knowledge and engagement. His book *Australian TV News: New Forms, Functions and Futures* (Intellect, 2013) focusses on emergent news formats, and their potential to generate public knowledge and deeper levels of audience engagement. He is currently studying how Twitter and other online communication platforms affect or shape the audience experience for traditional media forms and events (e.g., TV, sport), and how to use social media data capture, analysis, and mapping methods as an audience research technique.

CLAUDIA HAUFF (@charlottehase) is a Postdoctoral Researcher in the Web Information Systems group at Delft University of Technology, The Netherlands, working in the areas of Information Retrieval and User Modelling & Personalisation. Claudia received her PhD from the University of Twente, where her research focussed on system-oriented Information Retrieval, in particular query performance prediction and retrieval system evaluation.

ALFRED HERMIDA (@hermida) is an award-winning online news pioneer, digital media scholar, and journalism educator. He is an Associate Professor at the Graduate School of Journalism at the University of British Columbia, Canada. His research focusses on changes in journalistic practices, social media, and emerging genres of journalism, with his work appearing in *Journalism Practice* and *Journalism Studies*. He co-authored *Participatory Journalism: Guarding Open Gates at Online Newspapers* (Wiley-Blackwell, 2011). Hermida was a BBC journalist for 16 years, including four as a correspondent in the Middle East, and was a founding member of the BBC News website in 1997. A regular media commentator, his work has appeared in *The Globe and Mail*, *PBS*, *BBCNews.com* and the Nieman Journalism Lab.

TIM HIGHFIELD (@timhighfield) is a Research Fellow with the ARC Centre of Excellence in Creative Industries and Innovation, and a Sessional Academic at Curtin University, Perth, Australia. He was awarded his PhD in 2011 from Queensland University of Technology, Australia. His PhD thesis studied political blogging in Australia and France, while his current research interests include examining the uses of social media, such as Twitter, within discussions and commentary around political debates and popular culture. His recent publications include co-authored journal articles in *Media International Australia* and *Social Science Computer Review*.

GEERT-JAN HOUBEN (@gjhouben) is Professor of Web Information Systems at the Software Technology department at Delft University of Technology, The Netherlands. His main research interests are in Web Engineering, in particular the engineering of Web information systems that involve Web and Semantic, Web technology, and User Modelling, Adaptation and Personalisation. He is Managing Editor of the *Journal of Web Engineering;* Chair of the Steering Committee for ICWE, the International Conference on Web Engineering; and member of the Editorial Board of ACM TWEB, ACM Transactions on the Web.

NINA KRÜGER (@NinaKrger) is a Research Assistant in the research group of communication and collaboration management at the Institute of Information Systems at the University of Münster, Germany. She studied communication science, ethnology, and psychology, and completed her Magister in 2011. As grounding for her studies, she underwent job training in an IT enterprise, where she gained practical experience in dealing with social media and their use for collaborative purposes in business. Nina's research focusses on the internal and external use of social media for corporate communication in all its facets.

ANDERS OLOF LARSSON (@a_larsson) is a Postdoctoral Fellow at the Department of Media and Communication, University of Oslo, Norway. He is also associated with the Swedish Research School for Management and Information Technology. His work has been published in journals such as *New Media and Society, Convergence, The Information Society*, and *Journal of Information Technology and Politics*. Larsson's Web site can be found at http://andersoloflarsson.se/

ALEX LEAVITT (@alexleavitt) is a PhD student and Researcher in the Annenberg School for Communication & Journalism at the University of Southern California, USA, where he is advised by Professor Henry Jenkins. Alex studies participation and online communication across networked technologies, particularly social media platforms, emergent online communities, and the information practices of media subcultures. More information about his research is available at http://alexleavitt.com/

MERJA MAHRT is a media and communication scholar currently working at Heinrich Heine University, Düsseldorf, Germany. Her research focusses on the audience of media and the functions media fulfil for their users and society, from traditional mass media like television, newspapers, and magazines to Web 2.0 applications. She is the spokesperson of the interdisciplinary researchers group, Science and the Internet (http://www.nfg-win.uni-duesseldorf.de/en/). Within this project, her research is concerned with academic blogs, their readership, and the different uses bloggers and readers make of them.

AXEL MAIREDER (@axelmaireder) has been a Research Assistant and Doctoral Student at the Department of Communication, University of Vienna, Austria, since 2009. After graduating from this department in 2006, he has been working as Researcher for projects on the Internet use of teachers and students, funded by the Austrian Ministry of Education. His current research focusses on practices, networks, and dynamics of social media communication within public discourses, and particularly in political communication.

YANA MANYUKHINA has a BA in International Relations from Baku State University, Azerbaijan (2009), and an MA in Mass Communications from the University of Leicester, UK (2011), both with distinction. Her research interests include: research methods for the social sciences, especially applied to issues around health communication, nutrition, and public health; food advertising and food media; food policies; consumer research; brands and culture. She commenced a PhD in this area in 2012. She was a Researcher on the "Reading the Riots on Twitter" investigation, from which her interest for researching social media stems.

ALICE MARWICK (@alicetiara) is an Assistant Professor at Fordham University in the Department of Communication and Media Studies, and a Research Affiliate at the Berkman Center for Internet and Society. Her work looks at online identity and consumer culture through lenses of privacy, consumption, and celebrity. She is currently working on two ethnographic projects—one examining youth technology use, and the other looking at femininity and domesticity in social media such as fashion blogs, Tumblr, and Pinterest. Her book *Status Update: Celebrity, Publicity and Self-Branding in Web 2.0* is under contract with Yale University Press. Alice has a PhD from New York University's Department of Media, Culture and Communication, and was previously a Postdoctoral Researcher in social media at Microsoft Research New England.

DIANA MAYNARD (@dianamaynard) is a Research Fellow at the University of Sheffield, UK. She has a PhD in Automatic Term Recognition from Manchester Metropolitan University, and has been involved in research in NLP since 1994. Her main interests are in information extraction, opinion mining, terminology, and social media. Since 2000, she has led the development of USFD's open-source multilingual IE tools, and has led research teams on a number of UK and EU projects. She is Chair of the annual GATE training courses, and leads the GATE consultancy on IE and opinion mining. She has published extensively, organised a number of national and international conferences, workshops, and tutorials, given invited talks and keynote speeches, and reviews project proposals for RNTL. She is currently joint Coordinator of the Semantic Web Challenge.

HALLVARD MOE (@halmoe) is Associate Professor of Media Studies at the University of Bergen, Norway. In 2011, he was a visiting scholar at the ARC Centre of Excellence for Creative Industries and Innovation at Queensland University of Technology. Moe's research interests include media policy, democratic theory, and the use of new media platforms in the public sphere. His recent work on Twitter, published in journals such as *New Media & Society*, *International Journal of Communication*, and *Nordicom Review*, focusses on their uses in public debate in the Scandinavian countries.

MIRANDA MOWBRAY is a Senior Researcher at HP Labs, Bristol, UK, where her research interests include big data for security, and online communities. Her recent publications include "Enhancing Privacy in Cloud Computing via Policy-Based Obfuscation", *J. Supercomputing*, 61 (2012): 267–291, with Siani Pearson and Yun Shen; "Business-Driven Short-Term Management of a Hybrid IT Infrastructure", *JPDC* 72.2 (2012): 106–119, with Paolo Ditarso Maciel Jr. et al.; "Efficient Prevention of Credit Card Leakage from Enterprise Networks", CMS 2011, *LNCS* 7025: 238–240, with Matthew Hall and Reinoud Koornstra; and "A Rice Cooker Wants to Be My Friend on Twitter", *Proc. Ethicomp 2011*, 322–329.

CHRISTOPH NEUBERGER is a full Professor at the Department of Communication Science and Media Research (IfKW) at the Ludwig-

Maximilians-University Munich, Germany. His research interests include media change, online journalism, activities of press and broadcasting on the Internet, social media, journalism theory, and media quality. Recent book publications include: Christoph Neuberger, Hanna Jo vom Hofe, and Christian Nuernbergk, *Twitter und Journalismus: Der Einfluss des "Social Web" auf die Nachrichten* [*Twitter and Journalism: The Influence of the Social Web on News*] (3rd edition, Düsseldorf, Germany: Landesanstalt für Medien Nordrhein-Westfalen (LfM), 2010); and Klaus Meier and Christoph Neuberger (eds.), *Journalismusforschung* [*Journalism Research*] (Baden-Baden: Nomos, 2012).

TANYA NITINS (@DrTNitins) is a Lecturer at Queensland University of Technology, Australia, in the area of Entertainment Industries. Her previous research has focussed on product placement, brand development, and new media. Her book *Selling James Bond: Product Placement in the James Bond Films* was published in 2011. Dr. Nitins has also been intrinsically involved in various research projects focussed on new media services and applications, locative media, and building online user communities.

CHRISTIAN NUERNBERGK (@nuernbergk) is a Postdoctoral Researcher and Lecturer at the Department of Communication Science and Media Research (IfKW) at the Ludwig-Maximilians-University Munich, Germany. His research interests include political communication, online journalism, networked public sphere, online social networks, the blogosphere / blogging and microblogging, and alternative journalism. Recent book publications include: Christoph Neuberger, Hanna Jo vom Hofe, and Christian Nuernbergk, *Twitter und Journalismus: Der Einfluss des "Social Web" auf die Nachrichten* [*Twitter and Journalism: The Influence of the Social Web on News*] (3rd edition, Düsseldorf, Germany: Landesanstalt für Medien Nordrhein-Westfalen (LfM), 2010).

KATY PARRY (@reticentk) is a Lecturer in Communication Studies at the Institute of Communication Studies at the University of Leeds, England. Her research interests include war and media; photojournalism and visual culture; and political communications and political culture across media genres. Prior to joining Leeds, she worked on an AHRC-funded project with Kay Richardson and John Corner at the University of Liverpool, exploring the ways in which media formats other than journalism portray politics; now published as *Political Culture and Media Genre* (Palgrave, 2012). In addition to publishing articles from this project, she continues to write on media visualisation of conflict.

JOHANNES PASSMANN (@J_Passmann) is a PhD candidate at the DFG Locating Media Graduate School at the University of Siegen in Germany. His PhD thesis is an ethnography of the Favstar scene on German-speaking Twitter, which he conceptualises as a gift economy. From this perspective, the history of the social web is described as a history of the accountability of gifts. Johannes has worked as a Lecturer in the New Media and Digital Culture Master's programme at Utrecht University in the Netherlands. His most recent publication is "Beinahe Medien: Die medialen Grenzen der Geomedien", with Tristan Thielmann, in R. Buschauer and K. S. Willis (eds.), *Locative Media: Multidisciplinary Perspectives on Media and Locality* (Bielefeld, 2012).

ISABELLA PETERS (@isabella83) is a Researcher at Heinrich Heine University Düsseldorf, Germany, and holds a PhD in information science. Her book *Folksonomies: Knowledge Representation and Information Retrieval in Web 2.0* was published in 2009. Peters's research priorities include folksonomies in knowledge representation, information retrieval, and knowledge management, as well as scholarly communication on the web and altmetrics.

WIM PETERS (@wilhelmus101) is a Senior Research Scientist in the Department of Computer Science at the University of Sheffield, UK. He has been active in the field of computational linguistics for 16 years, and has participated in various EU and national projects covering multilingual thesaurus creation, corpus building and annotation, lexical tuning, information extraction in various domains, semantic resource analysis, and ontology creation and evaluation. Some of the projects he has been involved with are EuroWordNet (multilingual resource creation), DotKom (adaptive information extraction), LOIS (legal wordnet building), DALOS (knowledge acquisition from legal texts), NeOn (life cycle of ontology networks), and CLARIN (the creation of a grid-based research infrastructure for the humanities and social sciences). For most of these projects, Wim coordinated the University of Sheffield's efforts. Presently, Wim is Coordinator of the FP7 IP Arcomem (http://www.arcomem.eu/), which addresses the needs of memory institutions in the age of the Social Web by creating a social- and semantic-aware Web preservation system that transforms archives into meaningful collective memories.

NICHOLAS PROFERES (@moduloone) is a PhD student at the University of Wisconsin-Milwaukee's School of Information Studies (USA). He holds a BS in Information Technology from George Mason University and an MA in Communication, Culture and Technology from Georgetown University. His research interests include Internet research ethics, big data, privacy, and information policy.

CORNELIUS PUSCHMANN (@coffee001) is a Postdoctoral Researcher at Humboldt Universität zu Berlin's School of Library and Information Science (Germany) who studies computer-mediated communication and the Internet's impact on society. His current project "Networking, Visibility, Information: A Study of Digital Genres of Scholarly Communication and the Motives of their Users" investigates the use of (micro)blogs in academia, combining qualitative social research with language analysis. His other interests include language-based approaches to CMC (stylistic analysis, pragmatics) and corpus linguistics. Cornelius holds a PhD in English Linguistics from Heinrich-Heine-Universität Düsseldorf, and is the author of *The Corporate Blog as an Emerging Genre of Computer-Mediated Communication*.

THOMAS RISSE is the Deputy Managing Director of the L3S Research Center in Hannover, Germany. He received a PhD in Computer Science from Darmstadt University of Technology, Germany, in 2006. Prior to joining the L3S Research Center in 2007, he led the intelligent information environments group at Fraunhofer IPSI, Darmstadt. He was the Technical Director of the European-funded project BRICKS, which explored decentralised digital library infrastructures, and Coordinator of FP7 Living Web Archive (LiWA) project. Currently, he is the Technical Director of the FP7 ARCOMEM project on Web archiving using social media information. Thomas Risse's research interests are Semantic Evolution, Digital Libraries, Web Archiving, Data Management in Distributed Systems and Self-Organising Systems.

RICHARD ROGERS is University Professor, and holds the Chair in New Media & Digital Culture at the University of Amsterdam, The Netherlands. He is Director of Govcom.org, the group responsible for the Issue Crawler and other info-political tools, and the Digital Methods Initiative, dedicated to developing methods and tools for online social research. Rogers is the author of *Digital Methods* (MIT Press, 2013).

MIRKO TOBIAS SCHÄFER (@mirkoschaefer) is Assistant Professor of New Media and Digital Culture at Utrecht University, the Netherlands, and Research Fellow at Vienna University of Applied Arts, Austria. Mirko studied theatre, film and media studies, and communication studies at Vienna University, and digital culture at Utrecht University. He obtained a Master's degree in theatre, film, and media studies from the University of Vienna in 2002, and a PhD from Utrecht University in 2008. Mirko's research interest revolves around the socio-political impact of media technology. His publications cover user participation in cultural production, hacking communities, the politics of software design, and communication in social media. He is co-editor and co-author of the volume *Digital Material: Tracing New Media in Everyday Life and Technology* (published by Amsterdam University in 2009), and author of *Bastard Culture! How User Participation Transforms Cultural Production* (published by Amsterdam University Press in 2011).

JAN-HINRIK SCHMIDT (@janschmidt) is Senior Researcher for Digital Interactive Media and Political Communication at the Hans-Bredow-Institute for Media Research in Hamburg, Germany. His research interests focus on the practices and consequences of the social Web, mainly the structural changes in identity management, social networks, the public sphere, and privacy. His most recent monograph *Das neue Netz* (*The New Web*) was published in an updated second edition in 2011. More detailed information on other publications, research projects, and activities can be found on his blog http://www.schmidtmitdete.de/

PIERRE SENELLART (@pierresenellart) is an Associate Professor in the DBWeb team at Télécom ParisTech, France, the leading French engineering school specialising in information technology. An alumnus of the École normale supérieure, he obtained his PhD (2007) in computer science from Université Paris-Sud under the supervision of Serge Abiteboul, and his Habilitation à diriger les recherches (2012) from Université Pierre et Marie Curie. His research interests focus around theoretical aspects of database management systems and the World Wide Web, and more specifically on the intentional indexing of the deep Web, probabilistic XML databases, and graph mining.

STEFAN STIEGLITZ (@wikuk) is Assistant Professor of Communication and Collaboration Management at the Institute of Information Systems at the University of Münster, Germany. He is founder and Academic Director of the Competence Center Smarter Work at the European Research Center for

Information Systems (ERCIS). His research focusses on economic, social, and technological aspects of social media. Of particular interest in his work is to investigate the usage of social media in the context of enterprises as well as politics. Stieglitz studied business economics at the universities of Cologne, Paderborn, and Potsdam. He published more than 60 articles in reputable international journals and conferences. He is also a reviewer for international journals and conferences in the field of information systems.

KE TAO (@taubau) is a PhD student working in the Web Information Systems Group, Delft University of Technology, The Netherlands. He holds Bachelor and Master degrees, majoring in Computer Science and Technology, from National University of Defense Technology in Changsha, China. His current research focusses on search in the Social Web, User Modelling, Personalisation, and Linked Data.

MIKE THELWALL (@mikethelwall) is Professor of Information Science and leader of the Statistical Cybermetrics Research Group at the University of Wolverhampton, UK, and a Research Associate at the Oxford Internet Institute. Mike has developed a wide range of tools for gathering and analysing Web data, including hyperlink analysis, sentiment analysis, and content analysis for Twitter, YouTube, blogs, and the general Web. His publications include 152 refereed journal articles, including D. Wilkinson and M. Thelwall, "Trending Twitter Topics in English: An International Comparison", *Journal of the American Society for Information Science and Technology*, 63(8) (2012), 1631–1646; as well as seven book chapters and two books, including *Introduction to Webometrics*. He is an Associate Editor of the *Journal of the American Society for Information Science and Technology*, and sits on four other editorial boards.

CAJA THIMM (@CThimm) is Professor for Media Studies and Intermediality at the University of Bonn, Germany. Her main research interests are online communication theory, social media, and organisational and political communication online.

TIMO VAN TREECK (@timovt) is a Research Associate in the team for educational development at Cologne University of Applied Sciences, Germany. As a member of the interdisciplinary researchers group Science and the Internet, he analysed educational beliefs of teachers and decision-makers in universities, and organised media training and a seminar for doctoral students. He has worked in different projects relating to blended learning, especially in academic staff development and controlling. His current research and implementation activities focus on educational beliefs, eportfolios, and diversity management in (online) teaching and learning. He is a member of the research commission of the German Association for Educational and Academic Staff Development in Higher Education (dghd).

FARIDA VIS (@flygirltwo) is a Research Fellow in the Social Sciences in the Information School at the University of Sheffield, England. Her work is centrally concerned with researching social media, crisis communication, and citizen engagement. She led the social media analysis on an academic team that examined 2.6 million riot tweets as part of *The Guardian*'s groundbreaking "Reading the Riots" project, which won a Data Journalism Award for showing the ways in which rumours spread on Twitter during the riots. Her textbook for Sage, *Researching Social Media* (2014), is written with computer scientist Mike Thelwall, highlighting the need for such interdisciplinary work in this area.

HANNA JO VOM HOFE works as a Communication Consultant at Media Authority of North Rhine-Westphalia (LfM) Nova in Düsseldorf, Germany. Prior to this, she was a Research Assistant at the Department of Communication Science and the Faculty of Educational and Social Sciences at the University of Münster, after graduating with a thesis on political communication on Twitter in 2010. Her research interests include political communication, online journalism, and social networks. Recent book publications include: Christoph Neuberger, Hanna Jo vom Hofe, and Christian Nuernbergk, *Twitter und Journalismus: Der Einfluss des "Social Web" auf die Nachrichten* [*Twitter and Journalism: The Influence of the Social Web on News*] (3rd edition, Düsseldorf, Germany: Landesanstalt für Medien Nordrhein-Westfalen (LfM), 2010).

KATRIN WELLER (@kwelle) is an Information Scientist working as a Post Doctoral Researcher at GESIS Leibniz Institute for the Social Sciences, Germany. Until December 2012, she worked at Heinrich Heine University Düsseldorf, Germany. She has been involved in different research projects on social media and their role in e-learning, knowledge sharing, and collaborative knowledge management. As a member of the interdisciplinary researchers group Science and the Internet, she has investigated how novel Internet technologies change scientists' work environments, with a particular focus on informetric indicators for Twitter communication. She is